Progress in New Cosmologies
Beyond the Big Bang

Progress in New Cosmologies
Beyond the Big Bang

Edited by

Halton C. Arp
Max-Planck Institute for Astrophysics
Garching bei München, Germany

C. Roy Keys
Publisher, *Apeiron*
Montréal, Québec, Canada

and

Konrad Rudnicki
Jagiellonian University Astronomical Observatory
Cracow, Poland

Plenum Press • New York and London

Library of Congress Cataloging in Publication Data

Progress in new cosmologies: beyond the big bang / edited by Halton C. Arp, C. Roy Keys, Konrad Rudnicki.
        p.   cm.
    Proceedings of the Thirteenth Cracow Summer School of Cosmology on Progress in New Cosmologies, held September 7–12, 1992, in Łódź, Poland.
    Includes bibliographical references and index.
    ISBN 0-306-44635-9

    1. Cosmology—Congresses. 2. Astrophysics—Congresses. I. Arp, Halton, C. II. Keys, C. Roy. III. Rudnicki, Konrad. IV. Cracow Summer School of Cosmology on Progress in New Cosmologies (13th: 1992: Łódź, Poland)
QB981.P863 1993   93-42612
523.1—dc20   CIP

Proceedings of the Thirteenth Cracow Summer School of Cosmology on Progress in New Cosmologies, held under the auspices of the Omega Foundation and *Apeiron*, September 7–12, 1992, at the Physics Institute, University of Łódź, Łódź, Poland

ISBN 0-306-44635-9

©1993 Plenum Press, New York
A Division of Plenum Publishing Corporation
233 Spring Street, New York, N.Y. 10013

All rights reserved

No part of this book may be reproduced, stored in a retrieval system, or transmitted in any form or by any means, electronic, mechanical, photocopying, microfilming, recording, or otherwise, without written permission from the Publisher.

Printed in the United States of America

Dedication

In memory of Fritz Zwicky

> Some examples of Fritz Zwicky's approach to scientific ideas and hypotheses are given.

I had the opportunity to present the philosophical bases of the scientific methodology of Fritz Zwicky during the XIth Krakow Summer School of Cosmology "Morphological Cosmology" devoted to the 90th anniversary of his birth (Flin & Duerbeck 1989). On that occasion I discussed the specific Zwickean "Morphological Method" or "Morphological Approach" on the general basis of the Goetheanistic Theory of Knowledge. This paper has been published (Rudnicki 1989), and consequently I shall not go into great detail here. Nor do I wish to tell the life-story of this great astronomer and cosmologist, since this has been done in a voluminous book by Roland Mueller (1986) and all can read it there. Rather, I wish to mention some selected characteristics of his practical approach to the realm of scientific ideas and hypotheses.

A key to this approach is given in his *Morphological Astronomy* (Zwicky 1957), where we read the following:

> If rain begins to fall on previously dry areas on the earth, the water on the ground will make its way from high levels to low levels in a variety of ways. Some of these ways will be more or less obvious, being predetermined by pronounced mountain formations and valleys, while others will appear more or less at random. Whatever courses are being followed by the first waters, their existence will largely prejudice these chosen by latter floods. A system of ruts will consequently be established which has a high degree of permanence. The water rushing to the sea will sift the earth in these ruts and leave the extended layers of earth outside essentially unexplored. Just as the rains open up the earth here and there, ideas unlock the doors to various aspects of life, fixing the attention of men on some aspects while partly or entirely ignoring others. Once man is in a rut he seems to have the urge to dig even deeper, and what often is most unfortunate, he does not take the excavated debris with him like the waters, but throws it over the edge, thus covering up the unexplored territory and making it impossible for him to see outside his rut. The mud which he is throwing may even hit his neighbors in the eyes, intentionally or unintentionally, and make it difficult for them to see anything at all.

Zwicky, in his astronomical activity, never followed the "predetermined ways." When, according to general opinion, all the single stars, even novae at maximum brightness, were much less luminous than galaxies, he, together with Walter Baade, examined the old observations and organized a systematic search for very luminous objects, finally proclaiming a new type of single objects—supernovae, which could have the luminosity of entire galaxies. When Edwin Hubble and his followers talked about the uniform background of galaxies and some exceptional agglomerations of them—clusters of galaxies, Zwicky wanted to see whether such uniform background really existed. To check it, he introduced Schmidt cameras into astronomy (the first 8-inch version was in Pasadena on the roof of Robinson Building, followed in 1936 by the 18-inch version on Palomar) and discovered that all the galaxies take part in the clustering. Questioning the prejudice that a galaxy always consists of a nucleus and something around it (a shell, disk or halo), he made a systematic survey of galaxies and discovered compact galaxies, consisting of nucleus alone.

We should also not forget that while most extragalactic astronomers considered the redshifts of extragalactic objects as due to Doppler effect, Zwicky together with Milton Humason always examined other possibilities. To express the possibility of alternatives, he systematically used the expression "symbolic recession velocity" V_s, instead of commonly used "radial velocity" V_r. He never fell into another extreme prejudice, which holds that redshifts are not Doppler in nature. He simply wanted to explore all possibilities first, so long as they were not excluded by valid scientific arguments.

In general, in his morphological approach he never stuck to one hypothesis (even if it was his own hypothesis), but always worked with entire sets, with continua of hypotheses, and considered all possibilities as long as they remained feasible.

And this is also the aim of our present School of Cosmology. We have to present as many observational facts and as many theoretical hypotheses as possible, and not stick to any single one, as long as alternatives still exist. Of course it is much easier to do research by "digging even deeper" in a wide, "generally accepted" scientific trend. It is easier to get grants, to obtain scientific degrees and honors; yet, in this case, we would merely be following scientific fashion instead of seeking the truth.

References

Flin, P, and Duerbeck, H.W., ed., 1989, *Morphological Cosmology* (Proceedings, Krakow, Poland 1986) Springer-Verlag.
Mueller, R., 1986, *Fritz Zwicky*, Verlag Baeschlin.
Rudnicki, K., 1989, in: *Morphological Cosmology*, Flin, P, and Duerbeck, H.W., ed., (Proceedings, Krakow, Poland 1986) Springer-Verlag.
Zwicky, F., 1957, *Morphological Astronomy*, Springer-Verlag.

Konrad Rudnicki

Foreword

The Hidden Hypotheses Behind the Big Bang

It is quite unavoidable that many philosophical *a priori* assumptions lurk behind the debate between supporters of the Big Bang and the anti-BB camp. The same battle has been waged in physics between the determinists and the opposing viewpoint. Therefore, by way of introduction to this symposium, I would like to discuss, albeit briefly, the many "hypotheses", essentially of a metaphysical nature, which are often used without being clearly stated.

The first hypothesis is the idea that the Universe has some *origin, or origins*. Opposing this is the idea that the Universe is eternal, essentially without beginning, no matter how it might change—the old Platonic system, opposed by an Aristotelian view! Or Pope Pius XII or Abbé Lemaître or Friedmann versus Einstein or Hoyle or Segal, *etc*.

The second hypothesis is the need for a "minimum of hypotheses"—the *simplicity* argument. One is expected to account for all the observations with a minimum number of hypotheses or assumptions. In other words, the idea is to "save the phenomena", and this has been an imperative since the time of Plato and Aristotle. But numerous contradictions have arisen between the hypotheses and the facts. This has led some scientists to introduce additional entities, such as the cosmological constant, dark matter, galaxy mergers, complicated geometries, and even a rest-mass for the photon. Some of the proponents of the latter idea were Einstein, de Broglie, Findlay-Freundlich, and later Vigier and myself.

Very similar to the argument—or rather the postulate—of simplicity is the principle of *beauty, a* typical Pythagorean concept. A theory of the universe is not adopted because it is "true", but because it fulfills some religious views about the universe, or simply because it is beautiful. Such motivations are not considered scientific, and hence are usually disguised. When this is done, we pay less attention to saving the phenomena, and only concern ourselves with the internal coherence of the theory. One view often expressed by authors is that modern cosmology demands primarily coherence, not proof. We would instead say that coherence is necessary, but not sufficient. Examples of this approach are recent efforts to introduce the superstring theory of space, or to achieve a grand unified theory of forces, which is required not by facts, but by a quest for beauty.

A third hypothesis is the *reduction* of phenomena in order to save only some facts, which are deemed more important or more cosmologically significant. This hypothesis springs naturally from the belief that all facts in disagreement with a theory are of secondary importance—epiphenomena—that can be easily forgotten. In cosmology, this has led to a neglect of solar physics, even though we know that solar physics has much to say about, for example, neutrinos, gravity waves, element abundances, *etc*. This reductionism also leads to *ad hoc* hypotheses, such as the "cosmological principle", according to which the universe is homogeneous and isotropic—a welcome notion when one is looking for "simple" or "economical" models. Even worse, assessing the "importance" of facts becomes a very *subjective* exercise, which can lead to a choice of fundamental data that differs from one author to another. For scientists devoted to the "old" classical Big Bang theory, the only items that count are the Hubble shift (which is interpreted as "expansion"), the background radiation (which is called "cosmological") and the abundances of light elements (which are called "primordial"). To scientists dedicated to what I shall call, for the sake of simplicity, the "new Big Bang", the inhomogeneous grouping of galaxies in the universe, the minute traces of inhomogeneity in an almost isotropic background radiation and the obvious young age of many galaxies are arguments that cannot be forgotten, any more than the age of the globular clusters. And finally, of primary importance to the anti Big Bang camp are the evidence of discordant redshifts, the hierarchical distribution of matter in space and the lack of any direct evidence for secular evolution over the last few billion years. Critical writers often assign such evidence greater weight than the three standard arguments offered in support of the Big Bang. Likewise, they tend to regard the background radiation field as not originating from the single, most distant visible shell in the Universe, and reject the notion that element abundances are in any way primordial.

In conclusion, I would issue yet another warning against any type of dogmatism that might give authors who adopt the above hypotheses (or at least some of them) a sense of security, or even certainty, justified only by coherence. As I pointed out earlier, coherence is not sufficient. There is *some* truth to each of these hypotheses: this is my own working hypothesis! Nevertheless, in cosmology, we are very far from penetrating to the depths of this truth. We must allow ourselves to be guided by more observations—especially of the phenomena now recognized as contradictory—for only observations can illuminate the way for the theoreticians as they strive to achieve a better description of the universe and to learn more of its secrets.

Jean-Claude Pecker
Collège de France

Preface

The present volume contains the proceedings of the 13th Krakow Summer School of Cosmology, which was held in Lodz, Poland from September 7 to 11, 1992. The School was attended by more than 60 astronomers, physicists, students and amateurs. Krakow Summer Schools of Cosmology were originated in Poland by Professors Jan Jerzy Kubikowski, Konrad Rudnicki and Andrzej Zieba in 1968. They take place every second year and are organised in different locations. Their aim is to convey up-to-date, first-hand information about observational and theoretical cosmology to young scientists, graduate students and teachers, as well as to serious amateurs of astronomy, physics and philosophy. The Schools aspire not only to show the current, generally accepted views, but also alternative ones as long as they are based on reliable premises. The first Schools were local in character. Since 1978 they have become international. Each School is devoted to one selected topic. The 13th edition of the School was held under the auspices of Omega Foundation of Lodz and the astronomy and physics journal *Apeiron*.

To fully appreciate the contents of this volume, we must look back at least to the turn of the century, when a trend began in physics to invest the mathematical laws in which the book of nature had been written since the time of Galileo with an absolute stature, and to restrict the definition of physical reality to laboratory measurement and observation—in short, to reduce the objective world to a series of sense impressions. This trend, of which, ironically, Mach and his followers were among the chief exponents (it will be recalled that Einstein sought to distance himself from Mach's philosophical views), advanced to the point that physics soon found itself relegated to the status of a dialect of mathematics. It became fashionable to seek a single, revelatory equation to describe the Universe, an equation in which, inevitably, time-dependent terms would play a predominant role. Thus, the infinite and eternal universe of the presocratic Greek philosophers, in particular Anaximander, is supplanted by a mathematical cipher wherein the construct "space" is endowed with plastic qualities and mysteriously allowed to expand (or contract) the river of time is made to spring from a source outside of time (the primordial singularity in "spacetime"). So profound has the divorce between the current cosmological model and the empirical evidence become that modern cosmologists must invoke increasingly contrived epicycles ("something foreign and wholly irrelevant" in the immortal words of Copernicus) in order to preserve the Big Bang dogma.

In most textbooks on astronomy, the possibility that the universe might not have had a beginning is excluded by the very definition given of cosmology. It is hardly surprising that the literature of modern cosmology, which presents the numerous helper hypotheses that have been grafted onto the Big Bang as the quintessence of theoretical acumen, has sought to depict efforts to frame an alternative to the expanding universe model as a succession of farcical episodes: Einstein retracts his cosmological constant with an embarrassed *mea culpa*; Milne's kinetic cosmology is dismissed a curious fit of whimsy; Hoyle's early investigation of steady state cosmology earns him unrelenting ridicule. Virtually absent from the historical record are the flashes of brilliance that shine like beacons through the darkness of subsequent decades: de Sitter's prediction of "spurious Doppler shifts" in 1917; MacMillan's qualitative description of an energy-conserving static universe (ca. 1920); Eddington's near exact prediction of the 3° K background without expansion (in 1926, and similar work by Nernst, Regener and Finlay-Freundlich); Zwicky's gravitational drag redshifting mechanism (1929), favoured by Hubble as the most plausible explanation of his "apparent velocity" shifts; and de Broglie's quantum tired-light redshift proposal (1962), based on an interaction between the photon's "pilot" wave and the medium of its propagation—to name but a few. The present volume, dedicated to the memory of Fritz Zwicky, is the fruit of efforts to revive this tradition. It may be seen as a continuation of work published in two earlier collections: New Ideas in Astronomy (proceedings of a symposium held in honour of Halton Arp in 1987), and Festschift Vigier (papers presented at a workshop held in Paris in 1990, and published as a special issue of *Apeiron*).

Many of the papers presented at the 13th Krakow Summer School of Cosmology represented synthetic approaches to a new cosmological model. However, if a few general themes could be distilled from these contributions, the list might include the following: observational contradictions to the Big Bang and the expanding universe, alternatives to the Doppler interpretation of "cosmological" redshifts, problems associated with astrophysical sources and the intergalactic medium and alternatives to conventional theories of gravity.

Exhaustive treatments of observational contradictions associated with the Big Bang are given by Halton Arp, who raises the issue of redshift discrepancies in nearby galaxies and young stars (which show systematically higher redshifts than their older counterparts), Victor Clube, who argues that a newly discovered relationship between apparently diverse classes of object, *viz.* spiral arms and giant comets, poses grave problems for expansion; and Eric Lerner, whose paper focusses on the incompatibility between the Big Bang and current observations of light elements, as well as the ever-recurring "age of the universe" problem and the near-perfect isotropy of the microwave background. Two relatively recent—yet by no means insignificant—problems are raised by William Napier and Jack Sulentic. Napier presents a detailed account of his tests of the Tifft effect, which demonstrate beyond doubt that redshift periodicities are real, while Sulentic examines the incidence of discordant redshifts in compact groups of galaxies. Miroslaw Zabierowski investigates a hitherto unsuspected case of redshift quantization within the Local Group, and finally, Fred Walker notes that a small sample of spiral galaxies of the same morphological type and absolute magnitude might have the same apparent rotation velocity, contrary to what would be expected if galactic redshifts are due to expansion.

Alternative explanations of the Hubble law are put forward by Amitabha Ghosh, who introduces a velocity-dependent term in the gravitational force law; Eugene Shtyrkov, who suggests that the velocity of light changes as light waves travel through intergalactic space; and Thomas and John Miller, whose investigation of the de Sitter redshift-magnitude relation brings quasars in as close as intermediate-redshift objects. Discussions of specific astrophysical objects are found in papers by Peter Browne, who applies a magnetic vortex tube model to the problem of beams of ultrarelativistic charged particles in active galactic nuclei, and William

Peter, E. Griv and Anthony Peratt, who indicate that the Alfven plasma model may explain radio lobes and extragalactic jets from classic double radio sources. Wieslaw Tkaczyk finds that the inhomogeneous spatial distribution of gamma-ray bursts may fit an unconventional cosmology; Svetlana Triphonova and Anatoly Lagutin present a unified cascade model for producing gamma-radiation in active galactic nuclei; and Bogdan Wszolek proposes that dust in the South Coma void may cause additional redshifting in light originating from galaxies beyond the void.

In addition to phenomenologically motivated modifications to the gravity laws proposed by Andre Assis and Amitabha Ghosh (by the addition of a velocity term to the Newtonian force law), David Roscoe discusses a Galilean invariant formalism for gravitational action. Franco Selleri interprets the equations of special relativity in the presence of a fundamental frame, while Henrik Broberg shows that particles can be described as locally confined systems of vacuum energy. Finally, comprehensive new cosmologies are presented by Toivo Jaakkola, whose equilibrium model takes as its point of departure a coupling between gravitation and electromagnetism, and Tom Van Flandern, who describes a universe model deduced from first principles.

Exciting progress is now being made in the subjects of episodic matter creation, understanding of the physics of non-velocity redshifts, the significance of quantized redshifts, more general solutions of the equations of general relativity and the workings of Machian physical laws. Perhaps we have reached the point where we can take on the ultimate challenge of relating the mysterious world of quantum mechanics to the realm of classical physics. It is clear by now that these seminal investigations will be carried out by a small number of researchers working outside the constricting assumptions of conventional astronomy and physics.

The crucial discoveries needed to break away from current dogma will only be communicated in alternative journals, conferences and books such as the present one, where investigators can speak frankly about the fundamental issues. The dissemination of such publications is important, and the future will show, we believe, that real progress depends on a more general and enlightened participation by a large number of people who are concerned with achieving better understanding of our universe.

The editors, on behalf of all the participants, wish to thank the local organizers Wieslaw Tkaczyk of the Physics Department, University of Lodz, and Mieczyslaw Borkowski, Director of Omega Foundation, for making all arrangements for the conference on the ground. We also wish to acknowledge the efforts of local organizing committee members Marcin Tkaczyk, Radoslaw Borkowski, Wojciech Blonski, Tomasz Gierka, Mariola Kubiak, Grzegorz Kociolek, Bozena Mirys and Leszek Wojtczak, as well as the staff of the Physics Institute at University of Lodz, whose dedicated work was crucial to the success of the conference. The material for the proceedings was prepared with invaluable assistance from Svetlana Triphonova. Finally, we would like to express our gratitude to Omega Foundation for generously providing funds for the event, to the participants for delivering excellent manuscripts and to Konrad Rudnicki for forging ahead against all odds.

Halton C. Arp
C. Roy Keys
Konrad Rudnicki

Contents

Fitting Theory to Observation—From Stars to Cosmology 1
    Halton Arp

Redshift Periodicity in the Local Supercluster 29
    W.M. Napier and B.N.G. Guthrie

Compact Groups of Galaxies 49
    Jack W. Sulentic

Is there Matter in Voids? 67
    Bogdan Wszolek

Redshifts and Arp-like Configurations in the Local Group 71
    M. Zabierowski

Are the Galaxies Really Receding? 81
    Fred L. Walker

The Case Against the Big Bang 89
    Eric J. Lerner

De Sitter Redshift: The Old and the New 105
    John B. Miller and Thomas E. Miller

Equilibrium Cosmology 111
    Toivo Jaakkola

A Steady-State Cosmology 153
    A.K.T. Assis

Cosmological Principles 169
    Konrad Rudnicki

The Meta Model: A New Deductive Cosmology from First Principles 177
    T. Van Flandern

Dark Matter, Spiral Arms and Giant Comets 187
    S.V.M. Clube

Active Galactic Nuclei: Their Synchrotron and
    Cerenkov Radiations 205
    P.F. Browne

Computer Simulations of Galaxies 237
    W. Peter, E. Griv and A.L. Peratt

Are there Gamma-Ray Burst Sources at Cosmological Distances? 249
    W. Tkaczyk

Gamma-Ray Emission Regions in AGNs 259
    Svetlana Triphonova and Anatoly Lagutin

On the Meaning of Special Relativity
    if a Fundamental Frame Exists 269
    F. Selleri

Galilean Gravitation on a Manifold 285
    D.F. Roscoe

Astrophysical and Cosmological Consequences
    of Velocity-Dependent Inertial Induction 305
    Amitabha Ghosh

A New Interpretation of Cosmological Redshifts:
    Variable Light Velocity 327
    Eugene I. Shtyrkov

Quantized Vacuum Energy and the Hierarchy of Matter 333
    Henrik Broberg

Index 353

Fitting Theory to Observation—From Stars to Cosmology

Halton Arp

Max-Planck-Institut für Astrophysik
Garching bei München, Germany

A review of observational contradictions to the conventional interpretation of redshifts as velocities includes new discrepancies in nearby galaxies and also in the young stars they contain. It is established that the strongest empirical relation is between the redshift and age of an object, and that velocities are not significant factors in cosmic redshifts. Expanding universes are, therefore, excluded.

As an example of how a successful theory must incorporate all observational facts, I describe a more general solution of the equations of general relativity. In this solution we do not make the special case assumption that the mass of elementary particles is constant. We solve the more general case where mass is variable, the case which, *a priori*, is more likely to apply to the universe as a whole. We discuss how this theory predicts all the currently accepted astrophysical observations as well as the "forbidden" data which violate current theory.

Nearby Groups of Galaxies

The current theory of the Big Bang rests entirely on the interpretation of displacements of lines of elements in the spectra of small fuzzy spots in the sky called galaxies. But if we investigate the nearest galaxies to us, the ones we can resolve in detail (and, therefore, know the most about), we immediately encounter irreconcilable contradictions with the expanding universe theory. Figure 1 shows the redshifts of all the major companions relative to the redshift of their dominant galaxy in two of the nearest groups of galaxies. If the redshifts of the companions are due to orbital velocity around the central galaxy, there should be as many negative as positive spectral shifts. Instead, 21 out of 21 of these companions are redshifted positively. This has one chance in two million of happening accidentally and, therefore, proves that these companion galaxies have intrinsic, excess redshifts. (Arp 1987).

There is surprising news, however. A new member has been recently added to our Local (M31) Group. After a lengthy delay, it has been decided that IC342, seen through considerable absorbing material at low galactic latitudes in our own galaxy, is really within the confines of our Local Group. Our Milky Way galaxy is about 0.6 Mpc away from M31. On roughly the other side of M31 from us, IC342 lies about 1.2 Mpc distant from M31. (Madore and Freedman 1992, McCall 1989). But this galaxy counted among the M31 Group, as shown in Figure 1, has the highest redshift of all presently accepted Local Group members: $c\Delta z \sim +289$ km s^{-1}!

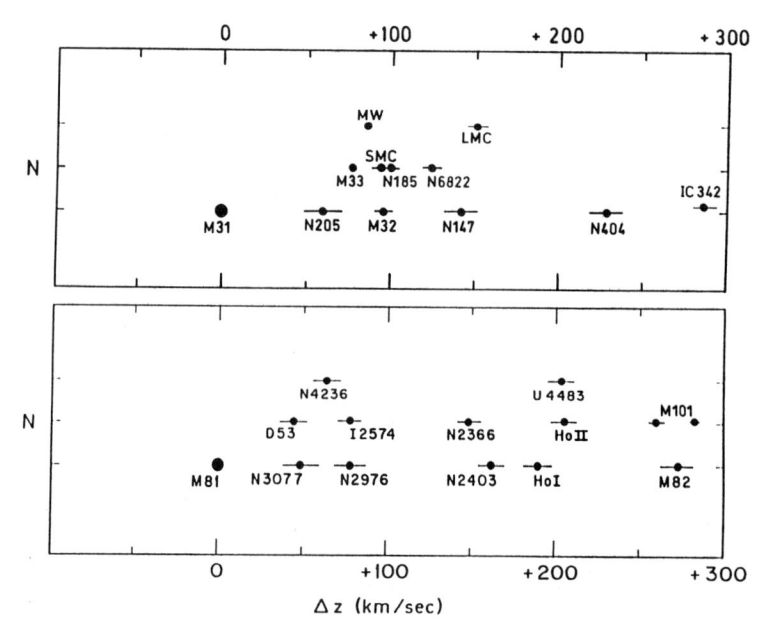

Figure 1. Relative redshifts (Δz = km s^{-1}) of all major companion galaxies in two nearest groups of galaxies. The dominant galaxies in the two groups are M31 (Local Group) and M81 in the next major group. N represents the approximate number of objects.

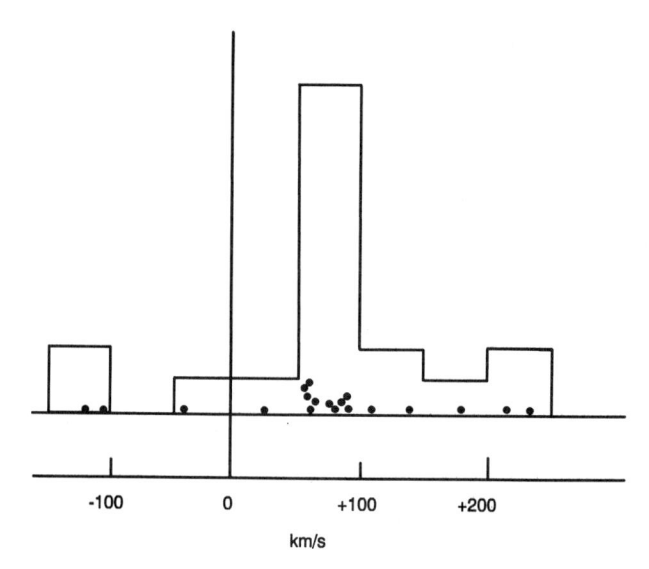

Figure 2. Residual redshifts of nineteen galaxies known to be companions of large galaxies. This announcement of the systematic redshift of companion galaxies appeared in *Nature* in 1970 (Vol. 25, March 14 1970). The significance then was about one part in a few thousand. Now that the significance has grown to about one chance in 4 million, the effect is not discussed in major journals. (Figure excerpted from Field, Arp and Bahcall 1973).

The chance now becomes 1 in 4 million that companion galaxies do not have intrinsic redshifts! It is important to note that in 1970, excess redshifts in companion galaxies were routinely announced in *Nature* magazine, as shown in Table 1 and Figure 2. The significance was then already a part in a few thousand. This result was confirmed by Bottinelli and Gougenheim (1973) and Jaakkola (1973). An independent investigation by Collin-Souffrin, Pecker and Tovmassian (1974) also supported the result, and yielded the important correlation that compact companions tended to be more systematically redshifted than lower surface brightness companions. Today, when the effect has grown to overwhelming significance, there is little likelihood the results and their implications will be discussed in the major professional journals. Establishment astronomers often argue that discordant redshift evidence has had the opportunity to be published and read, but the judgment of astronomers was that it was not significant and, therefore, not worth further discussion. The truth of the matter is that as soon as the consequences for conventional theory were realized it became rapidly more difficult to communicate the evidence. As the evidence grew to overwhelming proportions, discussion shrank to the vanishing point.

Table 1. Differential redshifts in companion galaxies (as of 1970)

Dominant galaxy	Companion galaxy	Differential redshift (km s^{-1})
M31	M32	+85
	NGC 205	+62
	NGC 185	+58
	M33	+57
M81	M82	+234
	NGC 185	+58
	NGC 3077	−104*
	IC 2574	+91
	HO II	+215
NGC 5128	NGC 5102	+77
	NGC 5236	+64
	NGC 5253	−42
	NGC 5068	+139
M51	NGC 5195	+109
Atlas 48	Companion	−120
Atlas 58	Companion	+60
Atlas 82	Companion	+90
Atlas 86	Companion	+23
Atlas 87	Companion	+180

* Corrected later to +57 (km s^{-1}).

As a result, younger astronomers are not aware of the fact that companion galaxies are systematically redshifted, and they try to apply completely incorrect dynamical calculations to their data (which show the effect also). One of many examples of this can be seen in a paper by Vader and Chaboyer (1992).

The systematic redshift of companions, of course, has now been established in every group tested out to the limit of applicable redshift surveys; Arp and Sulentic (1985) measured over 100 galaxies in more than 40 different groups with

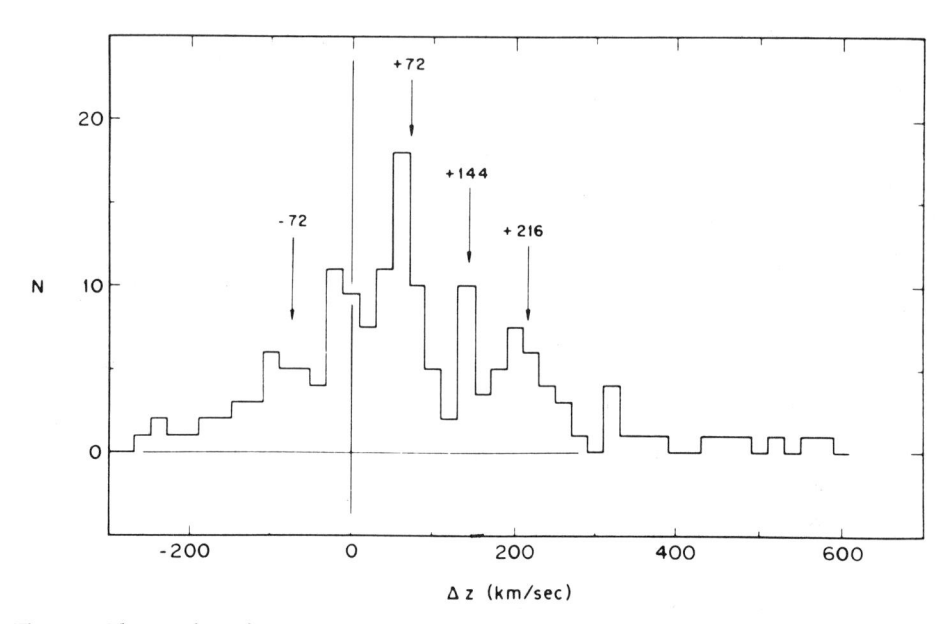

Figure 3. The number of companion galaxies as a function of their redshift with respect to the central galaxy (From Arp and Sulentic 1985). The preferred values of redshift previously found in independent galaxy samples by W. Tifft are marked by arrows.

the Arecibo radio telescope and tested a total of 160 galaxies in groups with dominant galaxies as faint as apparent magnitude 11.8. Overwhelming evidence was found that the companions are systematically redshifted, as shown in Figure 3. Later M. Geller and J. Huchra at the Center for Astrophysics (CFA) tested their surveys and reported no effect. R. Brent Tully also reported no effect in groups he had picked out. But when a team of astronomers from Trieste (Girardi *et al.* 1990) analyzed this same Geller-Huchra and Tully data, they found a strong confirmation of the systematic redshift excess of companion galaxies in groups. Figure 4 here shows the significance of their excess numbers greater than velocity difference zero for the Geller-Huchra (GH) data.

It is instructive to note that even though Girardi *et al.* proposed an explanation that avoided intrinsic redshifts, their paper still had enormous difficulty getting published. Theirs was one of the two "conventional" explanations advanced for the phenomenon. One proposed model had companions expanding away from the central galaxy and our cone of vision intercepting more on the far receding side than on the near approaching side. The other model had companions falling toward the central galaxy and being hidden by dust on the far, approaching side. Aside from the disaster for the observed stability of groups, these models were obviously hopeless from the outset, in the sense that they predicted large numbers of hitherto undiscovered Local and nearby group members. It was indeed ludicrous to consider that *any* large, low redshift galaxies had remained undiscovered.

Of course, to accept the obvious conclusion from these empirical results meant that the redshift of the companion galaxies was an intrinsic property which overwhelmed the spectral shifts due to true, Doppler velocities. In this case there would be even less than zero evidence for systematic *velocities* of galaxies in general, and hence no evidence that the universe was expanding. The Big Bang

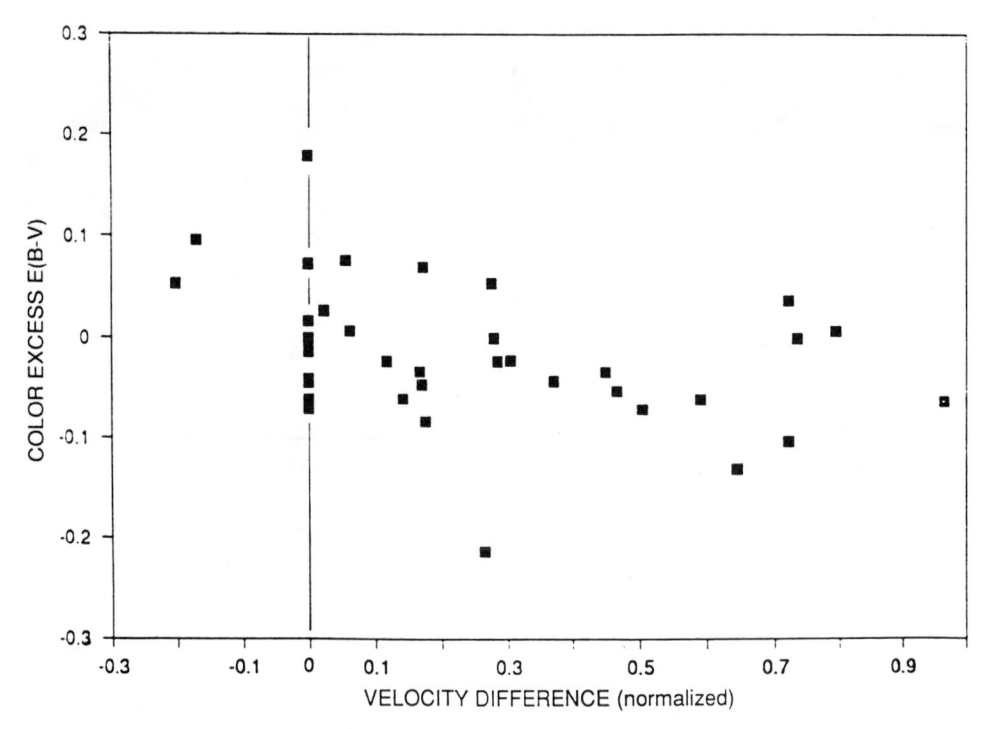

Figure 4. Geller-Huchra groups. The blueness of the companion galaxy is shown to be correlated with the amount of its excess redshift in this diagram from Girardi *et al.* (preprint and 1992).

would be devastated. Clearly this latter the consequence has caused the data to be swept firmly under the rug.

A particularly striking example of how the observational data is misinterpreted is provided by a recent paper by Zaritsky (1992) and particularly a paper by Dennis Zaritsky, Rodney Smith, Carlos Frenk and Simon White (1993). The last author is a researcher on the hypothesis of unseen matter working at the Institute of Astronomy in Cambridge, England. In this latter paper, they survey what they call satellites around isolated spirals of type Sb–Sc and perform dynamical analyses which lead to the conclusion "... that isolated spiral galaxies have massive halos that extend to many optical radii."

Although a total of 16 Figures were presented in their paper, the critical data which showed the differential redshift of the companion galaxy as a function of its relative faintness was not plotted. This figure shown here as Figure 5 was plotted from the data which these authors tabulated. It is immediately apparent that, excluding the obviously distinct set of very faint $4.5 < \Delta m_b < 7.2$ mag companions, the remaining companions of appreciable mass and luminosity are systematically shifted to higher redshift. This is conspicuous even though they have arbitrarily excluded companions brighter than $\Delta m_b < 2.2$ mag, a class of companions in which objects like M82, connected to M81 with a hydrogen bridge, has an excess redshift of +288 km s^{-1}. Of course, doing dynamical analyses with redshifts which do not represent velocities does not lead to meaningful results.

The most startling aspect of this study is, however, the fact that the authors do not reference even one of the eighteen independent studies from 1970 onward that

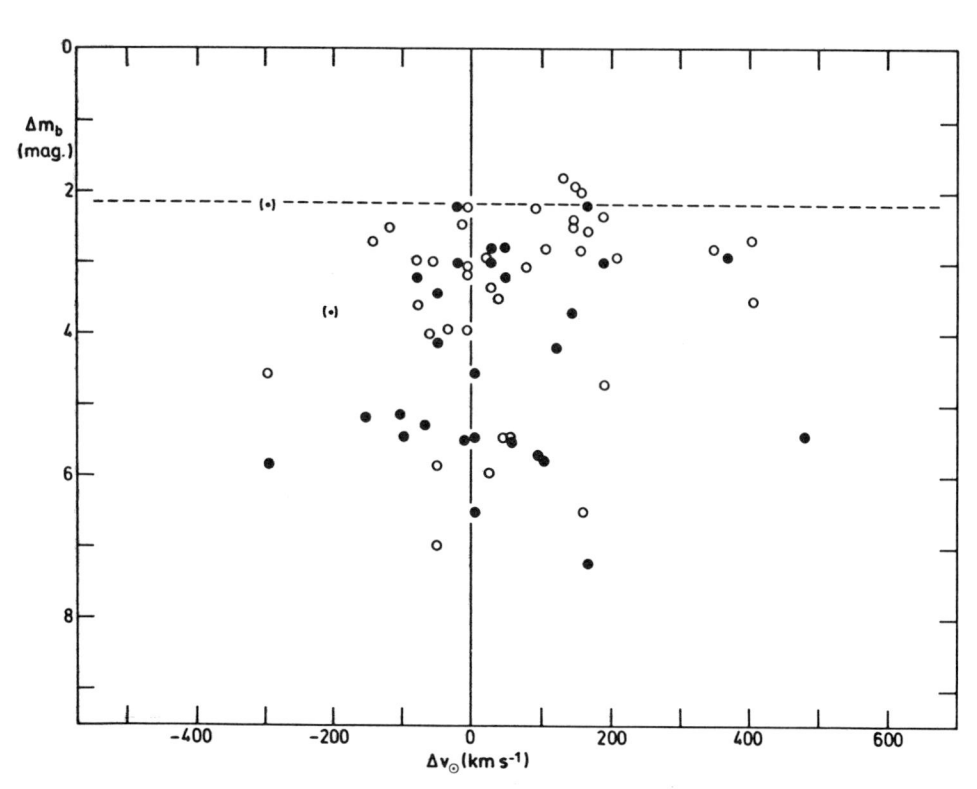

Figure 5. Data on "satellite" galaxies from Zaritsky et al. (1992). Filled circles are low surface brightness dwarfs of de Vaucouleurs Type 9–11. Two extreme and undefined objects classified $T = 15$ are indicated by parenthesis. The authors arbitrarily excluded objects classified companion galaxies brighter than $\Delta m_b < 2.2$ mag. One can again clearly confirm, however, that the *major* companion galaxies are systematically redshifted relative to the dominant galaxy in the group.

established the systematic redshift of companion galaxies. (Most of these references are available, for example, in Girardi et al. 1992). Their new data again confirms the effect but they ignore this confirmation as well as all the previous documentation of the excess redshift phenomenon in companion galaxies.

What appears to be the actual situation is that younger galaxies are created in or near the centers of older galaxies and are expelled as intrinsically higher redshift galaxies into the near neighborhood of their parent. As the companions emerge, they entrain some pieces of the parent galaxy with them, and these small pieces have the same age and intrinsic redshift as the older galaxy. These are the very low surface brightness dwarfs which are several *hundred* times less luminous than the parent galaxy. The actual, more recently created *major* companions (as they are always referred to in my published analyses) are of the order of 10% of the luminosity of the main galaxy and are not all in the class of tiny pieces torn off the original galaxy. Considering the small mass-to-luminosity ratios of these low surface brightness dwarfs, they become even more negligible compared to the mean mass exterior to the parent galaxy which shows such clear cut systematic redshift.

Physical Correlations with Excess Redshift

If we ask the natural question, "What is observationally different about the companions compared to the central, dominant galaxy?" we immediately realize that although they are both made up of the usual stars, gas and dust, many of the companions have a relatively larger percentage of younger, hotter stars. This can be strikingly seen in the extreme excess redshift companions NGC404 and M82. Their spectra are dominated by Hydrogen Balmer lines in absorption, characteristic of stars of evolutionary age 10^8 to 10^9 years rather than older stars of 10^{10} years age. Such stars are hotter and, therefore, bluer. This younger stellar content is confirmed by the effect seen in Figure 4 where the excess companion redshifts increase with blueness.

In fact, the blueness-redshift excess correlation is seen all the way from the smallest to the largest excess redshifts. Compact blue, peculiar and active galaxies show redshifts in the 1,000 to 10,000 km s^{-1} excess redshift range. In the case of the largest excess redshift quasars, they are so energetic and compact that they must expand quickly from their observed state and, hence, must be dynamically very young. The consequence of this is very important, namely: *empirically the cause of the intrinsic redshift is something related to age.*

Quantization of Redshifts

There is another clue to the nature of non-velocity redshifts, *viz.* the tendency for extragalactic redshifts to occur at certain preferred values. W.G. Tifft in 1972 pointed out the tendency for redshifts to be periodic or quantized. In spite of enormous disbelief and ridicule, every succeeding test has confirmed and strengthened the result. For example, Figure 3 in the present paper shows a test by H.Arp and J. Sulentic in 1985 of a completely different set of galaxies with the most accurate redshift measured to that date. The previously predicted peaks at 72 km s^{-1} intervals stand out conspicuously.

In 1967 G. Burbidge and E.M. Burbidge pointed out the existence of preferred redshifts for quasars. In 1971 K.G. Karlsson showed they obeyed the mathematical formula $\Delta \ln(1+z) = 0.206$. Many investigations confirmed this, but one of the most recent (Arp *et al.* 1990) found a confidence level of 99.97% for the existence of the periodicity.

It is instructive to note that a young postdoctoral student, then at the Institute of Astronomy in Cambridge, England where the director is Martin Rees, rapidly published a paper titled "Against the $\Delta \ln(1+z) \cong 0.205$ Periodicity in Quasar Redshifts" (Scott 1991). The most complete sample available, the 3CR radio quasars, he claimed, showed no periodicity. Yet both he and the editor of the journal were shown plots where three of the periodicities, z = .60, .96 and 1.41 were conspicuous. In this and other samples, he included the faintest apparent magnitude, most uncertain and probably the most distant quasars, the same quasars which Arp *et al.* had shown did *not* exhibit the periodicity. Finally, in a new sample of bright X-ray quasars he again found the periodicity, but ventured the opinion that it would go away with further measures (*i.e.* fainter quasars). He ended his paper with the statement: "The conclusion is, there is no evidence for such a periodic structure."

This is a useful example to cite because it well characterizes how observational evidence which is contrary to the current assumed model is misrepresented.

In the redshift ranges intermediate between the bright galaxies and the quasars, pencil beam surveys in various directions in the sky also reveal clumping of redshifts. (Broadhurst et al. 1990). The investigators, after considerable delay, rather nervously announced this result, but only after turning the redshifts into distances via the assumed redshift-distance relation. They concluded the galaxies were spaced periodically apart by about 128 Mpc. Now, it is a cardinal rule of observational science that one should report the observed quantities which, in this case, were redshifts. To obtain this primary data, however, I had to go to their graphs to read off the measurements, the preferred redshifts. It can be seen in Figure 6 of the present paper that two of the preferred redshifts agree with quasar peaks. The peak at $z = 0.3$ for galaxies is also conspicuous by casual inspection wherever one looks at galaxy redshifts in the literature. This is as it should be because quasars and galaxies are extragalactic objects which are continuous in their physical properties. What this agreement does, then, is give an independent confirmation of two of the quasar quantization peaks. It also demonstrates that the whole realm of measured redshifts—from the smallest to the largest—is quantized. Redshift is a physical property which comes in discrete, periodic quantities from the highest redshift, large periodicities to the lowest redshift, small periodicities. This property is schematically illustrated in Figure 6.

LARGE SCALE QUANTUM MECHANICS

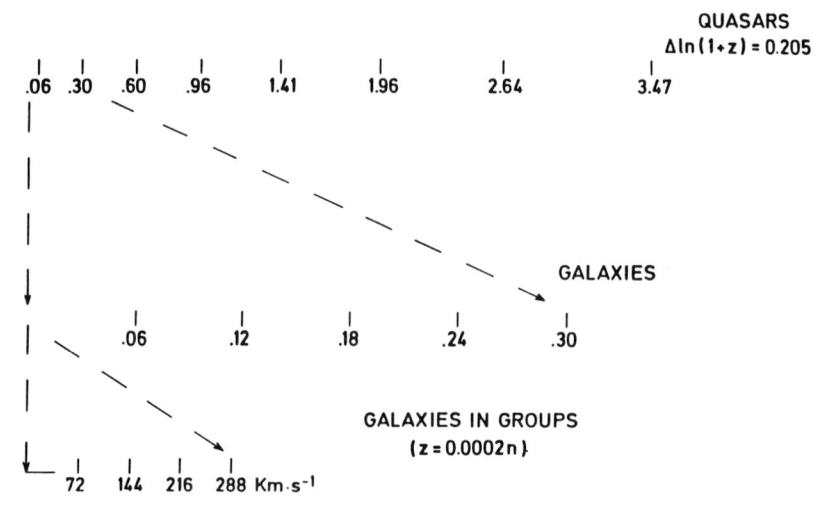

Figure 6. Periodized redshifts: Redshifts from the highest (quasars) to the lowest (differential redshifts between galaxies in groups) show preferred, *i.e.* quantized values of redshift.

Quasars: $\Delta \ln(1+z) = 0.205$ at .06, .30, .60, .96, 1.41, 1.96, 2.64, 3.47

Galaxies: .06, .12, .18, .24, .30

Galaxies in Groups ($z = 0.0002n$): 72, 144, 216, 288 km·s^{-1}

As for the low redshift, small periodicities, it is very important to note that analysis, first by Tifft and then by Arp showed that the redshifts of the galaxies in the Local Group were quantized in 72 km s^{-1} steps. In the Arp (1986) analysis of the redshift, the periodicity obtained was 72.4 km s^{-1}. Now four intervals of this period would be $4 \times 72.4 = 289.6$ km s^{-1}. But, astonishingly enough, the galaxy just added to the Local Group, IC342, has the largest excess redshift with respect to M31 and it is observed to be 289 km s^{-1}. In general, hydrogen redshifts are accurate to ± 8 km s^{-1}, but since IC342 is so large and nearby, its HI redshift is probably much more accurate, so the agreement with the predicted periodicity is even more impressive.

Finally, the evidence obtained by W. Napier and B. Guthrie on the smallest quantization interval of ≈ 37.5 km s^{-1} as reported by Napier in these proceedings is the most powerful of all the quantization evidence. In all directions in the sky up to redshifts of 2600 km s^{-1} he finds galaxies quantized in this small interval of redshift with enormous significance. The most important consequence of the observation is that real velocities of galaxies cannot average much more than about 20 km s^{-1} because, projected at random angles to the radial line of sight, they would wash out the 37 km s^{-1} quantization.

Again, we have evidence that extragalactic redshifts have negligible connection with velocities, and that we must look for a cause of intrinsic redshift which is physically related to the age of the object and which can be quantized.

Intrinsic Redshifts in Stars

If galaxies in our Local Group have excess, non-velocity redshifts, we should be able to ask whether some parts of these resolved galaxies show more excess redshift than others. In fact, we have predicted that excess redshift is correlated with younger age, so it is appropriate to look at the youngest stars in these nearby galaxies. This is easy to do because the youngest stars are the highest luminosity supergiants. (After a few million years they burn away to become much fainter stars.)

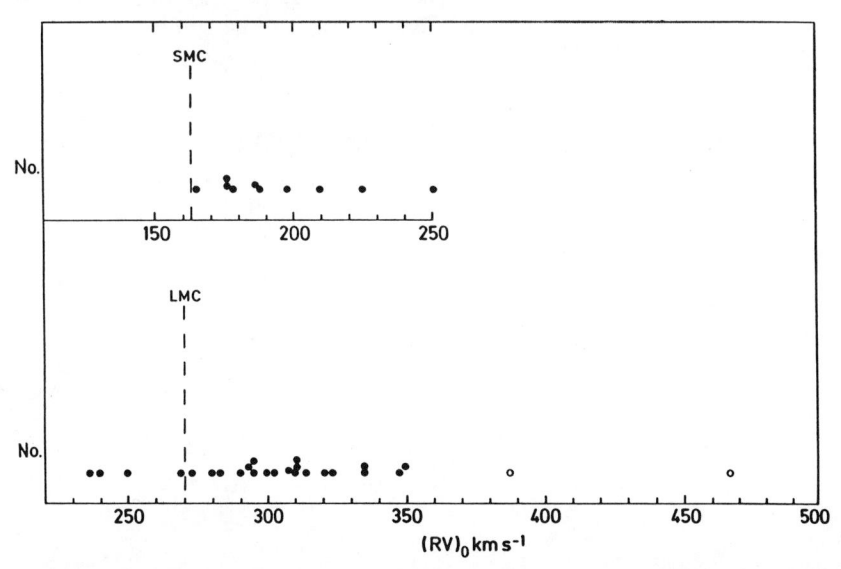

Figure 7. After correction for velocity of mass loss in each individual star, this best measured sample in the Magellanic clouds shows almost every supergiant has redshift higher than the mean redshift of its galaxy.

In the Large and Small Magellanic Cloud (LMC and SMC) individual supergiants have been well observed by now. Just looking at their redshifts shows that they are indeed systematically higher than the mean of the rest of the material in the galaxy. In fact, the confidence level of just this first order result is at the 99.8% level.

But now we must realize that supergiant stars have tenuous atmospheres in which the radiation pressure from the star propels mass outward in a so called "mass loss wind." This is standard knowledge in astrophysics and this negative velocity coming toward the observer in the absorption lines which overlie the bright photosphere must be corrected for in order to get the true systemic spectral shift of the star. In practice, the effect has been ignored because it makes the excess redshift of young stars even more excessive.

If we correct the reported spectral shifts (always incorrectly called radial velocities in the literature) for these negative outflow velocities, we find that *almost every supergiant star in both Magellanic Clouds has an excess redshift* relative to the well determined mean spectral shift of the remaining stars and gas in the galaxy. This is shown here in Figure 7.

Does this result hold for the nearest galaxy of all, our own Milky Way system? Indeed it does, and it has been known since 1911! But the hot, young stars in our own galaxy cannot all be exploding away from us in every direction we look. If this empirical evidence had been heeded, the assumption that redshifts meant only velocity would never have been made, and the conclusion that the universe was expanding would not have been promulgated to the public for more than 60 years.

How good was the contradictory evidence during all these years? After W.W. Campbell announced the "K effect" in 1911 from the first systematic spectroscopic measures of bright stars, the effect has been confirmed over and over again. A few sample determinations are shown in Table 2 here (Table excerpted from Arp 1992b).

Just to illustrate the effect in one case, I show here in Figure 8 the measured spectral shifts for the supergiants in the largest, most conspicuous young star clusters in our own Milky Way Galaxy, h + χ Persei. It is clearly evident that the excess

Table 2. Excess Redshift (K-Trumpler effect) in our Galaxy

Stars	$c\Delta z$	Reference
B stars	+4.7 ± 0.02	Smart and Green 1936
OB stars in Gould's Belt	+ 5	See Frogel and Stothers 1977
O stars in Clusters (re B's)	+10 ± 1	Trumpler (1935; 1956)
B stars in Orion	+11.4 ± .2	Findlay-Freundlich 1954
O stars in Clusters	+17.6 ± .5	
Supergiants in h + χ Per	+15	Present paper
Supergiants in Assn's	+15	
Single O stars	+0.6 ± 1	Conti *et al.* 1977
O-type binaries	+8.5 ± 1	
O-type binaries	+2.0	Gies 1987
Cluster/assn's O's	+1.8	
Field O's	+6.4	
Runaway O's	+21.9	

For references, see Arp (1992b).

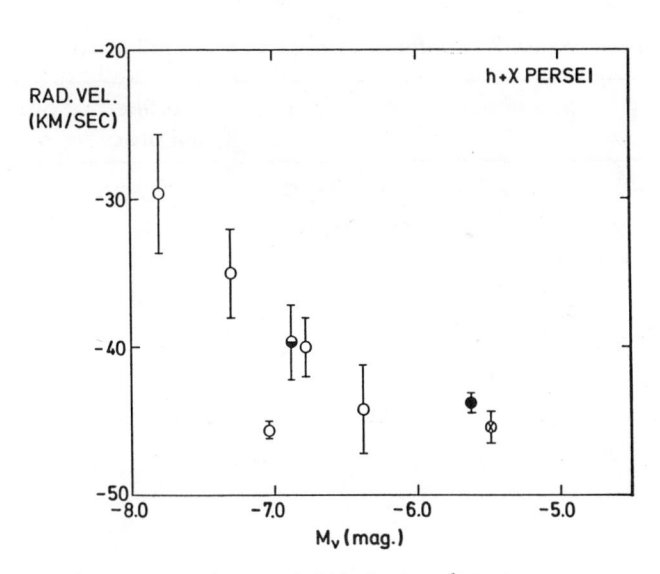

Figure 8. The points show measured spectral shifts (in km s^{-1}) in the supergiants in h + χ Persei, the largest, young star clusters in our own Milky Way Galaxy. The figure demonstrates the increasing redshift as the stars become more luminous (hence evolutionarily younger).

redshift increases with increasing luminosity for stars in the same physical cluster. Of course, the increasing luminosity is quantitatively calculable in terms of the younger evolutionary age of the stars*. So, again, we have the empirical result that the excess redshift increases as the age of the material decreases.

Roberta Humphries, whose data was plotted in Figure 8, was asked to comment on the excess redshift effect which it showed. She never answered. The detailed, documented analysis of all evidence for systematic redshift in young stars was submitted to the European Journal, *Astronomy and Astrophysics*. The editor, James Leqeux, himself an expert on bright stars in the Magellanic Cloud, forwarded a derogatory referee's report and refused to publish the data. He also did not allow the results to be orally presented at a nearby meeting specifically concerned with the Magellanic Clouds.

After some years' delay, however, the publication of the detailed analysis has now been entered in the record (Arp 1992b). The detailed data which shows that the most luminous stars in the three galaxies we know the most about—the Milky Way, Large and Small Magellanic Clouds—have excess intrinsic redshifts can now be referred to if anyone so desires. Naturally, we would like now to extend this a little further and ask what is revealed by the next nearest galaxies, where we can still tell

---

* Note on the ages of stars: Evolutionary age means time since the star ignited after, on the conventional view, collapsing from an extended gas cloud. The conventional assumption is that the prestellar material was all created at the same time. But if some material were created more recently, stars made out of the newer material by the usual process would be "younger" than the rest. Stars from the older material would have burned through the supergiant phase, on the average, already, and the brightest stars would, therefore, be preferentially from the younger material and, on our interpretation, of systematically higher redshift.

Table 3. Excess Redshifts of Luminous Stars in Nearby Galaxies

Galaxy	Kind of Stars	K-effect	Mass loss correction	Total excess redshift $c\Delta z$
Milky Way	h + χ Pers.	+(7) km s^{-1}	+29 km s^{-1}	(+36) km s^{-1}
LMC	supergiants	7	+22	+29 ± 6
SMC	"	17	+17	+34 ± 8
NGC 1569	A		—	>+36 ± 17
"	B		—	>+35 ± 22
NGC 277	early integ. spectrum		—	>+21 ± 8
NGC 4399	"		—	>+25 ± 15
M31	irreg. blue vars.		—	(+100)
M33	"		—	>+21

something about individual stars. The results of this investigation are given here in Table 3.

It is striking to note that even though we cannot make the corrections for mass loss in these more distant systems, the measured excess redshift is significantly positive and comparable to the amount measured in the three nearest galaxies. A pertinent comment is that stars with chemical compositions which contain fewer metals feel less radiation pressure and, therefore, have lower velocity mass-loss winds. This can be seen empirically in the fact that the SMC has a smaller mass-loss wind correction than the Milky Way and LMC. The four systems immediately following the SMC in Table 3 appear young, thereby, presumably, suffering less metal enrichment, and may turn out to have rather lower mass loss corrections when they are eventually measured. But they all, as they stand, represent lower limits to the total excess redshifts of the younger stars.

At the end of Table 3 are listed irregular blue variables in M31 and M33. These are very luminous, blue stars which have been known since Hubble's time. The absorption line redshifts (again from Roberta Humphries) are used to derive the excesses in Table 3. It is astonishing that such large redshift discrepancies have gone unremarked for such a long time. The redshifts used in Table 3 are absorption line redshifts and, therefore, have an unknown component of negative mass loss velocity in them. Spectral shifts derived from emission lines, however, should refer to regions outside the line of sight to the photosphere and be much less prone to approaching wind corrections. It is of great interest, then, that Dr. Otmar Stahl, Heidelberg, privately communicated to me his emission line spectral shifts for three of these irregular blue variables.

Table 4. Emission Line Redshifts of Some Irregular Blue Supergiant Variables

star	cz (emission)	Δz (excess redshifts)
M31 AE	−217 km s^{-1}	+80 km s^{-1}
M31 AF	−282	+15
M33 Var C	−117	+63

Again, the star average is surprisingly large compared to their galaxies. (They are not corrected for rotation of the galaxy to which they belong so we can only rely on their average.) In the case of M31, however, we should remember that M31 has an intrinsic blue shift of about -86 km s^{-1} when viewed from our own galaxy. Therefore, an M31 star of an age equal to the mean age of one of our own stars would have immediately a +86 km s^{-1} excess with respect to M31.

When all the measures that should be done on these systems are finally done, one could reasonably expect that, with the exception of M31, the excess redshift of the luminous stars would come out near 37 km s^{-1}. That is the empirical suggestion of the values so far in Table 3. This is an important number because, as the discussion earlier on quantized redshifts showed, the most significant, empirical peak in redshifts is close to 37 km s^{-1}. The appearance of this quantization value in galaxies reinforces the finding of the same period in stars. If we say that stars are created periodically at intervals of 3×10^6 years, then their redshifts should group 35 km s^{-1} apart (see later discussion on redshift-age formula).

The Case for Intrinsic Redshift-Age Relation

So far we have investigated the closest galaxies to us and the brightest stars in them. We have found incontrovertible evidence that very little of the galaxy redshift can be due to velocity. We have found that the intrinsic redshift of the galaxy increases as the proportion of young stars it contains increases. Then we have found the youngest, most luminous stars themselves, individually, have intrinsic redshifts which increase generally with the relative youth of the stars. These represent so many independent kinds of determinations on so many independent systems that there would seem to be scarcely any reasonable doubt left that a major cause of extragalactic redshifts is intrinsic and age-related. At this point, before we see whether such observations could be incorporated into the laws of physics as we presently understand them, we should say a few words about tired light.

Tired Light Mechanisms

The plausible arguments that photons on their voyage through space must lose at least some of their energy and suffer a redshift has been made by so many people for so long that I will not attempt to review the subject here. But I will argue the case that the *dominant* cause of systematic redshift we see in extragalactic objects must be due to some intrinsic property of the material in the object and not to intervening material.

The argument is simply that there are cases of high redshift galaxies and quasars seen interacting with low redshift galaxies (Arp 1987). Because both objects are at the same distance they are seen through the same optical path length and, therefore, their redshift difference must be intrinsic. Even if one postulates a special interacting medium around the high redshift object, one should see gradients of redshifts and occultation effects when different redshift objects intermingle (Arp 1990). Observationally, however, one always sees just discrete sets of redshifts.

In the end, we may eventually be able to observe tired light effects as a higher order effect once the large discordances, which, I believe, are caused by age differences, are corrected for. The most straightforward way of doing this would be to study the redshift behavior of similar kinds of objects as they are viewed through increasing amounts of galactic material at lower and lower galactic latitudes. Multiparameter analyses of catalogued redshifts, apparent magnitudes and diameters and object types with galactic and supergalactic latitudes should be able to demonstrate effects of tired light or set quantitative upper limits on the mechanism.

A Theoretical Basis for a Redshift-Age Formula

If we consider a homogeneous, Euclidean universe in which the mass of a subatomic particle, m_p, varies as the amount of "gravitons" it can exchange with particles within its light signal sphere, then:

$$m_p \propto \int_0^{ct} \frac{4\pi r^2 dr}{r} \propto t^2 \tag{1}$$

Since the energy of a photon emitted from an atom varies as the mass of the electron making the orbital jumps, then the wavelength λ varies inversely as the electron mass:

$$\lambda \propto m_e^{-1} \tag{2}$$

and the redshift, z, varies as:

$$\frac{1+z_1}{1+z_0} = \frac{t_0^2}{t_1^2} \tag{3}$$

where z_0 is the redshift of matter created t_0 years ago and z_1 is the redshift of matter created t_1 years ago.

From this basic formula (3), the redshift of any galaxy or quasar can be calculated from its age since creation (Arp 1991).

The fundamental step in this derivation is the assumption that inertial mass is induced by interaction with all other visible masses as a $1/r$ law. Qualitatively this is Machian physics. But does this satisfy Einsteinian physics? Following Narlikar (1977) we can write the GR field equations as:

$$\frac{1}{2}m^2\left(R_{ik} - \frac{1}{2}g_{ik}R\right) = -3T_{ik} + m\left(\Box m g_{ik} - m_{;ik}\right)$$
$$+ 2\left(m_{,i}m_{,k} - \frac{1}{4}m^{-1}m_{,l}g_{ik}\right) \tag{4}$$
$$\Box m + \frac{1}{6}Rm = N$$

We see straight away that a solution for these equations in flat space time is given by the Minkowski metric with the mass function:

$$m = at^2 \quad a = \text{constant} \tag{5}$$

Now, there is a great advantage of this solution over the usual solution. The usual solution is due to A. Friedmann in 1922, who assumed the mass of elementary particles was constant, i.e. $m \neq m(r,t)$. This assumption was natural in terrestrial laboratories and perhaps fine for the local solar system. But did it make any sense cosmologically? If there was only one particle in the universe what would its mass be? In any case the solution in which mass could vary as a function of spacetime coordinates is much more general and, therefore, to be preferred. Scientifically, it is much better to find the general solution first and later make simplifying assumptions if conditions warrant.

The Friedmann solution, of course, led to the expanding universe, as embraced by Einstein, Eddington, Le Maître et al., where the radius must change as a function of time. But our solution leads to a nonexpanding universe where, for galaxies born at the same time, the look-back time to distant galaxies reveals them at a time when they were younger than our own galaxy Therefore, they appear to have an intrinsic redshift which is proportional to their distance (Narlikar and Arp 1992). This gives an *exact* Hubble law—much better than the expanding universe solution in which currently supposed peculiar velocities could and should destroy the tightness of the observed redshift-distance relation. Therefore, our more general, non-expanding solution gives a better fit to the fundamental Hubble relation which originally gave rise to the conventional Big Bang interpretation.

In another important aspect our more general solution is vastly preferable, *viz.* the singular points in spacetime where physics completely breaks down in the usual $m = $ const. treatment of the field equations. For our $m \propto t^2$ solution the particle mass must pass through zero. There the $m = 0$ hypersurfaces, Kembhavi (1978) showed, are just the old spacetime singularities that embarrass the Friedmann solution. Moreover $m = 0$ just at $t = 0$, is the natural creation point for the mass particles. We no longer have an unexplained, *ad hoc* creation for all matter at an arbitrary instant as in the Big Bang—instead we have continuous creation possible throughout spacetime. Finally, when passing through $m = 0$ we pass from $m^2 < 0$ to $m^2 > 0$, that is, from the quantum mechanical realm to the classical physics realm (Khlatnikov 1992). We will comment later on the conclusion from the observation that the quantization in redshifts could only be accounted for by imprinting at the $m = 0$, quantum mechanical domain of the creation point.

Quantitative Predictions of the New Theory

From the standpoint of the universe, if we could watch the history of a galaxy run backward in time, we should see the masses of its constituent particles diminish as it approached its origin. Now the rate at which atomic time runs is dependent on the mass of its particles (m_p). Since m was smaller in the past in this galaxy, time ran slower. The amount of universal time (t) elapsed in an interval of the galaxy's time (τ) is:

$$\tau = \frac{t^3}{3t_0^2} \tag{6}$$

At the origin, $t = t_0$ and $\tau = \tau_0$, giving:

$$\tau_0 = \frac{t_0}{3} \qquad (7)$$

Differentiating our redshift-age formula of (3) with respect to Δt yields:

$$H_0 = \frac{2}{3\tau_0} \qquad (8)$$

where H is the Hubble constant at $z = 0$.

Details of this derivation can be consulted in Arp (1991). But the upshot is that for a flat spacetime, homogeneous universe (the most reasonable starting approximation) we obtain the same expression for the Hubble constant that was obtained in the conventional solution of the equations of general relativity. That is not surprising, since our solution with $m \propto t^2$ is simply a conformal transformation of the usual solution—i.e. it is mathematically the same—only the physical interpretation is completely different.

We can then compute what the Hubble constant, H_0, must be from the age of our galaxy (τ_0). As customary, the age of the oldest stars is assumed equal to the age of our galaxy. We obtain:

$17 > \tau_0 > 13 \times 10^9$ yrs.   observed age of galaxy
$39 < H_0 < 51$ km s^{-1} Mpc^{-1} H_0 predicted by (8)
$42 < H_0 < 56$ km s^{-1} Mpc^{-1} H_0 observed

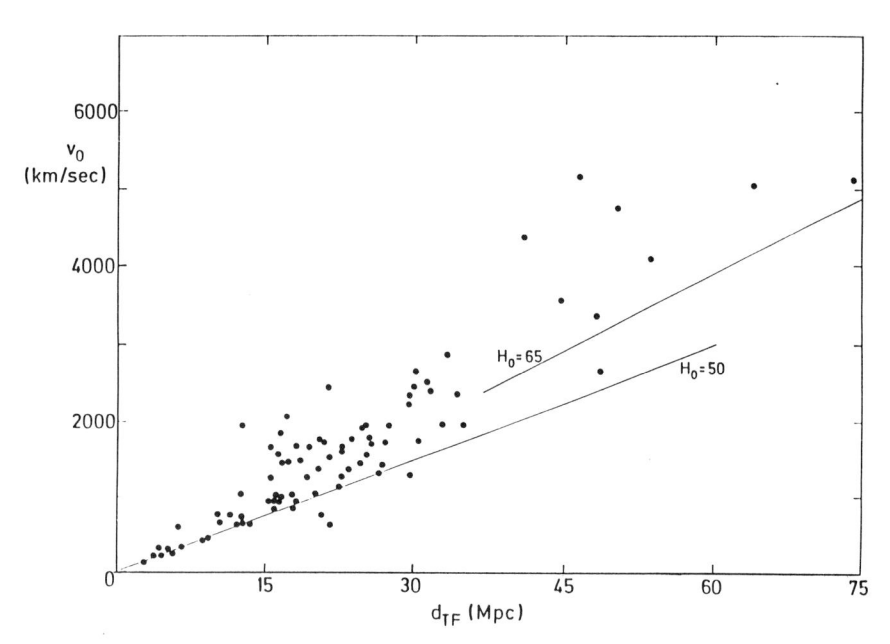

Figure 9. Distances (d_{TF}) from Tully-Fisher estimates of rotational mass, luminosity and apparent magnitude of spiral galaxies are plotted against measured redshift (in km s^{-1}). The redshift-distance relation for low redshift galaxies is accurately $H_0 = 50$. At higher redshifts younger galaxies are encountered which have intrinsic redshifts and give higher values of H_0.

The astonishing result is that the application of equation (3), required for galaxies whose intrinsic redshifts are a function of their age, leads to a Hubble constant which is, within the observational uncertainties, numerically equal to that observed. The advantage of our variable mass solution over the expanding universe solution is, however, that our formula predicts the observed Hubble constant with only one observational parameter and has no possibility of adjusting the predicted value, as the Big Bang does, by changing the geometry of the universe or introducing repulsive cosmological constants.

Discordant Values of the Hubble Constant

It is well known that a fierce struggle has been going on for many years between advocates of the $H_0 \approx 50$ and $H_0 \approx 80$ to 100. The evidence on both sides is so persuasive that one might suspect that both sides are correct, observationally speaking. How could this be?

Figure 9 shows the best redshift-distance relation available for spiral galaxies. Here the masses are estimated from rotation speeds of the galaxies, luminosities estimated from the masses and the distances estimated from the difference between apparent magnitude and absolute magnitude. These "Tully-Fisher" distances are, therefore, somewhat independent of the systemic redshifts of the galaxies to which they are being correlated. One sees immediately that nearby (say inside the distance of the Local Super Cluster, at about 15 Mpc from our Local Group of Galaxies), rather accurately, $H_0 = 50$. But at greater distances one derives greater H_0's.

In the Big Bang there is supposed to be only one expansion speed over such relatively small regions of the universe. Obviously, Figure 9 shows this is not true. What our redshift-age, equation (3), requires, however, is that all galaxies the same age as our own should give a linear redshift-distance relation with $H_0 = 50$ km s^{-1} Mpc^{-1}. This is true, as the measures in Figure 9 show, out to a distance of about 15 Mpc from our Local Group. Beyond this distance, however, if there are galaxies younger than our own, they will have excess redshifts; and this is exactly what is shown by the higher z spirals.

So we conclude that the faction claiming $H_0 \approx 80$ has more accurate distances and redshifts to galaxies in a larger volume of space than the side which claims 50 throughout. But both sides refuse to acknowledge that in reality, the value of H_0 increases from low redshift samples to samples which include high redshifts. Actually one can get any H_0 one wishes by choosing a particular kind of galaxy to measure it with. For example, in our Local Group IC 342 would yield $H_0 = 161$. The only solution to this absurd situation is the one given by our general solution to the Einstein field equations where the universe is not expanding and the redshift reflects the relative age of the galaxy we are viewing.

H_o and the Virgo Cluster

A key object in the determination of the value of the Hubble constant, and, therefore, a bone of contention between both sides, is the mean redshift of the Virgo

Cluster at its correct distance from us. Table 5 below is extracted from Arp (1988), and shows that in all three of the relevant parameters, the two sides have adopted values which either maximize or minimize their resultant H_0.

Table 5. Determination of H_0 from the Virgo Cluster

Mean redshift	Infall to Virgo	Distance	H_o	Author
967 km s^{-1}	220 km s^{-1}	21.6 Mpc	55	Tammann and Sandage (1985)
1165	250	15.0*	94	de Vaucouleurs (1982)

* Distance to S cloud, on de Vaucouleurs precepts the E "cloud" would be closer.

Rigorously, even if one believes redshifts should be interpreted as velocities, the above adopted mean redshifts are both incorrect. That is because the mean redshift should refer to the redshift of the average *mass* in the Cluster. The luminosity weighted mean is only 863 km s^{-1}. (This is still an overestimate because it is derived from the blue luminosity, not the red luminosity). What the conventional procedure does is overweight the galaxies of small or negligible mass in deriving the mean redshift. Naturally, this yields too high a redshift. It is the same result we got in 1970, as reviewed in the beginning of this paper; the companion galaxies have systematically higher redshifts than the dominant galaxies. The size of the effect is even the same, $100 < \Delta cz < 300$ km s^{-1}.

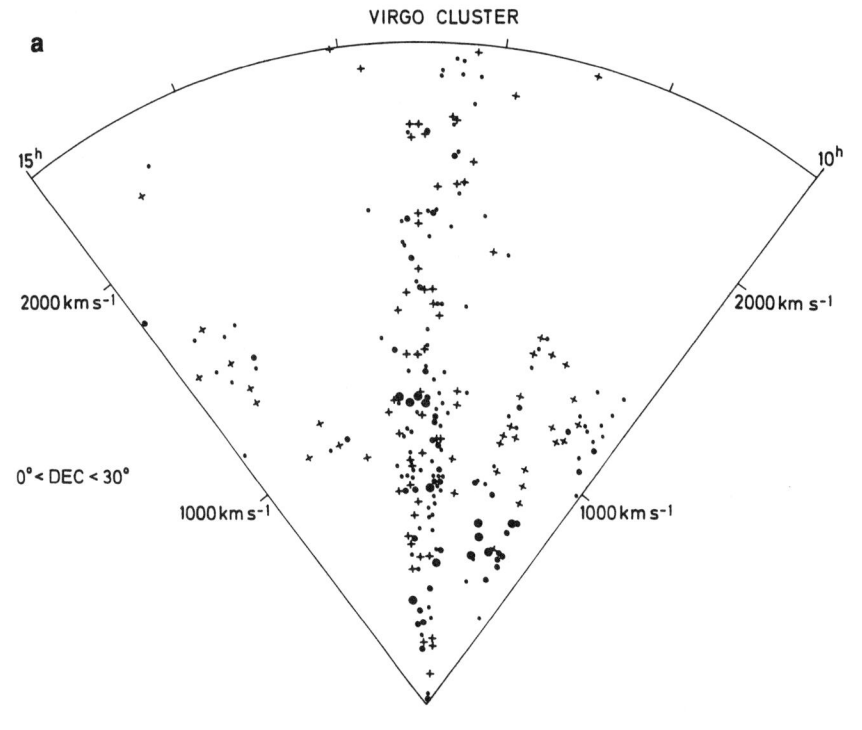

Figure 10a. The distribution of redshifts in the Virgo direction in the sky. Data from revised Shapley Ames Catalog (Sandage & Tamman 1981). Size of symbol according to apparent brightness of galaxy.

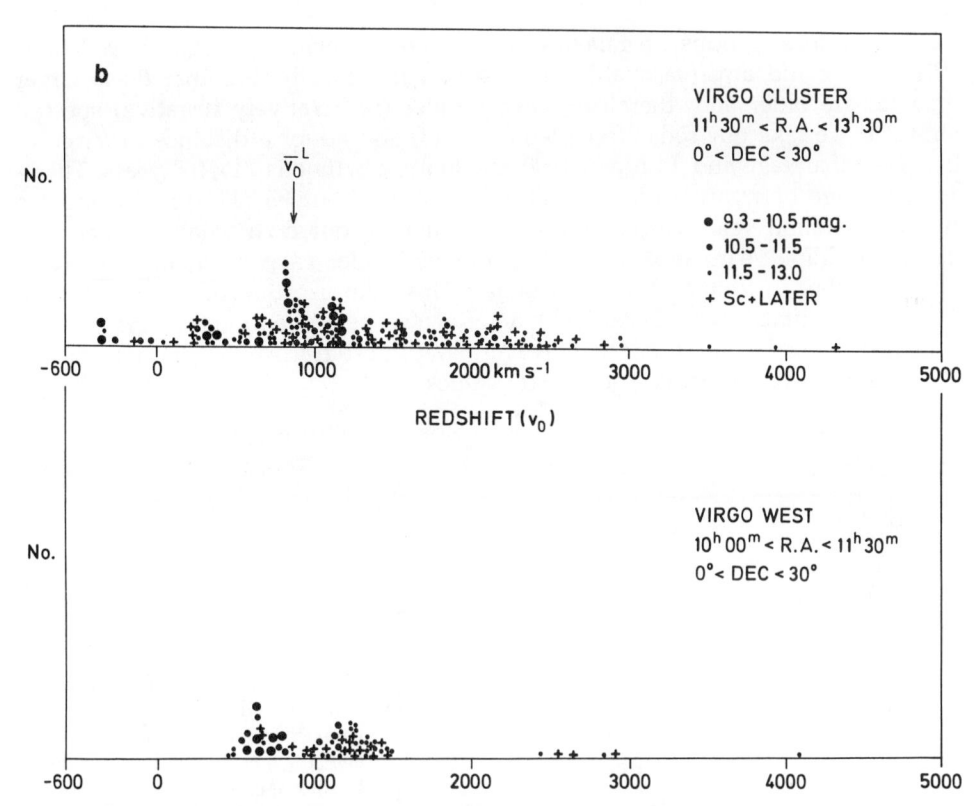

Figure 10b. Redshift distribution of galaxies in main Virgo Cluster and Virgo West. The blue luminosity weighted mean redshift is shown by the arrow (from Arp 1988). This demonstrates that fainter galaxies in the "Finger of God" configurations represent galaxies with intrinsic redshifts.

Of course, the situation is even much worse than that. Take any given class of galaxy in the Virgo Cluster, say the S_0's. One could obtain any mean redshift, from a few hundred km s^{-1} to almost 2000 km s^{-1}, depending on whether bright or faint Virgo members were selected (see Arp 1988, Figure 2).

An illustration of this chaos in the determination of H_0 can be seen in a recent paper by Sandage (1992). In this paper he claims that by using a supposedly very luminous type of spiral galaxy called an ScI, the very low value of $H_0 = 43$ km s^{-1} Mpc^{-1} is derived. (Actually this type of galaxy has been shown to be the least luminous kind of spiral galaxy: Arp 1988b, 1990c, 1991b). But in Sandage's Table 1 he lists "The Four Largest Virgo Cluster ScI Spirals." One can easily write down the measured redshifts as taken from Sandage's own *Revised Shapley Ames Catalog* and derive a mean redshift of $v_0 = 1747$ km s^{-1}. Using Sandage's own infall velocity to Virgo of 220 km s^{-1} and Sandage's own distance to Virgo of 21.6 Mpc one quickly calculates $H_0 = 91$ km s^{-1} for the ScI's which Sandage says are members of the Virgo Cluster! This is a far cry from $H_0 = 43$ km s^{-1}, but the higher figure is clearly the more correct one for these younger, intrinsically redshifted kinds of galaxies.

The only negative redshifts in the sky outside our Local Group are seven in number, and fall in the center of this populous cluster. Hence they must be members. But they are predominantly big S_a and S_b galaxies, the same kind that dominate more local groups of galaxies and also have relatively negative redshifts. What else could any reasonable judgment of the data be but that these larger galaxies are older and, therefore, have smaller (*i.e.* relatively negative) spectral shifts? If we take our redshift-age equation (3) and adopt a distance to Virgo of 21.9 Mpc (Sandage and Tamman 1990) the look-back time is 71×10^6 years. Therefore, the age of creation of these Virgo giants is about 95×10^6 yrs, or about a hundred million years earlier than the creation of our own galaxy, which was about 15 billion years ago. As we might expect, older galaxies created relatively close together in time look rather similar. This interpretation of the redshifts in Virgo is quantitatively consistent with an $H_0 \approx 50$. The conventional analyses which give both $H_0 = 50$ and $H_0 = 80$, however, simply arbitrarily pick the parameters which give their respective "observed" values.

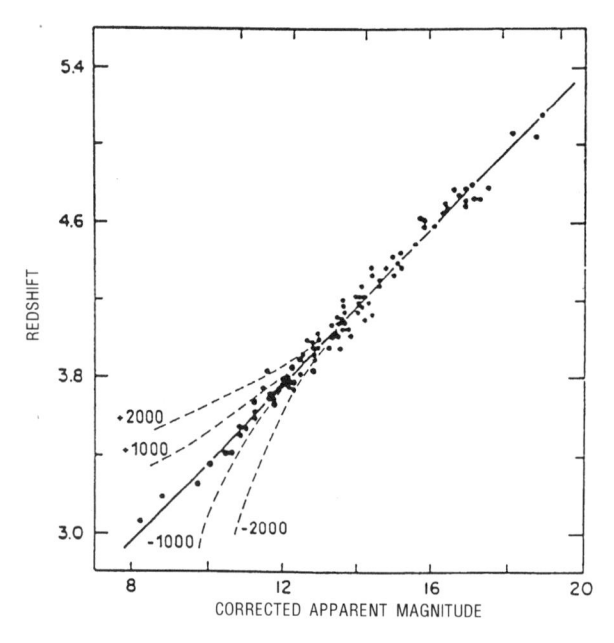

Figure 11. The redshift-apparent magnitude diagram for clusters of galaxies as measured by A. Sandage (1975) is shown. Peculiar velocities of clusters are reported $\simeq 2000$ kms^{-1} and should have dispersed the lower 1/3 of the Hubble diagram as shown by the dashed lines in the figure.

Another way of illustrating this problem is shown in Figure 10. If we use the redshifts as a measure of distance in the Virgo region of the sky, the galaxies all stretch out in a line just in the direction we are looking. This "Finger of God" stretches as far as we have redshifts (even to negative z's). Figures 10a and b are derived from data in the *Revised Shapley Ames Catalog* (Sandage and Tamman 1981) which is a complete sample to a limiting magnitude near $B_{ap} \approx 13.0$ mag. Essentially the same diagram is obtained from deeper surveys (Zucca *et al.* 1991). As can be seen from both Figures 10a and b, however, the brightest (most massive) galaxies are not at the mean redshift of the Virgo Cluster, as they must be on conventional assumptions! The fainter galaxies, therefore, cannot represent peculiar orbital velocities around the massive galaxies. Instead, as in the nearby groups, the smaller galaxies have excess, intrinsic redshifts of varying amounts.

If researchers plotted the wedge diagrams with galaxy symbols proportional to the brightness of the galaxy, as in Figure 10, they would immediately see the velocity interpretation was untenable. Even as they are customarily plotted, without brightness differentiation, one should see a thickening or "knot" at the true redshift of the cluster representing the massive galaxies at the centre, plus orbiting galaxies far out and moving slowly, plus orbits transverse to the line or sight which do not give appreciable radial peculiar velocity. Instead, the diagrams show the galaxies evenly spread out along the "Fingers of God". These "Fingers of God" show up all over the sky whenever there are concentrations of galaxies, and should tell astronomers, at a glance, that velocity differences cannot be responsible for the observed effect.

The Virgo Cluster is just like every other cluster that has been studied in detail in that each type of galaxy within the cluster has its own systematic redshift value. The younger Sc's have the highest. Of course the first discovered, most famous, brightest apparent magnitude quasar, 3C273, also falls in the projected confines of the Virgo Cluster. Recent spectra of 3C273 in the ultraviolet show about 10 times more absorbing clouds than normal between the redshift of 3C273 and the Virgo Cluster. Are absorbing clouds just accidentally piled up in excess just in the line of sight behind the Virgo Cluster? Or should we instead believe the numerous independent studies which show quasars of various redshifts are actually physically associated with the Virgo Cluster (Arp 1992a)?

The Hubble Relation at Larger Redshift

Figure 11 shows an adaptation of the famous redshift-apparent magnitude diagram for clusters of galaxies as reported by A. Sandage (1975). It is often stated that the relation is so tight that it precludes any other explanation than an expanding universe. But, in fact, the relation is much too tight. Various observers have reported peculiar velocities of clusters and groups of galaxies of 2000 km s^{-1} or more (see Narlikar and Arp 1992). The dashed lines in Figure 11 show how the existence of real velocities of this size would blow up the lower third of the pictured Hubble relation. How can the observational conclusions of these two conventional analyses be reconciled? Only if the dispersion observed in redshifts of clusters is not velocity but age differences and, further, if Sandage has measured only clusters of galaxies which have similar age characteristics.

It is interesting to contrast the claims of the observers who report peculiar velocities of galaxies exceeding 2000 km s^{-1} with the published results of Yahil, Sandage and Tamman (1980) that the Hubble flow is probably quieter than 50 km s^{-1}. Those claims too can only be reconciled if spectral shifts generally do not indicate velocity differences.

Evolution Away from the Hubble Relation at High z

Spinrad and Djorgovski (1987) report measures of radio galaxies which deviate from the Hubble relation by 5–6 mags at $z \approx 1.5$. This is conventionally attributed to evolution, but it requires these galaxies to be 100–240 times brighter in the past than at present. Naturally this requires "star bursts" of unprecedented scale

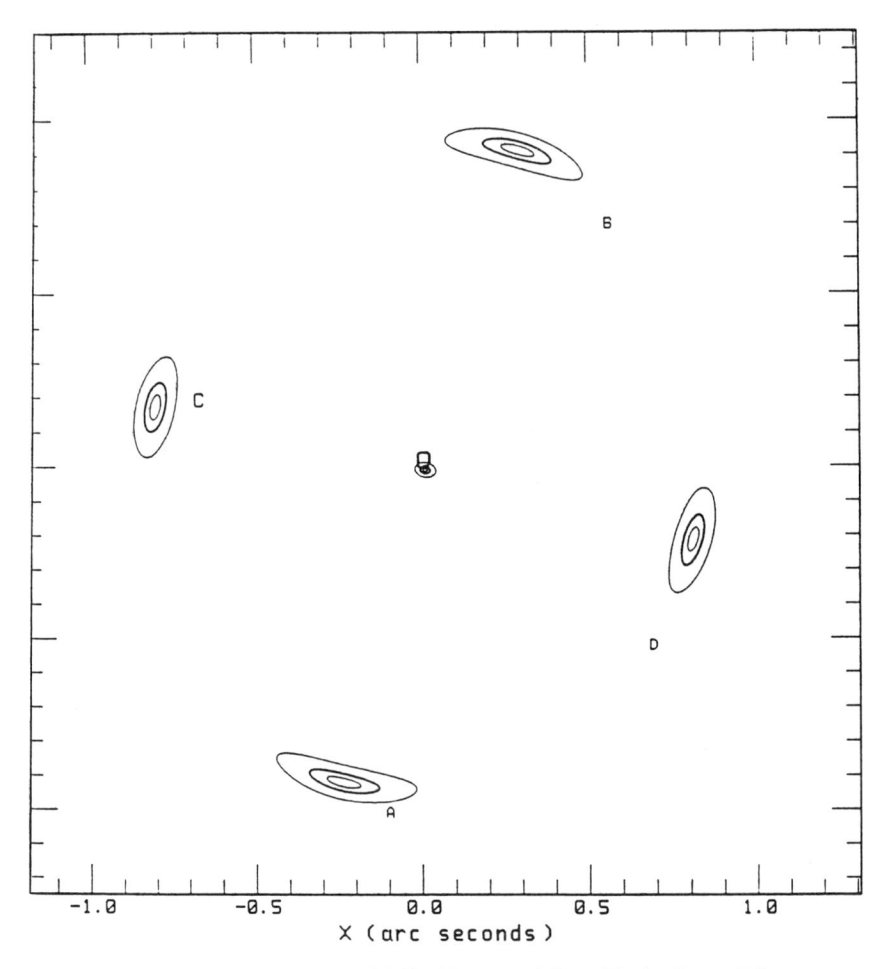

Figure 12. High resolution images in redshifted Lyman alpha with the Space Telescope are shown for the system of four quasars surrounding the low redshift spiral galaxy G2237 + 0305. The lower diagram shows that the predictions of gravitational lens theory are exactly opposite to the observations. The observations indicate physical ejection from the galaxy (See Arp and Crane 1992).

and would make it necessary to observe hydrogen dominated precursor galaxies which have not been seen. The observational fact that these precursor galaxies are not seen means that apparently young galaxies really are young in the sense that they are more recently created. This observation, by itself, then rules out the Big Bang, where all galaxies are supposed to have been formed in the beginning.

If, however, these active galaxies and the material in them have been created more recently, we would expect by our precepts to have them deviate to higher redshift from the Hubble line. In this respect, the radio galaxies should show a deviation due to intrinsic redshift intermediate between quasars and normal galaxies. This would agree with their generally intermediate physical properties.

More normal E galaxies, however, can be measured out to redshifts $z \approx 1$. For observations in the infrared where young stars hardly affect the magnitude, we see deviations of about 2 mag. brightward from an unevolved Hubble line of $q_0 = 0$. It is interesting to note, however, that our predicted value of H_0 pertains only locally, for $z \to 0$. At $z = 1$ we predict $H = 2.8 H_0$ (see Narlikar and Arp 1992). If this were interpreted as a deviation from the Hubble relation in an expanding universe, it would require a normal galaxy to be 2.3 mag. more luminous in the past. But, in fact, as we see in the conventional analyses of the evolution of stellar assemblages, this is just about the 2 mag. deviation from the Hubble line which is required for normal E galaxies. The point is that the additional epicycle of systematic evolution which is needed in the Big Bang theory to reconcile theory with observations is not needed in the flat spacetime, continuous creation cosmology discussed here.

High Redshift Quasars Associated with Nearby Galaxies

The evidence that high redshift quasars, which on the Big Bang hypothesis should be out at the edge of the universe, are, in fact, associated with nearby galaxies has been growing since 1966. Reviews of this extensive evidence are available in Arp (1987) and (1992a).

Of course, the quasars nearest in angular separation to the low redshift parent galaxy are the most disturbing to advocates of conventional distance criteria. One famous example is the bright apparent magnitude quasar Markarian 205 which is only 40" south of the extremely disturbed, radio ejecting galaxy NGC4319. In a study with the Hubble space telescope, Bahcall *et al.* (1992) found that the quasar had much less absorption in front of it than would be expected due to the intervening galaxy NGC4319 (about 10% of what was expected). They then concluded that this was "consistent with the cosmological interpretation." Of course, a more accurate statement would have been, "this is most consistent with the quasar being 90% in front of the galaxy."

X-ray extensions from the quasar to the nucleus of this active galaxy were uncovered in the Einstein satellite archives (Arp 1990b). After much resistance, the object has finally been assigned observation time with the German X-ray satellite, ROSAT.

But one of the associations that conventionalists could not ascribe to chance was the so-called "Einstein Cross", which consisted of four quasar images of

$z = 1.7$ all within one arc sec of the nucleus of a galaxy of redshift $z = 0.039$. It was explained by gravitational lensing and splitting of a fainter, accidentally aligned background quasar. By consulting the space telescope archives, however, Arp and Crane (1992) were able to add and image process all the best resolution images of the configuration. Figure 12 here shows that the predictions of the gravitational lens imaging are completely and orthogonally violated.

Instead there are luminous, filamentary connections back to the nucleus in just the fashion expected if the quasars arose from an ejection episode in the nucleus of the galaxy. It seems difficult to conceive of any more direct observational evidence against the gravitational lens model and for the ejection model.

Of course, this observational evidence from the world's most expensive telescope was refused publication in the journals which routinely publish space telescope data. We can comment here that gravitational lensing, which is used regularly as a *deus ex machina* to explain a host of discordant observations, depends on very massive galaxies to create the needed effects. We have seen earlier that galaxy masses have been customarily overestimated by one to two orders of magnitude. For realistic galaxy masses, gravitational lens effects may some day show up, but on a much smaller scale than currently claimed.

Gravitational lens interpretations which *are* rapidly published, however, claim the data confirms the standard theories. One example of such a paper is Rix, Schneider and Bahcall (1992). They compute that lensing in the Einstein Cross requires a mass of $1.1 \times 10^{10} M_\odot$ inside the very small radius of 0.29 where the quasar images are located. This computation is essentially no different from Howard Yee's (1988) analysis which derived a mass of $1.2 \times 10^{10} M_\odot$. As Yee stated, this leads to a mass-to-light ratio of $M/L \geq 13$, which "...is near the high end of that of large spiral galaxies... but is entirely acceptable." Well not quite, if you consult the measured M/L ratios for galaxies (Kormendy 1988) you see that this M/L is completely above bulges of spiral galaxies, even above that for ellipticals and that is for a *lower limit* on the required M/L. In fact, if you compute the luminosity required for an elliptical to have this M/L ratio it comes out $M_\beta \approx -25$ mag! How bright is $M_\beta \approx -25$ mag? Well, quasars are defined as starting at $M_\beta = -23$ mag. So this is a galaxy 2 magnitudes brighter than a quasar! And this is just a lower limit!!

This so-called confirmation of the lens theory by the observations rests on assumptions that the redshift dispersions in galaxy interiors represent both velocities and velocities in equilibrium, which, as we have seen, the observations show is incorrect on both counts and generally overestimates the masses in galaxies. Furthermore, the mass needed for the lensing is underestimated by assuming it is all concentrated at a point in the center of the galaxy—hence the lower limit. In addition, a disk galaxy is called an elliptical, and mass-to-light ratio comparisons are shuffled between blue, visual and red. Even after all this, the result is that one requires an extraordinary and unprecedented galaxy to satisfy the lens requirements.

This is called a rigorous and exact confirmation of the lens predictions, and it is there for all to read in the professional journals. Direct images of the object with the best resolution telescope available directly contradict the predictions of lens theory, but are refused publication in these same journals. It would seem that if the theory had not failed, there would be no need to censor the contradictory data.

Cosmic Background Radiation

In April 1992, enormous publicity was given to the announcement that a satellite observing in the microwave region of the energy spectrum had detected irregularities in the sky. These were interpreted as "fingerprints of primordial galaxy formation" and said to have (once again) proved the correctness of the Big Bang theory.

The claim of "proof" of Big Bang was later severely criticized by rival establishment astronomers. Nevertheless, there was never any discussion of how the evidence actually is very difficult to reconcile with a Big Bang model. The point is that in a universe expanding faster at each further distance observed, the 2.7° K black body energy curve would be smeared out unrecognizably by Doppler recession velocities. Therefore, one must only observe a single, thin shell. In the Big Bang model, it is assumed this is the shell at which radiation decoupled from matter in the beginning. *But* the irregularities which were forming galaxies should be seen! The really astonishing fact of the background observations is that they are smooth to about one part in a hundred thousand—not that they are irregular (on too large an angular scale) to about one part in a million.

In the nonexpanding universe, however, the intergalactic medium can be observed from here to the edge of the visible universe with no velocity smearing. The integration through this maximum distance is most capable of smoothing out all fluctuations in background radiation received from all depths in the universe. In the nonexpanding universe an obvious, and much simpler, explanation of the observation is that we are simply seeing the temperature of the underlying extragalactic medium.

It is customarily stated that the Big Bang *predicted* the correct temperature of the cosmic background. But, as A. Assis discusses in this volume, G. Gamow (1961) predicted T = 50° K. It was static, tired-light models which predicted values around 2.8° K. As early as 1926, A. S. Eddington calculated the temperature of interstellar space as ~3° K. Many investigators have since pointed out that if one takes the ambient starlight and thermalizes it to lower energy photons, one gets closely the observed microwave background temperature. F. Hoyle and C. Wickramasinghe have suggested that iron whiskers blown out of supernova explosions could effect this thermalization (see Arp *et al.* 1990). Narlikar and Arp (unpublished) have considered whether $m = 0$ electrons, which are extremely efficient thermalizers, could be the agents responsible. Creation events, in the variable mass theory, could supply abundant low mass particles, either from active galaxies or in smaller events spread throughout the intergalactic medium. Recently, Burbidge, Hoyle and Narlikar (in press) have summarized the evidence from recent gamma ray observations which can be interpreted as direct detection of creation events.

Mass Creation and Quantum Mechanics

One of the great searches in modern physics has been to connect the realm of the submicroscopic quantum mechanics to the macroscopic world of classical mechanics. There are, however, some classical formulae which work in the quantum domain if $m^2 < 0$ (see for eqs. I. Khalatnikov 1992). It is very provocative, therefore, when we obtain a general solution for the classical equations of general relativity, equation (5) here, which gives $m \to 0$ when $t \to 0$. In other words our creation events go exactly through the juncture between quantum and classical mechanics. It is, therefore, possible to conceive of new-born matter having been previously in a state where its constituent wave packets were unbounded and spread throughout the universe. The materialization of this matter at a certain point in space at $t \approx 0$ would then represent not matter from "outside" the universe but merely the rearranging of mass/energy within the universe. In this case conservation laws could hold in general, while at the same time opening vastly greater possibilities for morphological change within the universe.

Are there any observations which can give clues in the macroscopic domain to the nature of their connection to the quantum domain? There is one kind of observation which apparently does just that. That observation is the quantization of redshifts. As Figure 6 showed earlier, the whole of the measured redshift phenomenon is split into periodicities: large periodicities for the large redshifts and small periodicities for the small redshifts. It has been commented that the

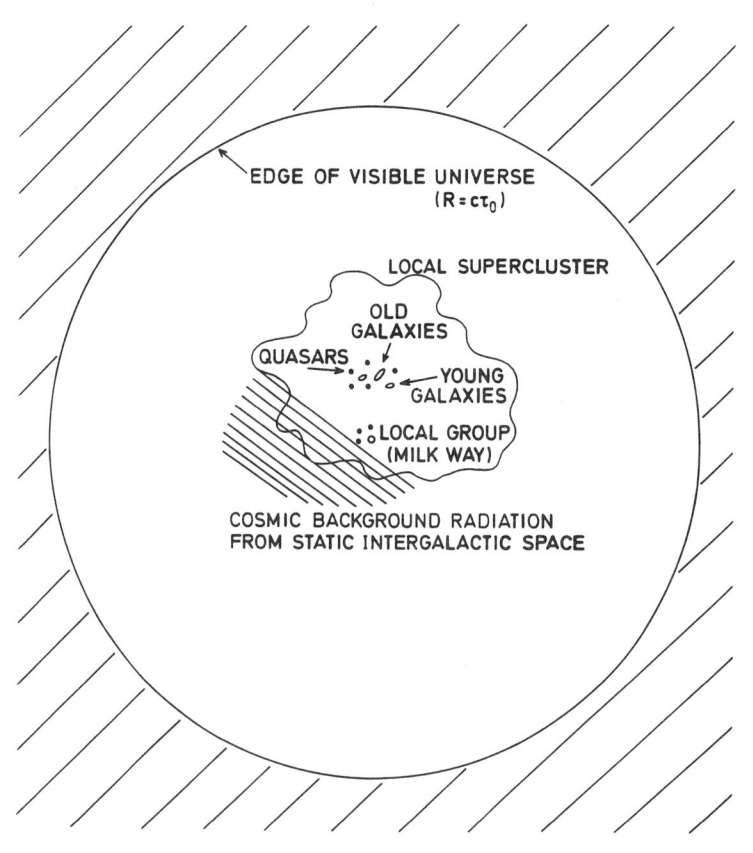

Figure 13. Schematic model of our nonexpanding universe is shown. The region inside an indefinitely large universe with which we communicate is the speed of light times the age of our galaxy, $r = c\tau_0$. Background radiation can be smooth and pervasive or very slightly non-homogeneous in direction of Local Supercluster.

presence of any appreciable peculiar velocities of these extragalactic objects is ruled out by this quantization. If the age is responsible for the spectral shift, however, the age of the extragalactic objects could be quantized. Since the material comes through the quantum domain of $m = 0$, periodicities of Δt could well be impressed at that stage which would remain imprinted, albeit growing relatively smaller as the material gained age and mass.

This cannot be the whole story, of course, because as more distant objects of our own age are considered, the look back time determines their apparent redshift. This should be continuous, but instead is implied to be some sort of cellular structure. Possibly coherence and interference effects will have to be considered.

However, whereas the pervasive quantization observed in redshifts seems possibly compatible with age and quantum effects, the current interpretation of redshifts as velocities offers no hope of either glimpsing the correct model of the universe, or gaining physical understanding of the physical laws which relate the macroscopic and microscopic scales. The discouraging situation is, of course, that so long as the validity of the observations is denied on the grounds they do not fit the current theory—just that long will it be impossible to correct the theory.

The Best Current Model of the Universe

Finally, considering *all* the observations, the model of the universe which *can* fit is shown in Figure 13. It is a universe which is not expanding and in which we, from our Milky Way Galaxy, see out to a radius of the speed of light times the age since our creation. Of course, galaxies could have existed before that, and if they were enormously brighter than galaxies we have any knowledge of, we could see out even further than this $r = c\tau_0 = 15$ billion light years. Therefore, we must always be open to the possibility of surprises as time passes and as signals from new events just penetrate to us or as our horizon expands to embrace older events.

The Local Supercluster is the largest aggregate of material about which we have certain knowledge. We are near the edge, at about 15–20 Mpc from the center. When we get through remapping all the young, high redshift objects like quasars and active galaxies to their correct distances rather than their redshift $\equiv$ velocity distances, then the Local Group and Local Supercluster will become much more populated relative to the more distant regions than is presently believed. In fact, it may be relatively quite empty beyond the confines of the Local Supercluster. In that case, a cap may be set upon the exchange of gravitons with distant matter and, with it an upper limit upon the blue spectral shift of the oldest galaxies which we can see.

The cosmic background radiation could also be fairly concentrated in the vicinity of the Local Supercluster, because there is no velocity restriction on the Black Body peak wavelength. Asymmetries and irregularities could be indicative of local features, whereas the utterly smooth part could represent integrations through more distant regions.

In summary, a true scientific process of inferring a theory which fits all the observations, instead of the usual inverse procedure yields a consistent, physically plausible picture which opens the possibility of understanding much more about the universe in which we live. We have not dared to invent "new physics"; only carefully, empirically, earned our way to more correct physics.

References

Arp, H., 1986, *Astron.Astrophys.* 156:207.
Arp, H., 1987, *Quasars, Redshifts and Controversies*, Interstellar Media. Berkeley.
Arp, H., 1988, *Astron.Astrophys.* 202:70.
Arp, H., 1988b, in: *New Ideas in Astronomy*, Cambridge Univ. Press, eds. F. Bertola, J.W. Sulentic, B.F. Madore, p.161.
Arp, H., 1990a, *IEEE Trans. on Plasma Sci.* 18:77.
Arp, H., 1990b, *Phys.Lett.A.* 146:172.
Arp, H., 1990c, *Astrophys. Space Sci.* 167:183.
Arp, H., 1991, *Apeiron* 9-10:18.
Arp, H., 1991b, *Sky and Telescope* 81:373.
Arp, H., 1992a, Observational Problems in Extragalactic Astronomy, in: *IAU Highlights of Astronomy*, Vol. 9:43.
Arp, H., 1992b, *Mon. Not. Roy. Astr. Soc.* 239:800.
Arp, H., Burbidge, G.R., Hoyle, F., Narlikar, J.V. and Wickramasinghe, 1990, *Nature* 346:807.
Arp, H. and Crane, P., 1992, *Phys. Lett.A* 168:6.
Arp, H. and Sulentic, J.W. 1985, *Ap. J.* 291:88.
Arp, H., Bi, H.G., Chu, Y. and Zhu, X., 1990, *Astron.Astrophys.* 239:33.
Bahcall, J.N., Junuzzi, B.T., Schneider, D.P., Hartig, G.F. and Jenkins, E.B. 1992, *Ap. J.* submitted.
Bottinelli, L. and Gougenheim, L., 1973, *Astron. Astrophys.* 26:85.
Broadhurst, T.J., Ellis, R.S., Koo, D.C. and Szalay, A.S., 1990, *Nature* 343:726.
Collin-Souffrin, S., Pecker, J.-C. and Tovmassian, H.M., 1974, *Astron. Astrophys.* 30:351.
Field, G.B., Arp, H. and Bahcall, J.N., 1973, *The Redshift Controversy*, W.A. Benjamin Inc., Reading, Mass.
Jaakkola, T., 1973, *Astron. Astrophys.* 27:449.
Girardi, M., Mazzeti, M., Giurcin, G. and Mardirossian, F., 1992, *Ap. J.* 394:442.
Khalatnikov, I., 1992, *Phys. Lett. A.* 169:308.
Kembhavi, A.K., 1978, *Mon. Not. Royal Astr. Soc.* 185:807.
Kormendy, J., 1988, *Ap. J.* 325:128.
Madore, B.F., and Freedman, W.L., 1992, *P.A.S.P.* 104:362.
McCall, M.L., 1989, *A.J.* 97:1341.
Narlikar, J.V., 1977, *Ann. of Phys.* 107:325.
Narlikar, J.V. and Arp, H., 1992, *Ap. J.* in press (20 Feb 93).
Rix, H.-W., Schneider, D.P. and Bahcall, J.N. 1992, *A.J.* 104:959.
Sandage, A and Tamman, G., 1981, *Revised Shapley Ames Catalog of Bright Galaxies*, Carnegie Institution of Washington.
Sandage, A. 1992, $H_0 = 43 \pm 11$ km s^{-1} Mpc^{-1} Based on Angular Diameters of High Luminosity Field Spiral Galaxies. Preprint, The Observatories of the Carnegie Institution of Washington.
Sandage, A., 1975, *Ap. J.* 202:563.
Scott, D. 1991, *Astron. Astrophys.* 242:1.
Spinrad, H. and Djorgovski, S., 1987, *IAU Symposium* No. 124:29 (Dordrecht: Reidel).
Vader, J.P. and Chaboyer, B., 1992, *P.A.S.P.* 104:57.
Yee, H.K.C., 1988, *A.J.* 95:1331.
Zaritsky, D., 1992, *Ap. J.* 400:74.
Zaritsky, D., Smith, R., Frenk, C. and White, S.D.M., 1993, *Ap. J.* 10 March issue.
Zucca, E., 1991, *Mon. Not. Roy. Astr. Soc.* 253:401.

Redshift Periodicity in the Local Supercluster

W.M. Napier and B.N.G. Guthrie

Royal Observatory, Blackford Hill
Edinburgh EH9 3HJ, Scotland, U.K.

Persistent claims have been made over the last ~15 years that extragalactic redshifts, when corrected for the Sun's motion around the Galactic centre, occur in multiples of ~24, ~36 or ~72 km s^{-1}. A recent investigation by us of spiral galaxies out to 1,000 km s^{-1} gave evidence of a periodicity ~37.2 km s^{-1}. Here we extend our enquiry out to the edge of the Local Supercluster (~2600 km s^{-1}). We confirm that, when corrected for the Sun's galactocentric motion, the redshifts are strongly periodic ($P \sim 37.5$ km s^{-1}). The periodicity is coherent over the entire Supercluster. The formal confidence level of the result is very high, and the phenomenon apparently cannot be ascribed to observational artefact or group membership. Various types of 'oscillating physics' are considered, such as variations in the fundamental constants, or the coupling of light to a universal, coherently oscillating scalar field. These run up against geological constraints or other difficulties. The possibility that massive galaxies comprise a conformally expanding lattice is discussed.

1. Introduction

The term 'quantized redshifts' encompasses a set of claims which are surely amongst the most bizarre to have been made in modern astrophysics. The first claims under this category were made by Tifft, who in 1976 stated that the redshifts of galaxies in the Coma cluster were preferentially offset from each other in multiples of 72.46 km s^{-1}. These redshifts were determined optically, and were of relatively low accuracy. However, within a few years, more accurate extragalactic redshift determinations were becoming available from 21 cm determinations: the HI line profile of a spiral galaxy may be several hundred km s^{-1} wide, but if it is fairly symmetric, the systemic redshift of the galaxy may be consistently determined to within a few km s^{-1}. Making use of these more accurate redshift data, it was claimed that binary galaxies were similarly offset (Tifft 1980), and that this redshift periodicity (~72 km s^{-1}) occurred also within groups and associations of galaxies (Arp and Sulentic 1985).

In 1984, Tifft and Cocke (1984: TC hereinafter) claimed that the redshifts of galaxies occurring anywhere on the sky are periodic when a suitable correction for the solar motion is made. That is, there exists a global periodicity, and not one confined to the differential redshifts of adjacent galaxies. This global periodicity, however, was not 72.46 km s^{-1} but one half (36.3 km s^{-1}) if the galaxies were broad-lined, and one third (24.2 km s^{-1}) if they were narrow-lined. These periodicities emerged when the same solar motion was subtracted from each of the redshift sets.

The phenomenon, if real, is unrelated to known physics and inexplicable in terms of current cosmological paradigms. Further, the entities for which the periodicities are claimed are not exotic objects which might conceivably be the sites of unfamiliar physics, but ordinary galaxies. It is perhaps not surprising that few astronomers have taken these claims seriously. On the other hand, there is abundant historical precedent for the discovery of phenomena which were regarded as bizarre and inexplicable at the time. Implicit in our examination of the question, then, is a rejection of the philosophy that physics and cosmology are at the stage where unexpected new phenomena can be confidently excluded; we have adopted instead the view that the existence or otherwise of redshift 'quantization' is a matter for empirical enquiry. In recent years there has been a large increase in the numbers of redshift determinations using 21 cm line profiles; because of this greatly enhanced dataset, it should now be possible to settle the question through rigorous statistical analysis.

A previous analysis by us (Guthrie and Napier 1991) of field galaxies seemed to support the claims of redshift periodicity amongst spiral galaxies out to 1000 km s^{-1}. In that study, we found no periodicity when the redshifts were corrected for a solar vector with respect to the Local Group or the microwave background, or when it was allowed to vary with distance as described by de Vaucouleurs and Peters (1984). Although the mean solar apex may vary by ~30° over the distance range investigated (~15 Mpc), the periodicity emerged only when a local solar vector (*i.e.* one close to the Sun's motion around the Galactic centre) was subtracted from the heliocentric redshifts. It appears that the redshifts have to be corrected for a unique, local solar vector irrespective of the dimensions of the region being explored. This result, while unexpected, implies that a periodicity of ~24 or ~36 km s^{-1} in galactocentric redshifts might exist and be detectable to much greater distances, and so prompted us to investigate the question out to a much larger volume of space, namely the whole of the Local Supercluster (LSC). Galaxies belonging to the LSC have corrected redshifts ranging up to ~2600 km s^{-1} (Flin and Godlowski 1989). The outcome of this enquiry is described in the present paper: our earlier conclusions are confirmed and strengthened.

2. The Methodology

In testing a statistical hypothesis, the following recipe is generally followed:

(i) The hypothesis is set up, derived from a theory, an early set of data or whatever.
(ii) From this initial hypothesis, a specific prediction is made concerning the behaviour of new, independent data. The prediction should ideally contain no unspecified parameters, but if it does, the freedoms they introduce must be accounted for in the reckoning, for example by estimating the effective number of independent trials that they represent.
(iii) The prediction having been made, it is then tested on a new dataset. This new

set must be (a) independent of the initial one from which the hypothesis was formulated, and (b) unbiased. The latter condition requires, *inter alia*, that any culling of the data from a larger set should be done once and for all, prior to the analysis, in a way which will not alter the outcome of the testing. In evaluating the outcome of a test, other questions may reasonably be asked of the investigators, such as: did they play around with other datasets, other selections and values of parameters? Was there a 'termination bias', the tendency to stop when a 'success' has been attained?

(iv) It will often be found that, as new or better data accumulate, some modification of the original hypothesis gives a better fit: probably all statistical hypotheses are destined to fail in their original form (they are only models). Such modifications may carry with them the suspicion that the investigator is 'shifting the goalposts'; but equally, one expects the original perception of a phenomenon to be sharpened up as the data available expand or improve in quality. The test of whether a given 'improvement' of the original result is genuine is to apply the improved hypothesis to a new, independent dataset. Thus, exploratory and confirmatory phases intermingle until a precisely formulated hypothesis is reached or one runs out of data.

(v) Multiple hypothesis testing is important: periodicity might fit a redshift dataset better than the null (random) distribution; but clustering might fit better still.

3. The Technique

The technique most commonly applied in testing for periodicity is power spectrum analysis, in which a given set $\{V_i\}$ of N numbers is circularly transformed with respect to a trial period P, and a statistic $I = 2R^2/N$ is calculated. Here R represents the magnitude of the vector sum of the unit vectors

$$\mathbf{e}_i = \mathbf{e}_x \sin\left(\frac{2\pi V_i}{P}\right) + \mathbf{e}_y \cos\left(\frac{2\pi V_i}{P}\right) \tag{1}$$

Essentially a string, along which the signals $\{V_i\}$ are marked, is wrapped around a drum of circumference $2\pi P$, and the corresponding unit vectors in the plane are added. Thus in Figure 1, $R = OP$ is the distance moved from the origin by the legendary drunk man performing a random walk of N unit steps.

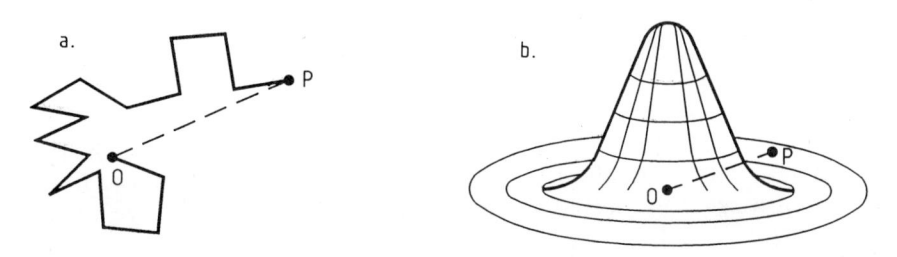

Figure 1. The connection between PSA and the drunkard's walk. Linear data are converted to circular by wrapping them around a drum of circumference $2\pi P$.

In such a random walk $R \propto \sqrt{N}$, and the statistic I represents a normalized distance whose behaviour is understood from the theory of the random walk in two dimensions. A power spectrum or periodogram is a plot of $I(v)$, where $v = 1/P$. A periodicity in the data may be observed as a peak at the relevant frequency; the higher the peak, the straighter the walk and hence the greater the probability of a real periodicity in the data. For large N, for random, uniform and independent data, and neglecting edge effects (non-integral wraps around the drum), the probability p of obtaining a value $I \geq I_{max}$ by chance in a single trial is

$$p = \tfrac{1}{2}\exp\left(\frac{-I_{max}}{\bar{I}}\right) \qquad (2)$$

and the mean value $\bar{I} = 2$.

3.1 Against PSA

Although power spectrum analysis in some form goes back to the 19th century, it was first rigorously discussed by Bartlett (1955), who found substantial problems with the method and quickly abandoned it. The apparent limitations of PSA have recently been discussed by Newman et al. (1992).

First, the statistic I is a biased estimator. A departure from non-uniformity in the redshift distribution will in general yield $\bar{I} \neq 2$ and also a departure from the exponential distribution (2). Only in the limit of high frequencies are these formulae applicable. Bias is also created if, as is always the case, the dataset N is finite.

Second, I is inconsistent, a fact which manifests itself in the 'grassy' or noisy appearance of a periodogram, the relative variance $\sigma(I)/\bar{I}$ remaining of order unity even as $N \to \infty$. Should a large number of trials be involved, even random data may yield surprisingly high values for I: for n_T independent trials, the expectation value of I_{max} is $E(I_{max}) \sim 1.2 + 2\ln n_T$. The inconsistency can be greatly reduced by the use of smoothing techniques (there is a large literature on 'window carpentry'), but at the cost of increasing the bias and introducing further degrees of freedom.

Finally, even when bias is not a problem (say for high frequencies), Newman et al. (1992) find on empirical grounds that, as N increases, convergence towards the form (2) is slow for high I. But of course, it is precisely the high peaks which are of interest in testing for periodicity. Newman et al. conclude that PSA is 'limited and possibly dangerous in that quantitative assessments in the sense of hypothesis testing are not meaningful'.

3.2 In Defence of PSA

In spite of these disadvantages, some of which are more widely known than others, PSA remains the most widely used period-hunting tool in the physical sciences (for discussions of the technique in an astrophysical context, see Scargle 1982, Horne and Ballunas 1986, Stothers 1991). Its usefulness lies in the increase of signal strength with N, and in its simple limiting statistical behaviour as described by (2). It is also a sensitive technique, signals too weak to be seen by eye often being detectable by it.

The bias due to the finiteness of N is, for a truncated series of data, of order $\ln N/N$ and so tends to zero as $N \to \infty$. It is not important, in probability terms, for the present problem. In the datasets analyzed so far (Guthrie and Napier 1990, 1991), the redshift distributions were fairly uniform, and the distribution of peaks in the power spectra, other than those under test, showed no significant departures from the expected $\bar{I} = 2$ and exponential decline. In any case, the periodicities under test are of high frequency in relation to those of trends and range of the dataset and so bias due to these factors is not expected.

The convergence of I towards the exponential formula has been examined by Buccheri and de Jager (1989) through numerical experiments. For samples of random data with $N \sim 100$, of the order discussed in this problem, the cumulative probability distribution of I closely follows the form (2) down to probabilities $p \sim 10^{-6}$, corresponding to $I \sim 28$.

Finally, the inconsistency of the statistic I (the variance equals the mean) does not prevent the estimation of the significance of a high peak if n_T can be assessed or comparison is made with suitable synthetic data (Guthrie and Napier 1991b).

In the present paper, the significance of the power $I(v)$ in the actual redshifts is assessed by a ranking procedure, simply comparing I(real) with I(synthetic) derived from artificially generated redshifts constructed so as to closely simulate the actual redshifts (including any non-uniformities in distribution). This procedure is robust, and questions of bias, consistency and convergence are simply bypassed: indeed almost any measure of periodicity would do! Analyses of the same data using absolute rather than ranked values of I, and using a linear rather than circular analytic technique (Stothers 1991), have yielded essentially the same results (Guthrie and Napier, unpublished).

4. The Database

4.1 The Galaxy Sample

In our previous study, mentioned above, we used the database for 6439 galaxies compiled by Bottinelli *et al.* (1990). Galaxies with the most accurately determined HI redshifts (*i.e.* those with database errors $\sigma_{cz} \leq 4$ km s^{-1}) were taken from the list. From these were eliminated possible members of the Virgo cluster, non-spirals and galaxies previously used by TC in their study. Of the remaining spirals, 89 had redshifts < 1000 km s^{-1} after correction for the Sun's motion around the Galactic centre (taken as $V_\odot = 252$ km s^{-1}, $l_\odot = 100°$, $b_\odot = 0°$, although the precise solar vector is not critical here). These constituted an independent dataset of the nearby field spirals, culled once and for all in an objective, reproducible manner from the catalogue. The data were then divided into 40 'more accurate' ($\sigma \leq 3$ km s^{-1}) and 49 'less accurate' galaxies (the remainder) and analyzed, separately and together.

For the present analysis, the culling was carried out in similar fashion, with the exception that the corrected redshift limit was extended from 1000 km s^{-1} to 2600 km s^{-1}, the recognized limit of the LSC. Thus, the expanded database incorporates our earlier one and so conclusions drawn from it are not to be read independently of our earlier ones. The present study may rather be regarded as a simpler statistical treatment of an extended redshift sample. With the extension to 2600 km s^{-1}, and after eliminating Virgo members and galaxies previously used by TC (34 of them), the dataset expands to 247 galaxies, of which 97 are 'more accurate'

($\sigma \leq 3$ km s^{-1} and redshift calibrators adopted by Baiesi-Pillastrini and Palumbo 1986); this compares with 89 galaxy redshifts in our earlier sample, of which 40 were 'more accurate' ($\sigma \leq 3$ km s^{-1}). Two large clusters (UMa and Fornax with 61 and 49 members respectively) are now encompassed; whether these should for consistency have been excluded from the sample as were the probable Virgo spirals is perhaps arguable. However because of the large computational effort involved in the study (~10^8 PSAs!) we generally operated only on the 97 very accurately measured redshifts, exploiting the full sample only when its behaviour as a function of accuracy was under consideration.

4.2 The Solar Vector

The circular velocity of the solar neighbourhood has been determined from HI data (Gunn et al. 1979), yielding $\Theta = 220 \pm 10$ km s^{-1}, and from an examination of the solar motion relative to the nearest members of the Local Group (Einasto 1979, Haud et al. 1985), yielding 221±5 km s^{-1} and 218±5 km s^{-1}. Taking account only of the solar peculiar motion relative to the local standard of rest (Delhaye 1965), a resultant galactocentric solar vector $V_\odot \simeq 232$ km s^{-1}, $l_\odot \simeq 88°$, $b_\odot \simeq 2°$ is found. However, the presence and orientation of a strong bar in the central regions of the Galaxy (Blitz and Spergel 1991) are consistent with a local expansion velocity in the range 20–40 km s^{-1} (Combes and Gerin 1985 and Napier, unpublished), while evidence has been presented (Clube and Waddington 1989) that the local standard of rest has an outward motion of ~40 km s^{-1}. Additionally, the Galactic disc may possess a warp, introducing some uncertainty (which can only be guessed at) in the local perpendicular velocity relative to the mass plane of the Galaxy. Allowing for these factors, the galactocentric solar vector may be taken roughly as $V_\odot \simeq 233 \pm 7$ km s^{-1}, $l_\odot \simeq 93 \pm 10°$, $b_\odot \simeq 2\pm10°$ where the uncertainties are largely in the modeling. This is very close to the $V_\odot$ for which TC claim the periodicities emerge (233.6 km s^{-1}, 98.6°, 0.2°), and justifies the formulation that: *the periodicities emerge when the heliocentric redshifts are corrected for the galactocentric solar motion*. The alleged periodicities are therefore nucleus to nucleus between galaxies. Thus, we are testing whether extragalactic redshifts tend to occur in multiples of ~24, ~36 or ~72 km s^{-1}, when corrected for the Sun's galactocentric motion $V_\odot$, against the null hypothesis that there is no periodicity.

5. Is the Hypothesis Reasonable?

A hunt for a signal within the range of uncertainty of $V_\odot$ is in effect an exercise in optimization of the statistic I, equivalent to carrying out a number of independent trials. The 'correction' for $V_\odot$ thus introduces three additional degrees of freedom, which might cause an otherwise insignificant signal to be artificially boosted.

This additional freedom can readily be allowed for by 'optimizing' artificial redshift data in identical fashion to the real data. However a potentially more serious consequence of having $V_\odot$ as a free parameter is that TC might at the outset have been guided to search for periodicity only in the local neighbourhood of an astrophysically significant solar vector and around a previously suspected period (or a simple fraction thereof). But one can imagine that, if one were to vary the solar vector over the entire celestial sphere and search for periodicity over a wide range, peaks would arise in all sorts of directions and for all sorts of 'periodicities'. In that case the ($V_\odot,P$) derived would measure nothing more than the bias of the original investigators, and its proximity to a solar vector of consequence would have no significance.

To investigate this possibility, the solar vector $V_\odot$ was varied in direction over the whole sky and in speed over the wide range 140 to 300 km s^{-1} (in steps of 2° in $b_\odot$, 2 or 3° in $l_\odot$, and 5 km s^{-1} in $V_\odot$). For each solar vector a set of 97 corrected redshifts was derived and a periodogram constructed in 490 equal frequency steps over the period range 20–200 km s^{-1}; thus in all, for two interlaced grids, about a million periodograms were constructed, for each of which the maximum power I was recorded.

Figure 2 reveals the outcome of the exercise, in which the ($V_\odot$, $l_\odot$, $b_\odot$) box has been collapsed on to ($l_\odot$, $b_\odot$) and the ten highest peaks have been plotted.

The presence of multiple peaks illustrates the importance of $V_\odot$ in this problem. One might have supposed that, for a real periodicity, only a single peak would have emerged, but (for reasons to be discussed) this is not the case. Five of the ten peaks correspond to periods 24±3 km s^{-1} and seem to lie roughly on a band just south of the galactic equator. The other five have essentially the same period (37.5±0.2 km s^{-1}), and three of them lie close to the current best estimates of the galactocentric solar motion, within its uncertainties. Keeping in mind that the dataset is completely independent of that employed by TC, this analysis indicates that the hypothesis (a redshift periodicity ~36.3 km s^{-1} after subtraction of the galactocentric $V_\odot$) is a reasonable one to test, in the sense that the whole sky is not filled with high peaks at all sorts of frequencies. Indeed, only the TC solution stands out.

The absolute deviations from periodicity of the corrected redshifts were evaluated for the ten peaks, and the inter-peak Spearman rank correlation coefficients were obtained. The three peaks marked (206, 217, 223) were found to be strongly correlated, as were (258, 292). There was also cross-correlation between the members of these two sets, although at a weaker level. The five peaks at ~37.5 km s^{-1} therefore appear to be caused by a single underlying phenomenon. As to why there should exist multiple peaks as well as signals at ~24 km s^{-1} in other directions of the sky, some more or less speculative comments are made later.

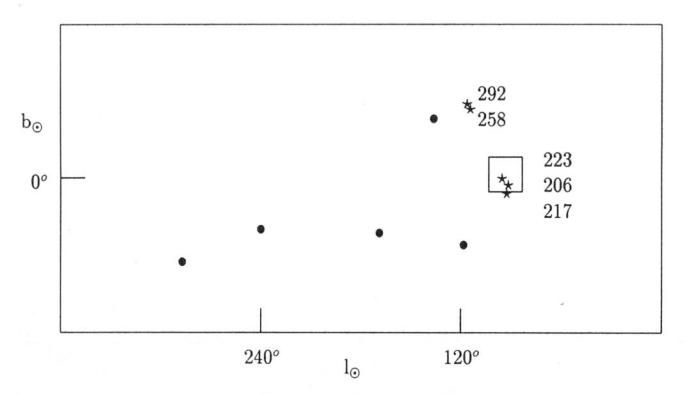

Figure 2. Ten highest peaks (out of ~10^6) in a whole-sky search (140 ≤ $V_\odot$ ≤ 360 km s^{-1}), over 20 ≤ P ≤ 200 km s^{-1}. ● = 24±3 km s^{-1}, ✻ = 37.5±2 km s^{-1}. The formal error box of the solar galactocentric motion is shown.

6. The Significance of the Signal

The next step in the analysis was to assess the significance of this redshift structure, in the sense of asking whether a random redshift distribution might yield it. TC claimed a periodicity of 36.3 km s^{-1} for a solar vector $(V_\odot, l_\odot, b_\odot)$ = (233.6 km s^{-1}, 98.6°, 0.2°) on the basis of 40 galaxies with broad HI lines. The proximity of three out of the ten highest peaks to this solution has probability $\leq 10^{-4}$, smaller if comparison is instead made with the best estimate of the Sun's galactocentric vector; however this calculation does not yet take account of the absolute heights of the peaks ($I \sim 38$).

First, the power structure in the neighbourhood of the region under investigation was examined: a coarse-grid exploration was carried out over the region shown in Table 1.

Table 1. Search range for the 97 spirals. P was searched in equal frequency steps. $(V_\odot, P)$ in km s^{-1}, $(l_\odot, b_\odot)$ in degrees.

parameter	min	max	step nos.
$V_\odot$	203	263	6
$l_\odot$	78	108	15
$b_\odot$	-13	+17	15
P	34	39	8

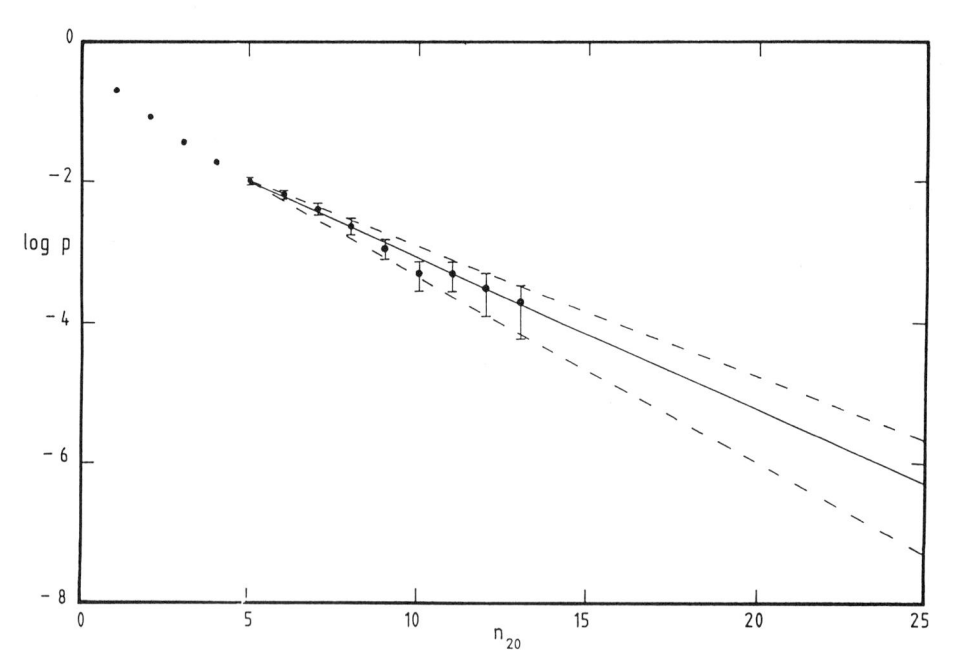

Figure 3. Probability that a set of randomized redshifts, constructed and analyzed as described in the text, would yield more than n_{20} spectral peaks.

Thus a generous allowance was made for uncertainties in the galactocentric solar vector and the predicted period, and 10,800 I-values were obtained. Of these, 12 were > 25, 25 were > 20 and 76 were > 15, i.e. $n_{25} = 12$, $n_{20} = 25$ and $n_{15} = 76$. These high I values are not in general independent.

These numbers were then compared with the corresponding values obtained by constructing synthetic datasets and analyzing them in identical fashion to the real data. The positions of the galaxies on the celestial sphere were preserved, and each real redshift was randomized by adding to it the difference of two random numbers in the range (0, 50) km s^{-1} (S.E. 20 km s^{-1}), large enough to smear out the periodicities under test but small enough to preserve the overall redshift distribution. The real and synthetic datasets were, therefore, identical in all respects except for the deletion of this fine structure, and so any significant difference between the real and synthetic data could only be ascribed to the presence of redshift structure coherent on a scale ≤ 50 km s^{-1} say.

Ten thousand sets of 97 randomized redshifts were so constructed, and grid searches were carried out on each set. No values of $n_{25} \geq 12$, $n_{20} \geq 25$ or $n_{15} \geq 76$ were found (the highest values obtained were $n_{25} = 5$, $n_{20} = 13$ and $n_{15} = 52$). The cumulative distributions of n_{20} and n_{15} were used to give the single-trial probability p of exceeding any prescribed value of n_{20} or n_{15}: in Figure 3, log p is plotted against n_{20}; an extrapolation yields the 1σ range for the probability of obtaining $n_{20} > 25$ in a single trial as $5 \times 10^{-8} \leq p(n_{20}) \leq 2 \times 10^{-6}$; for $n_{15} \geq 76$, $2 \times 10^{-6} \leq p(n_{15}) \leq 1.5 \times 10^{-5}$. Note that comparison is being made with a part of the 'whole-sky' box which is of significance in its own astrophysical right and not simply because it contains strong peaks. We see, therefore, that when the observed LSC redshifts are referred to the nucleus of the Galaxy, <u>strong redshift structure, centred on ~37.5 km s^{-1}, emerges.</u> We can make this statement at a confidence level of, say, a million to one. At this stage of the argument, we are not describing the structure as a periodicity. However already we see that there is a suggestion of something 'bizarre': why should strong redshift structure emerge with respect to the Sun's galactocentric motion, which can have little relevance to the redshift behaviour of galaxies scattered throughout the LSC?

7. Statistical Behaviour

Good statistical behaviour is expected of a real phenomenon: if the strength of the signal failed to increase with increasing quality of the data, or with increasing sample size, or if the derived parameters were unstable, the 'best' values hunting around without convergence as the dataset increased, then one would suspect that some gremlin lurked in the analysis.

In the present case, the signal strength appears to concentrate in the best data. In our initial analysis of galaxies out to 1000 km s^{-1}, we randomly extracted 40 galaxies from the 89 in the set, and computed the signal strengths; we found that the peak derived from the 40 'accurate' galaxies was exceeded by chance only with probability ~0.03 (the figure is ~0.002 for the 97 out of 247 LSC galaxies, but there may be a complication, as discussed below). The signal strength also increases with sample size: taking the 'best' galaxies in each case, $I \sim 29$ for the 40 nearby galaxies and $I \sim 38$ for the 97 LSC galaxies. Finally, the derived parameters are remarkably stable: as shown in Table 2, the signal holds steady at ~37.5 km s^{-1}, for a solar vector varying by only 3 km s^{-1} in speed and ~2° in direction, as the sample is extended from the best 40 to the best 103 redshifts (Section 8.3).

Table 2. Optimized solar vector for the most accurate available data, as a function of increasing sample size. Note (a) the stability of the peak, and (b) the progressive increase of signal strength. $N = 40$ refers to the original sample; $N = 97$ to the set described herein; and $N = 103$ to the 'best' dataset, constructed as described in the text.

N	I_{max}	P	$V_\odot$	$l_\odot$	$b_\odot$
40	29	37.5	215	94	−12
97	38	37.5	217	95	−12
103	52	37.5	218	96	−12

8. Gremlins

8.1 False Peaks

The various analyses carried out by Tifft and colleagues have involved small samples, often (in the earlier papers) of relatively low accuracy, while the solar vector was varied over a very limited range. However our analysis reveals that in finding and assessing any periodicity, $V_\odot$ cannot be treated as a 'fudge factor'. It is in fact an optimizing parameter which, uncontrolled, could mislead the investigator into deducing the existence of spurious periodicities.

A number of trials were carried out in which synthetic periodic data (say 100) were split into small groups (say 10) of adjacent redshifts, a random redshift was added to each group, and PSA was then carried out on the whole jumbled dataset. Quite frequently, the dominant peak was found at some simple fraction (½, ⅓ or ⅔ say) of the basic periodicity, depending on how the random runes were cast.

This result leads us to conjecture that the ~24 km s^{-1} peaks which turned up far from the galactocentric vector in our whole-sky study (Figure 2) may result from an analogous effect: essentially a solar motion correction far from the true one may stretch and distort the real redshift periodicity, resulting in a 'best fit' on to some fractional one. We also consider that the similar periodicity which TC claimed to have found for the narrow-line galaxies, which we cannot find, may be an artefact due to the smallness of their sample and their binning technique, which can settle on a fraction of any basic period. However, we emphasize that, at the time of writing, these comments are conjectural.

8.2 Radio Telescopes

Several investigations have been undertaken on internal errors within telescopes, and consistency of measurement between them (Rood 1982, Tifft and Cocke 1988, Tifft 1990, Tifft and Huchtmeier 1990). For the best data a consistency of ≤ 3 km s^{-1} is routinely obtained in the determination of systemic velocities, both with different profile reduction techniques and between telescopes. Each galaxy in the list of 97 has several line profile determinations, obtained from different telescopes, and a separate assessment of σ from each individual group of radio astronomers. It seems unlikely that an artefact, communal to all radio telescopes and capable of generating a false periodicity after subtraction of the solar galactocentric motion, could have gone unnoticed for ~30 years.

8.3 Foreground Contamination

Examination of the deviations of the redshifts of individual galaxies from the periodicity showed no obvious relation to sky position or morphological type. However, a few of the HI line profiles revealed evidence of asymmetry, as might arise from foreground contamination by our own Galaxy. An exploratory list of 115 redshifts was therefore compiled, comprising the 97 above enhanced by the 18 most accurate redshifts used by TC, and any galaxy whose profile (at the 20 percent level) was within 100 km s^{-1} of 0 km s^{-1} was rejected. This reduced the list to 103. The signal for this uncontaminated sample was very strong: $I_{max} = 52$ (see Figure 4) for $P = 37.45$ km s^{-1} and a solar vector (218.6 km s^{-1}, 95.8°, −11.5°). A histogram of the corrected redshift differences taken in pairs is shown in Figure 5 for these 103 redshifts. The periodicity—for Figure 5 clearly reveals it to be such—is coherent over the entire LSC (~63 cycles). Note that this is not a 'statistical result': when the most accurately observed redshifts of the LSC are corrected for the prescribed solar vector, the resultant $\{cz_i\}$ distribution is observed to be highly periodic, independently of any statistical analysis.

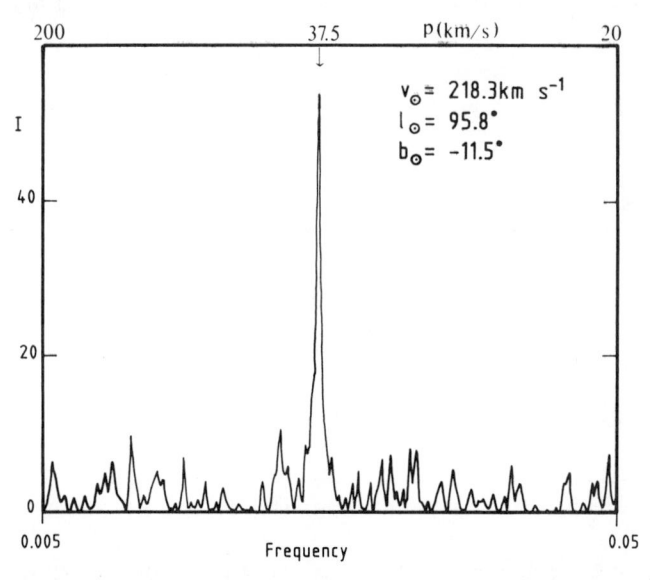

Figure 4. Power spectrum of the 103 most accurate, uncontaminated redshifts corrected for the optimum solar vector (218.6 km s^{-1}, 95.8°, −11.5°).

9. Group Membership

These results show that there is consistency with the hypothesis of a strong redshift periodicity between the nuclei of spiral galaxies in the LSC. The hypothesis that the redshift distribution is random gives an extremely poor fit by comparison and cannot be sustained. However we do not know, for example, that the solar vector has the precise parameters yielding the strong peak in Figure 4. It therefore remains to be seen whether some phenomenon other than periodicity might, with the search procedure used, be made to mimic a periodicity: after all, the procedure involves varying parameters to make the data appear as strongly periodic as possible!

Many of the 'field' galaxies in the sample in fact belong to loose groups and associations (catalogued by Fouqué et al. 1992) containing a few bright galaxies. Since the Monte Carlo simulations smear out redshift correlations $\lesssim 20$ km s^{-1} in the real data, it is conceivable that the difference between the real and the synthetic data is due, not to a periodicity, but rather to clustering, assuming the latter has redshift coherence on this scale.

9.1 Redshift Accuracy and Group Membership

As a prelude to this question, we examined the behaviour of the signal as a function of redshift accuracy and group membership. A sample of 261 spirals was taken, comprising those galaxies in the Bottinelli et al. (1990) catalogue with $cz < 2600$ km s^{-1} relative to the Galactic centre, with $\sigma_{cz} \leq 4$ km s^{-1}, and excluding members of the three large groups (Virgo, UMa and Fornax clusters). This sample includes a few galaxies previously used by TC but this may be methodologically justified as this part of our investigation is exploratory.

Power spectra were constructed over the ranges of period and solar vector shown in Table 3.

Table 3. Search range for the 261 spirals. P was searched in 24 equal frequency steps.

parameter	min	max	stepsize
$V_\odot$	200	260	2 kms^{-1}
$l_\odot$	60	120	1°
$b_\odot$	-30	+30	1°
P	37	39	

Two high peaks were found for this sample, namely:

$I = 38$ for $P = 37.8$ km s^{-1} when $(V_\odot, l_\odot, b_\odot) = (209.0, 94.5, -7.3)$; and
$I = 33$ for $P = 37.1$ km s^{-1} when $(V_\odot, l_\odot, b_\odot) = (219.9, 99.3, -4.1)$.

(These peaks are somewhat displaced from those previously derived, as expected since the overall accuracy of the sample is reduced.) The absolute deviations $|\epsilon_i|$ of the corrected redshifts from these two solutions are significantly correlated (Spearman's rank correlation coefficient $r_s \sim 0.33$). Thus the peaks are related, and the dependence of the mean absolute deviations $|\bar{\epsilon}_i|$ on redshift accuracy and group membership was studied.

The 261 redshifts were classed as 'accurate' ($\sigma_{cz} < 4$ km s^{-1} or calibrator: Baiesi-Pillastrini and Palumbo 1986) or 'less accurate' ($\sigma_{cz} = 4$ km s^{-1} and non-calibrator), and they were also designated as 'group' or 'non-group' according to the Fouqué *et al.* catalogue of groups. The values of $|\bar{\epsilon}_i|$ are on average smaller for the 111 'accurate' redshifts than for the 150 'less accurate' ones, the chance probability of this being $p \sim 0.014$. However the more accurately measured galaxies tend to belong to groups, a chi-squared test revealing that this correlation has chance probability $p \sim 0.002$ (Table 4).

Table 4. Correlation between accuracy of redshift measurements and group membership.

	group	non-group	
accurate	72	39	111
less accurate	66	84	150
total	138	123	261

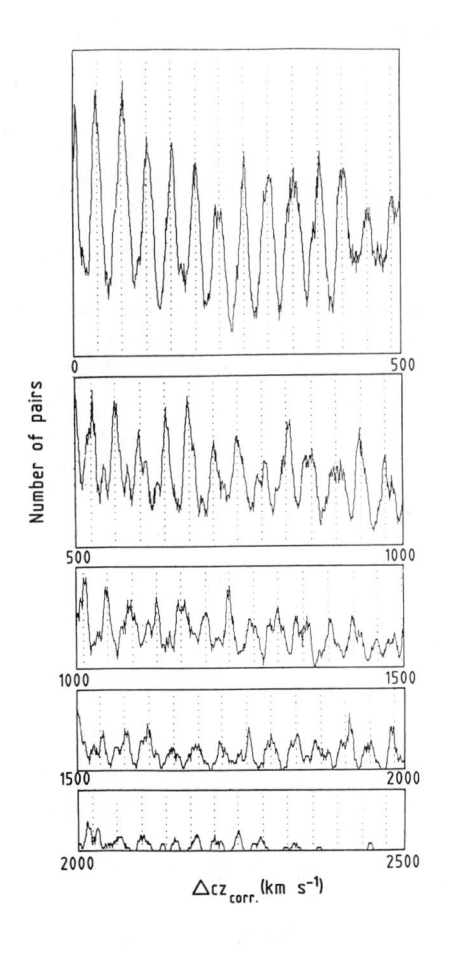

Figure 5. Two-point correlation function corresponding to the redshifts and optimum solar vector employed in the previous figure. Vertical dashed lines represent the best-fit periodicity, which seems to hold over the whole of the Local Supercluster.

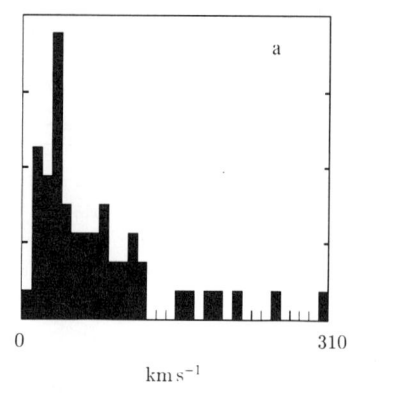

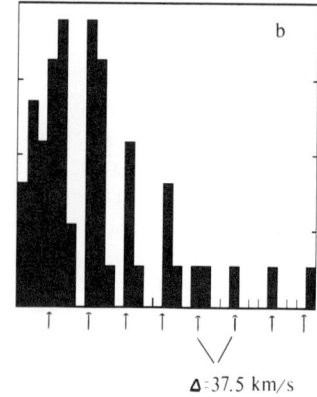

Figure 6. Histograms of differential redshifts dV for the 53 galaxies linked by group membership: (a) heliocentric redshifts, (b) redshifts corrected for $V_\odot = (216$ km s^{-1}, 93°, –13°). Bin-width is 10 km s^{-1}. Vertical arrows mark a periodicity of 37.6 km s^{-1} and zero phase.

Similar statistical exercises on the 261 galaxies revealed that there was a strong tendency for 'linked' galaxies (two or more measured in a group) to have more accurate redshifts than isolated galaxies or single representatives of groups ($p \sim 0.001$), and a further tendency ($p \sim 0.012$) for such linked, accurately measured galaxies to possess a stronger signal (grid searches yielding $I_{max} \sim 37$ for the 111 'linked' galaxies and ~22 for the 'unlinked' ones).

To determine whether the $|\bar{\epsilon}_i|$ are directly correlated with accuracy (rather than group membership), we examined the sets of 138 group and 123 non-group galaxies separately, before combining the probabilities. (The probabilities are derived using the one-sided Mann-Whitney test and are combined using the Fisher formula.) The upshot is that the signal does indeed seem to be stronger in the more accurately measured data ($p \sim 0.043$), consistent with a real signal, however caused.

These analyses confirm the tendency for the signal strength to reside in the best data; but they also reveal a complication, in that the best data tend to belong to small groups and associations. This illustrates the increasing difficulty of periodicity-testing as the search range extends beyond ~1000 km s^{-1} where only a small

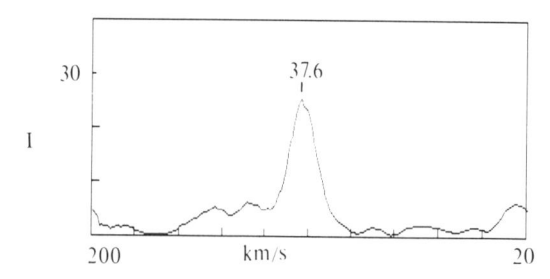

Figure 7. Power spectrum of differential redshifts in Figure 6b.

9.2 Local Redshift Periodicity

The 53 linked galaxies in the accurate sample of 97 comprise 9 doubles, 6 triplets, 3 quadruplets and 1 quintuplet. Subtracting each heliocentric redshift from every other, within each group, yields 55 local differential redshifts (of which 34 are independent). The sample is smaller and less accurate (the errors now being $\sqrt{2}$ times the previous ones), and the number of cycles within each group is small. Thus any periodicity will appear with less accuracy and reduced significance.

The uncorrected differential redshifts are plotted in Figure 6a. Although not readily evident to the eye, PSA of this sample yields a weak periodicity at ~ 38 km s^{-1}. Under the hypothesis that the corrected redshifts are periodic, it is not surprising that the raw redshifts should reveal one also, since the differential correction for the galactocentric solar vector, across a group of angularly close galaxies, is relatively small.

However, even this relatively small correction must be allowed for in the assessment of periodicity within a group. In Figure 6b are plotted the differential redshifts after subtraction of $V_\odot$ = (216 km s^{-1}, 93°, –13°). Since the solar correction is now second order, the precise choice within the accepted galactocentric range is not critical. A periodicity can now clearly be seen, and the corresponding power spectrum (Figure 7) peaks at 37.6 km s^{-1}, with I_{max} ~ 25.4. The significance of this peak value can be gauged by comparison with the I_{max} distribution obtained from synthetic data, equal to the real data except for the addition of random redshifts in the range ±50 km s^{-1} (S.E. 20 km s^{-1}). In these trials, signals anywhere in the range 20–200 km s^{-1} were recorded. The result of 1000 such trials is plotted in Figure 8. One peak of similar magnitude occurred, but at P = 20.5 km s^{-1} and ϕ ~ 66°, far from the ranges (24, 36 or 72 km s^{-1}) and zero phase being tested. Allowing for these latter factors, the periodicity seen in Figure 6b is found to have a chance probability $p \sim 10^{-4}$, to within a factor of a few.

Figure 6 reveals a difficulty with any hypothesis in which local redshift periodicity is ascribed to observational selection effects: any such factors would operate on the observed redshifts, whereas the periodicity emerges strongly only with respect to the corrected ones.

9.3 Local to Global

These results seem to indicate that there is indeed a periodicity within groups and associations of galaxies. It remains to be seen whether the fundamental process is this local periodicity, or whether galaxies in clusters are more or less test particles detecting a global one; the question reduces to whether these local periodicities are coherent in phase, from one cluster to the next. A grid search applied to the heliocentric (not differential) redshifts of the 53 group-linked galaxies reveals the presence of a sharp peak (I_{max} ~ 41) at a periodicity ~37.8 km s^{-1} and phase 27.5°, a remarkably strong signal for the size of the sample (Figure 9).

Its significance was assessed by shifting the groups randomly (within ±100 km s^{-1}) while preserving their internal relative heliocentric redshifts. This is equivalent to shifting the groups by 1–2 Mpc radially with respect to the Sun, destroying

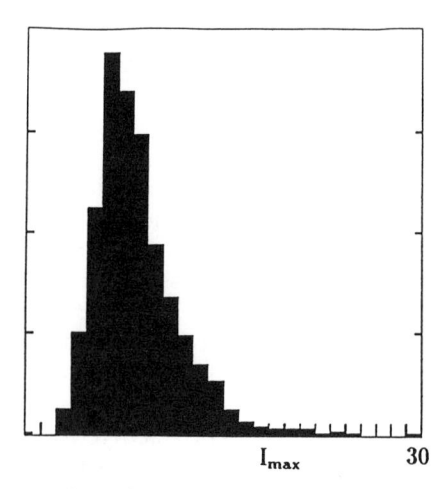

Figure 8. Distribution of I_{max} from PSAs of synthetic data simulating dV (real).

any global periodicity but preserving the internal ones. For each LSC so constructed, a grid search was carried out over a $20 \times 20°$ area, from 200 to 260 km s^{-1} in $V_\odot$, for periodicities between 34 and 39 km s^{-1}. The maximum peak found after 100 such trials was ~29.5 (Figure 10).

The significance of the $I \sim 41$ peak is hard to assess precisely from such a limited number of trials, but an exponential fit (theoretically expected) on to the high tail yields a probability $p \sim 10^{-4}$ that the real groups, by chance, would be so placed as to give the illusion of a strong global phenomenon. Thus, the evidence of these simulations is that (a) clustering cannot generate the observed signal; and (b) the periodicity is global in nature.

10. Summary

We have demonstrated the following:

(i) The data are consistent with the existence of a strong, coherent redshift periodicity of ~37.5 km s^{-1} extending over at least the Local Supercluster.
(ii) At a high confidence level, the observed peaks in the power spectra cannot be reproduced by a random redshift distribution.
(iii) Likewise, membership of groups and associations does not yield the observed signal.
(iv) The phenomenon is global, extending over at least the dimensions of the Local Supercluster.

The presence of an obscure observational artefact can never be absolutely excluded, but there is no independent evidence for its existence. The periodicity emerges only after correction for the galactocentric motion, and the phase coherence over the sky cannot readily be produced by, say, proximity effects. Thus, neither random nor clustered extragalactic redshifts, nor observational artefacts, seem capable of yielding the strong peaks observed, and we conclude that the periodicity is, in all likelihood, real.

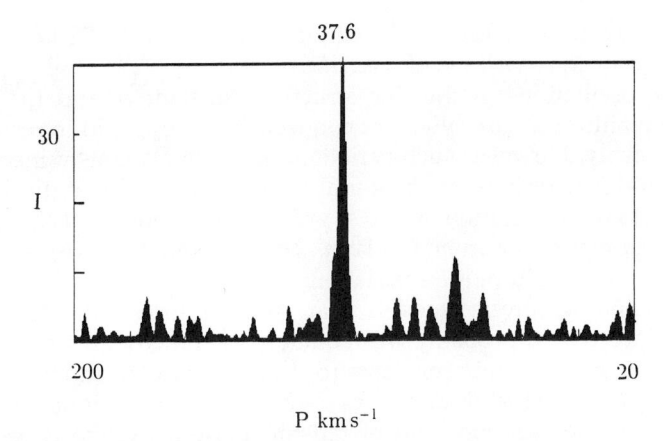

Figure 9. Power spectrum of the 53 linked galaxies, with $V_\odot$ correction.

11. Possible Physics

Until the phenomenon has been fully explored observationally, any discussion of the physics is likely to be premature. We may nevertheless consider a few hypotheses, with all due caution. In principle, redshift periodicity might arise from regularity in the structure of the LSC, or from oscillating physics, acting either on the atoms in the galaxies or on the photons along their flight paths.

If the redshifts were simply taken at face value, (i.e. as velocities) then we would have to suppose that the galaxies are arranged in a regular structure, have little or no peculiar motions, and that the whole rigid framework partakes conformally in the universal expansion. This simple notion has the advantage that it predicts the correct quantization interval Q to within a factor of two or three: it is given by $Q = H_o d$, where H_o represents the local Hubble constant and d is the projected scale length of the lattice. For $H_o = 75$ km s^{-1} Mpc^{-1} and $d = 0.5$ Mpc (if the mean distance between quantized galaxies in a group represents the lattice scale length), then $Q = 37.5$ km s^{-1}. A cubic lattice of side 37.5 km s^{-1} has r.m.s. dispersion, corner to corner, of ~10 km s^{-1}. Numerical trials are under way to determine whether the observations can, in fact, be reproduced by a lattice structure.

Assuming that a lattice exists and is not due to a multiply connected Universe (Fang 1990), how could it be maintained dynamically? A network of cosmic strings (Luo and Schramm 1992) is unlikely to have the required regularity. A gravitational pulse $\lambda \sim 1$ Mpc may be associated with the anisotropic collapse or explosion of a mass $M \sim c^2 \lambda / 2G \sim 10^{19}$ M$_\odot$, but would be too transient to generate periodic structure. If the nuclei of massive galaxies were formed, dissipatively, at the nodes of a system of standing waves, then the phenomenon might in principle be explained. The need for coherence of the periodicity over the LSC (at least) might imply inflation, and hence that such regularities existed within a microscopic horizon and were maintained during the inflationary expansion (cf. Buitrago 1988).

The hypothesis is outlandish, but only because the phenomenon is equally so. However, it is not clear that the hypothesis of lattice structure, whether imposed by 'quantum imprint' or otherwise, could account for the observed streaming motions;

and because of random projection, it cannot account for the ~72 km s^{-1} periodicity supposed to occur in binaries (Tifft *loc. cit.*).

Periodic oscillations in the fine structure constant α and the mass of the electron m_e (quantum electrodynamics connects the two) could in principle lead to redshift periodicity. However such variations are strongly constrained by geology: the Oklo natural reactor in west equatorial Africa, which went 'critical' ~2 Gyr ago, has been used to put a variation of $\leq 10^{-17}$ yr^{-1} on the proportional change in α, with a variation of similar order on m_e (Hill *et al.* 1990; Shlyakhter 1976). The periodicity, on the other hand, would require variations ~ 10^{-10} yr^{-1}.

Such oscillating physics models share the common difficulty that peculiar systemic motions would probably wash out any periodicity imposed through temporal variations in the atomic constants (or in the metric, or in Hubble's constant: Morikawa 1990). While the virial masses derived for galaxy groups may have little meaning, due to the superposition of redshift periodicity, the expected velocity dispersions based solely on the luminosity masses of galaxies are ≥ 20 km s^{-1} for a typical group, well in excess of the dispersion ~ 8 km s^{-1} observed for the 'best' solar vector. However, at the time of writing we have still to test the possibility that $\sigma \sim 20$ km s^{-1} (say), with a non-optimum vector, and is artificially reduced by the optimizing process.

Inflation, whether of the garden variety or of that originally envisaged by Hoyle and Narlikar (1966) in their C-field theory, might enter the story through the presence of a weak scalar field ϕ, oscillating in a harmonic potential due to its own mass $m^2\phi^2$ and interacting weakly with some other field or particle involved with light propagation (Hill *et al.* 1990, Crittenden and Steinhardt 1992) such fields are expected in some inflation-based cosmologies (Linde 1990). A rapidly oscillating vacuum energy and a finite photon rest mass ($\leq 10^{-43}$ g) may be implicit in such a photon/soft boson interaction. Finite photon mass can be made compatible with both Lorentz invariance and QED (Vigier 1990). The absence of peculiar motions is still problematic in such a scheme.

These highly speculative comments neglect the other redshift anomaly claims made by Arp and colleagues over the years. However, given the existence of one type of redshift anomaly, it becomes unsafe to ignore the others, and the true explanation may turn out to bear no relation to any of the above.

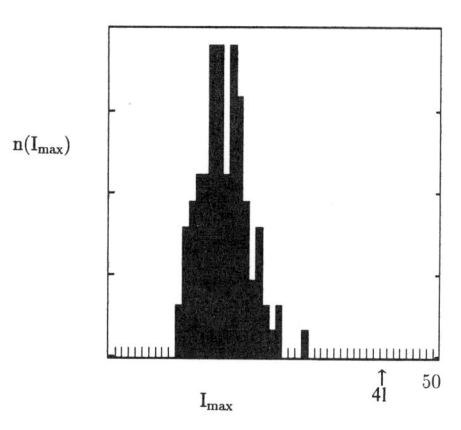

Figure 10. I_{max} distribution from 100 box searches on data simulating the 53 linked redshifts.

12. Conclusion

The falling of stones from the sky is physically impossible.
Paris Academy of Sciences (Memorandum, 1772).

Redshift anomalies have a long and contentious history in astrophysics. Our analysis of new data was intended to confirm or refute the existence of perhaps the most bizarre of such claims, namely redshift quantization. In the event, the phenomenon has been confirmed at a high confidence level.

An 'anomaly' in science, as defined by Lightman and Gingerich (1991), is an observed fact that is difficult to explain in terms of existing paradigms. Their historical analysis of past anomalies indicates that these are generally ignored until given compelling explanations within a new conceptual framework. Whether any of the currently fashionable concepts and models of cosmology can absorb this mysterious phenomenon remains to be seen. If not, the periodicity may indeed have to await the development of new perceptions before it can be assimilated into the mainstream of scientific thought.

References

Arp, H.C. and Sulentic, J.W., 1985, *Astrophys. J.* 291:88.
Baiesi-Pillastrini, G.C. and Palumbo G.G.C., 1986, *Astron. Astrophys.* 163:1.
Bartlett, M.S., 1955, *An Introduction to Stochastic Processes*, Cambridge University Press.
Blitz, L. and Spergel, D.N., 1991, *Astrophys. J.* 379:631.
Bottinelli, L., Gouguenheim, L., Fouqué, P. and Paturel, G., 1990, *Astron. Astrophys. Suppl.* 82:391.
Buccheri, R. and De Jager, O.C., 1989, in: *Timing Neutron Stars*, eds. H. Ögelman and E.P.J. van den Heuvel, p.95, Kluwer Academic.
Buitrago, J., 1988, *Astron. Lett.* 27:1.
Clube, S.V.M. and Waddington, W.G., 1989, *Mon. Not. R. astr. Soc.* 237:7P.
Combes, F. and Gerin, M., 1985, *Astron. Astrophys.* 150:327.
Crittenden, R.G. and Steinhardt, P.J., 1992, *Astrophys. J.* 395:360.
de Vaucouleurs, G. and Peters, W.L., 1984, *Astrophys. J.* 287:1.
Delhaye, J., 1965, in: *Galactic Structure, Stars and Stellar Systems*, eds. A. Blaauw and M. Schmidt, 5:61, University of Chicago Press.
Einasto, J., 1979, in *The Large-Scale Characteristics of the Galaxy*, IAU Symp. No. 84, ed. W.B. Burton, p.451, Reidel.
Fang, L., 1990, *Astron. Astrophys.* 239:24.
Flin, P. and Godlowski, W., 1989, *Sov. Astr. Lett.* 15:374.
Fouqué, P., Gourgoulhon, E., Charmaraux, P. and Paturel, G., 1992, *Astron. Astrophys. Suppl.* 93:211.
Gunn, J.E., Knapp, G.R. and Tremaine, S.D., 1979, *Astr. J.* 84:1181.
Guthrie, B.N.G. and Napier, W.M., 1990, *Mon. Not. R. Astr. Soc.* 243:431.
Guthrie, B.N.G. and Napier, W.M., 1991, *Mon. Not. R. Astr. Soc.* 253:533.
Haud, U., Joeveer, M. and Einasto, J., 1985, in: *The Milky Way Galaxy*, IAU Symp. No. 106, eds. H. van Woerden, R.J. Allen and W.B. Burton, p.85, Reidel.
Hill, C.T., Steinhardt, P.J. and Turner, M.S., 1990, *Phys. Lett. B* 252:343.
Horne, J.H. and Baliunas, S.L., 1986, *Astrophys. J.* 302:757.
Hoyle, F. and Narlikar, J.V., 1966, *Proc. Roy. Soc.* A290:162.
Lightman, A. and Shapiro, O., 1991, *Science* 255:690.
Linde, A.D., 1990, *Inflation and Quantum Cosmology*, Academic Press.
Luo, X. and Schramm, D.N., 1992, *Astrophys. J.* 394:12.
Morikawa, M., 1990, *Astrophys. J.* 362:L37.

Rood, H.J., 1982, *Astrophys. J. Suppl.* 49:111.
Scargle, J.D., 1982, *Astrophys. J.* 263:835.
Shlyakhter, A.I., 1976, *Nature* 264:340.
Stothers, R.B., 1991, *Astrophys. J.* 375:423.
Tifft, W.G., 1976, *Astrophys. J.* 206:38.
Tifft, W.G., 1980, *Astrophys. J.* 236:70.
Tifft, W.G., 1990, *Astrophys. J. Suppl.* 73:603.
Tifft, W.G and Cocke, W.J., 1984, *Astrophys. J.* 287:492.
Tifft, W.G. and Cocke, W.J., 1988, *Astrophys. J. Suppl.* 67:1.
Tifft, W.G. and Huchtmeier, W.K., 1990, *Astron. Astrophys. Suppl.* 84:47.
Vigier, J.-P., 1990, *IEEE Trans. Plasma Science* 18:64.

Compact Groups of Galaxies

Jack W. Sulentic

Department of Physics and Astronomy
University of Alabama, Tuscaloosa 35487 USA

We discuss the history and current status of the paradoxes associated with compact groups of galaxies. Dynamical theory requires that compact groups merge rapidly. Currently observed groups must, therefore, continually collapse out of the looser group environment. The observations do not support these model predictions nor do the models explain how the groups resist the onset of merger while in the process of formation. The observations are more consistent with the notion that the compact groups are long-lived or even primordial. At the same time, conventional theory completely ignores the large number of compact groups with at least one discordant redshift member. There are approximately nine times more such configurations than are expected by chance (37 *vs.* 4). Recognition of this fact may represent the greatest challenge to conventional ideas about this remarkable form of clustering.

Compact Groups and Clustering

Compact groups of galaxies (hereafter CGs) are a constant source of challenges to our ideas about galaxy evolution as well as the nature of the redshift. It can be said that the redshift controversy originated in the late 1950's with observations of the famous group called "Stephan's Quintet." Often apparent anomalies or perceived problems with a scientific paradigm do not stand the test of time. These false clues become weaker as more data is accumulated or some new theoretical idea shows them to not be the problem first imagined. Yet CGs have stood the test of time and remain a paradox. As additional data have accumulated, the problems posed by CGs have grown. This review the history of observations and ideas about compact groups will show that they pose some of the toughest questions for conventional ideas about galaxies. This is true whether or not the discordant redshift components are considered. Perhaps a resolution of the first problem lies with an understanding of the second.

It is now well established that we live in a universe characterized by clustering. Galaxies, the fundamental building blocks of the universe, are clustered on many scales from loose groups (*i.e.* the Local Group) to rich clusters. It is now realized that these groups and clusters exist in larger assemblages called superclusters. The concept of a more diffuse "field" population of galaxies has gradually

Figure 1. The environs of Seyfert's Sextet (Palomar 5m image). It is the most compact and one of the more isolated compact groups. There are no galaxies of comparable brightness within 45 arcmin.

faded over the past twenty years. This was perhaps most forcefully demonstrated by the discovery of large voids between the superclusters (see *e.g.* Sulentic 1980 for one of the first hints). Compact groups have emerged during the past decade as an "unexpected" form of galaxy clustering. CGs involve aggregates of from 4–10 galaxies with projected separations often on the order of the component diameters. Compact groups show surface density enhancements over their local environment on the order of 10^2 to 10^3 which imply space densities as high as 10^4 Mpc^{-3}. Accepted as physical aggregates, they are as dense as the cores of rich galaxy clusters. A most surprising aspect is that these dense cores are found in lower density regions of extragalactic space.

The reason that the compact groups have recently, in effect, thrust themselves upon us is historical. Some of the most famous ones were recorded in the peculiar galaxy catalogs of Vorontsov-Velyaminov (1959, 1977) and Arp (1966). The most famous of them, Stephan's Quintet, was discovered in the 19th century (Stephan 1877), and studied spectroscopically in the late 1950's. CGs were regarded by most astronomers as rare curiosities or even accidental projections of unrelated galaxies. Almost all of the papers about CGs published before 1985 focused on the three famous discordant redshift groups: Stephan's Quintet (SQ), Seyfert's Sextet (SS) and VV172 (VV). Early impressions and the focus of effort changed with the publication of a reasonably complete survey of these objects (Hickson 1982; hereafter HCG). The HCG contains a compilation of 100 compact groups found during a visual search of the Palomar Sky Survey. This survey indicates that approximately one percent of the nonclustered galaxies are members of compact groups or, alternatively, that one CG exists for every 50 loose groups (Mendes de Oliviera and

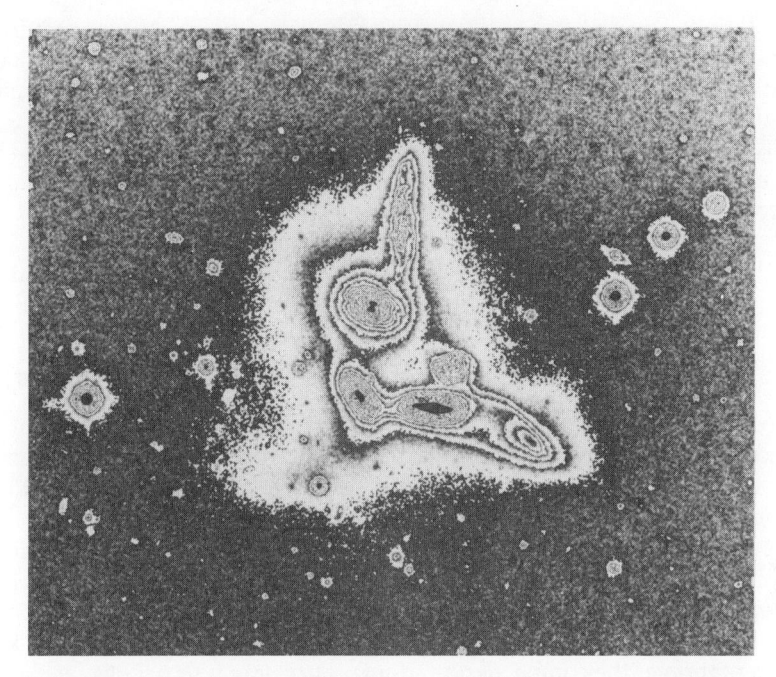

Figure 2. Computer processed image of Seyfert's Sextet. This 5m image was enhanced to show the luminous envelope surrounding the galaxies.

Hickson 1991). The realizations: a) that the compact groups are relatively common, and b) that they are difficult to explain and have generated considerable recent observational and theoretical interest.

Selection Criteria: The Uniqueness of Compact Groups

The selection criteria used in assembling the HCG are a reflection of the characteristics that had previously been noted about objects like SQ, SS and VV. They are compact aggregates of four or more galaxies located outside of clusters. Figure 1 shows an image of SS that was chosen to illustrate the relative isolation that is characteristic of most compact groups. There are no galaxies of comparable brightness within 48 group diameters (0.8 and 3.5 Mpc at the low and high redshift distances; $H_o = 75$). Figure 2 is a computer processed image of the same group that emphasizes the compactness and the signs of interaction between the galaxies (see Sulentic and Lorre 1983). The HCG selection criteria, consequently, involved richness, compactness and isolation.

Richness: The HCG consists of groups of four or more galaxies with a spread in apparent magnitude $m_F - m_B \leq 3.0$, where m_B and m_F are the magnitudes of the brightest and faintest components.

Compactness: Groups were required to have a mean surface brightness $\mu \leq 26.0$ mag. arcsec^{-2}.

Isolation: The distance to the nearest non group neighbor was required to be greater than three times the group radius ($R_N \geq 3 R_G$ where R_N is the distance to the nearest neighbor with apparent magnitude in the range of the group and R_G is the group radius). The final catalog contains 463 galaxies in 100 groups.

A very difficult question is the degree to which these selection criteria bias the HCG sample of compact groups. Certainly they insure that the HCG contains most of the compact groups with surface brightness as high as SQ, SS and VV ($\mu = 22, 20$ and 24 respectively). The HCG is estimated to be nearly complete for compact groups with $\mu \leq 24.0$ (Hickson 1982). A V/V_m calculation (Sulentic and Rabaça 1993) shows that the HCG becomes seriously incomplete for groups with a combined magnitude below $m_G = 13.0$. Where are these missing CGs and what are their properties? Will they change our basic description of compact groups and their relationship to other forms of clustering?

Possible evidence for the uniqueness of compact groups as a class of clustering (and, therefore, a negative answer to the latter question) comes from recent work towards a southern hemisphere catalog. Prandoni *et al.* (1992) are attempting an automated search of the southern sky for compact groups satisfying the HCG selection criteria. In the process of incorporating the selection criteria into their algorithm, they have graphically represented the selection domain of the HCG. It is possible to show the boundaries of the HCG selection domain in a plot of group radius versus magnitude of the brightest group member. Figure 3 shows a schematic representation of the compact group domain in relation to richness, isolation, and compactness. It is clear that compactness and richness are the defining criteria for most compact groups. The isolation criterion is redundant for groups with the brightest member above 16th magnitude. In other words, groups of four or more galaxies that have a combined surface brightness above 26.0 always satisfy the isolation criterion. Similar groups in or near clusters might not pass an isolation criterion. One candidate group was found near the Virgo cluster (Mamon 1989);

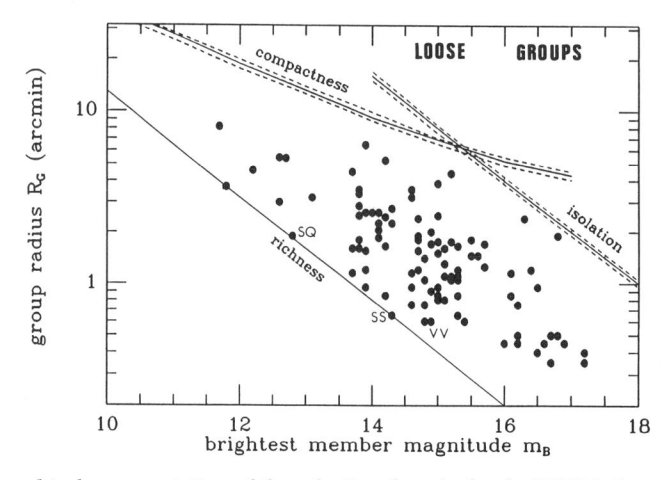

Figure 3. A graphical representation of the selection domain for the HCG (adapted from Iovino *et al.* 1993). Approximate boundaries of the richness, isolation and compactness criteria are indicated. The three famous quintets SQ, SS and VV are indicated.

but objects like SQ, SS and VV are not often seen near clusters. The HCG sample is plotted within the domain represented in Figure 3. If compact groups were an extension of the more general population of loose groups, we would expect the HCG points to cluster near the upper boundary of the domain. Looser groups would have larger size and lower surface brightness than compact groups like SS. Note that HCGs concentrate near the center of the selection domain. The most compact CGs fall near the lower richness boundary. The HCG was assembled with a visual search which will be most sensitive to the most obvious compact groups. Will the many CGs missed in the HCG survey fill in the upper part of the selection domain of Figure 3?

Recently Iovino *et al.* (1993) have published a list of the first 55 compact groups detected in the automated survey. Differences between brightest and faintest group member magnitudes were evenly scattered between 0 and 2.5 in the HCG while the new sample is strongly peaked between 2.5 and 3.0. Thus the new sample contains more low surface brightness groups, as expected from the V/V_m results. Despite the difference, however, the new groups do not concentrate at the upper boundary of the selection domain. They generally fall higher than HCG objects, but they do not appear to bridge the loose group domain. This suggests that compact groups may well represent a unique form of galaxy clustering. Images like Figures 1 and 2 certainly reinforce this impression.

Conventional Ideas About Compact Groups and Problems

Are compact groups physically dense systems and, if so, what are the implications of this? It may seem difficult to imagine that anyone would question the physical nature of groups like SQ and SS. The fact that this question has been raised, and for reasons beyond the issue of discordant redshift components, is a reflection of the difficulties that CGs pose for conventional ideas. Let us assume for the moment that most compact groups are real. Following the current fashion, we will ignore the discordant redshift components. Note that SQ, SS and VV are still compact quartets in this case. Components in the groups show very small projected separations, implying very high space densities. The velocity dispersion is usually quite small (SQ at 400 km s^{-1} is twice the mean for the HCG). Attempts to simulate the evolution of a compact group always lead to the same result: they merge very quickly into presumably elliptical-like remnants (Carnevali *et al.* 1981; Ishizawa *et al.* 1983; Mamon 1986; Barnes 1985, 1989). The exact time scale varies depending upon the exact recipe: *e.g.* distribution and amount of dark matter, velocity dispersion and group crossing time. The main result is that in timescales on the order of a few crossing times (typically a few times 10^8 years) the groups will have become triplets, binaries or single remnants. The final coalescence may take more than a Gyr, but the groups will disappear from a catalog like the HCG in much less time. This result implies that the compact groups now observed are all "young" systems. They are found in lower density regions of space, which implies that they continually form out of their looser group environment. Otherwise we should observe no compact groups at all. Note that attempts to prolong the lifetime of a CG (*e.g.* Governato *et al.* 1991) invoke special conditions that are not obviously applicable to most objects.

As is often the case, the observations appear to contradict the theoretical predictions. Let us count the ways: 1) The foremost difficulty is how to bring four or more galaxies very close together, allowing them time to dynamically evolve, before the onset of the first merger. 2) The ratio of early to late type galaxies in compact groups is different from their local environment. They show a distinct excess of E and S0 galaxies (Hickson *et al.* 1989, Mendes de Oliviera and Hickson 1991, Rabaça and Sulentic 1993). It seems unlikely that secular evolutionary effects would have time to account for these changes if most present groups are recently collapsed from the field. 3) The observations reveal little evidence for mergers in progress. 3a) The ellipticals in the HCG show no evidence of color anomalies expected if they are the first stage in the coalescence of currently observed groups (Zepf and Whitmore 1991). 3b) The ellipticals in compact groups have also been shown to be not sufficiently "first ranked" compared to the predictions of models that view them as partial mergers (Mendes de Oliviera and Hickson 1991). 3c) There is a surprisingly low level of optical, FIR and radio enhancement in compact groups considering that widely cited merger remnants like Arp 220 are so active at these wavelengths (Sulentic and Rabaça 1993, Sulentic and de Mello Rabaça 1993). In fact, long wavelength enhancements are generally interpreted as evidence for merger-induced starburst activity. 4) Searches for merger remnants in the environment where the compact groups are found reveal few candidates (Sulentic and Rabaça 1993). The models have suggested quite high densities of such remnants implied by the rapid coalescence timescales (*e.g.* 10^{-4} Mpc^{-3}, Barnes 1989). The few elliptical galaxies found in loose groups tend to be significantly lower in luminosity than would be expected for the remnants of compact groups. Claims to the contrary (*e.g.* Zepf and Whitmore 1991) involve comparison samples rich in luminous cluster (*e.g.* Centaurus A and Fornax A) elliptical galaxies (see Rabaça and Sulentic 1993).

It is the above impressive array of observational and theoretical contradictions that has motivated some to argue that compact groups cannot be real. The implications are otherwise too disturbing, for they suggest that the groups are long-lived, maybe even primordial configurations. We cannot rule out the possibility that the HCG represent the dynamically "lucky" part of a much larger primordial population of compact groups. Argument 4 above remains the strongest argument against this view since any existing remnant population is quite small. The published article that described the most sophisticated compact group simulation so far (Barnes 1989) was accompanied by an invited (or contributed) editorial. This accompanying piece, after summarizing the model predictions, contains the comment: "A quick check of recent observations shows no contradictions with the scheme!" It is left to the reader to decide whether the above discussion indicates a contradiction or not.

Compact Groups: Real or Imagined?

The strongest advocate of the chance projection hypothesis argues that 70% or more of the HCG represent various types of chance alignments or false groups (Mamon 1986, 1993). The strongest argument against the chance projection idea is the low surface density of galaxies found near the compact groups. Three independent estimates of the actual surface densities have been made (Sulentic 1987, Rood and Williams 1989, Kindl 1990). Each used slightly different procedures, but the studies all yield the same essential result. Figure 4 illustrates the results from

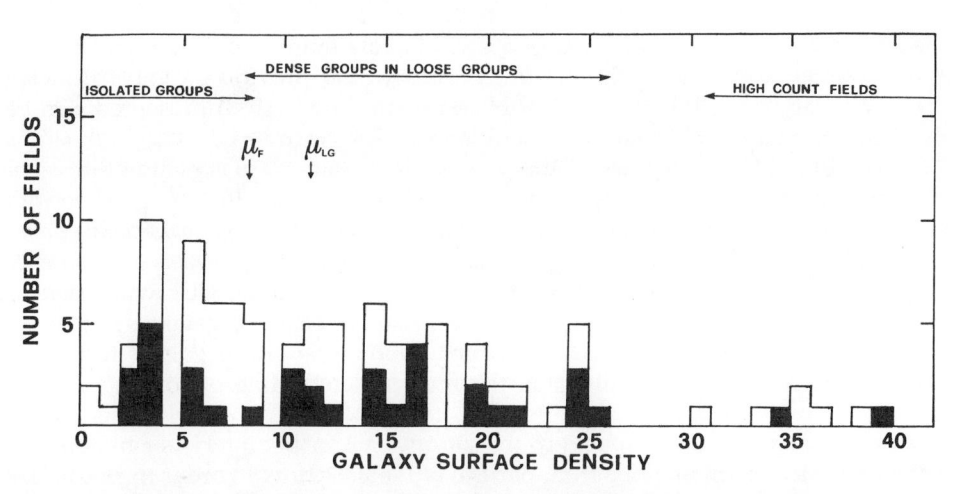

Figure 4. Distribution of galaxy surface densities for the HCG (adapted from Sulentic 1987). Values are given for a one half degree search radius in units of galaxies deg^{-2}.

Sulentic (1987) where the very isolated, loose group and clustered regimes are indicated. They suggest that compact groups are found in low surface density "loose-group" type environments. Is this a reasonable estimate? The Sulentic (1987) estimates were based upon galaxy counts out to a radius of one degree around each CG. A fixed search radius was chosen in order to avoid any dependence on the redshift. Following the conventional procedure, one degree at $V_o = 10^4$ km s^{-1} (and $H_o = 75$) corresponds to 2.3 Mpc which is a typical loose group diameter. This suggests that our procedure is also reasonable from the standard redshift-distance relation point of view. Estimates of the number of chance projections by accordant redshift interlopers using the above results yield very small numbers. The surface densities are simply too low to produce many false groups with accordant redshifts. Note that the fraction of compact groups decreases as the surface density increases in Figure 4. This emphasizes the point that CGs are typically found in nonclustered regions. If the opposite were true, then the number should increase with surface density until the effect of the isolation criterion becomes strong. Stated another way, if all isolated CGs are due to chance alignments, then we should see large numbers of CG-like configurations in clustered fields (even if they do not satisfy the HCG isolation criterion). My own observational experience suggests that this multitude of nonisolated false groups does not exist.

The other arguments against the chance projection hypothesis are multifold. They involve the observational evidence that the groups are physical systems. It is worthwhile to summarize here the principal elements: A) Compact groups tend to be deficient in neutral hydrogen by roughly a factor of two (Williams and Rood 1987). This probably indicates that the gas has been stripped in close encounters and collisions between the groups members. B) More than half of the spiral components that have been studied kinematically show rotation curve peculiarities (Rubin et al. 1991). C) Some of the groups are surrounded by a common luminous envelope (e.g. SQ, SS and VV172) (Sulentic 1987) or show bridges, tails and other deformations suggestive of interaction. D) Admittedly limited data show three out of five observed compact groups as X-ray sources, two of which are diffuse sources

(Bahcall *et al.* 1984). The detection of a considerable number of additional diffuse X-ray sources associated with compact groups would provide the most ironclad proof that they are physical systems. Hickson and Rood (1988) presented a more detailed summary of the evidence that compact groups are real.

The bulk of the evidence at the present time leads us to conclude that compact groups are real. Their properties are sufficiently unusual that we must consider them as a unique form of galaxy clustering: *aggregates of galaxies with implied densities like the cores of rich clusters but located outside of clusters; systems that appear to be much more long lived than theory predicts.* At the very least, CGs will teach us something fundamental and new about the dynamics of dense galaxy systems. At the outside, they may teach us new things about the entire galaxy formation process. We raise the possibility that the final understanding of compact groups may also involve the discordant redshift components. There are far too many discordant members of compact groups for any reasonable interloper hypothesis. This conclusion is not obvious from perusal of the literature. In order to emphasize the problem and provide the information upon which an opinion should be based, we review the history of discordant redshifts in compact groups. The history begins, appropriately enough, with a discussion of Stephan's Quintet.

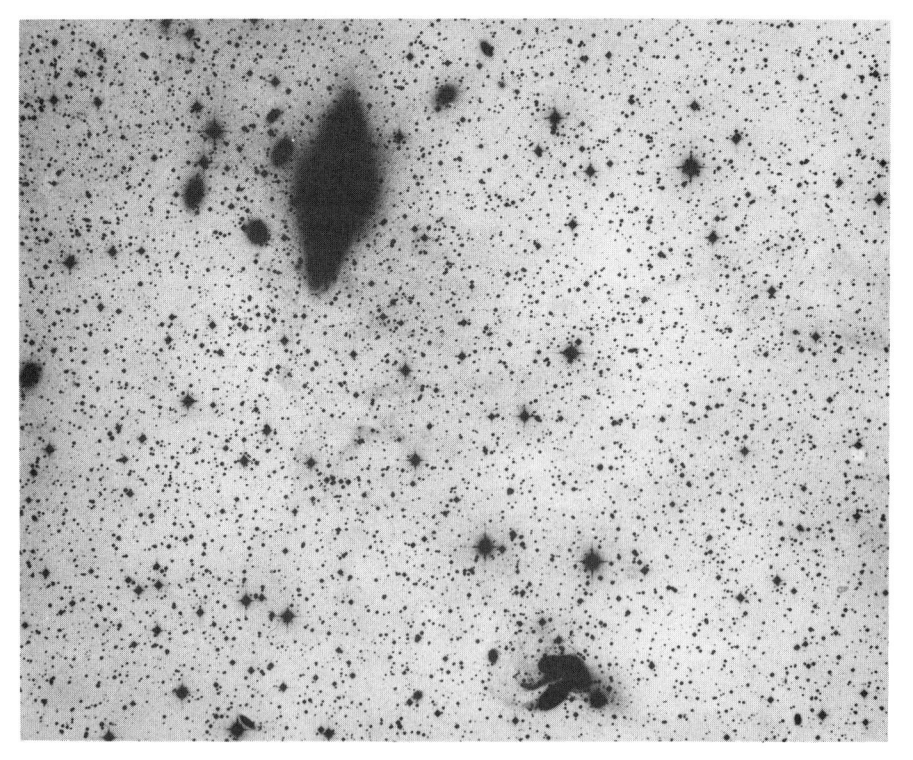

Figure 5. Image of the Stephan's Quintet region including NGC 7331 at the upper left (Palomar 1.2 m image). The galaxy images are overexposed to emphasize faint structure.

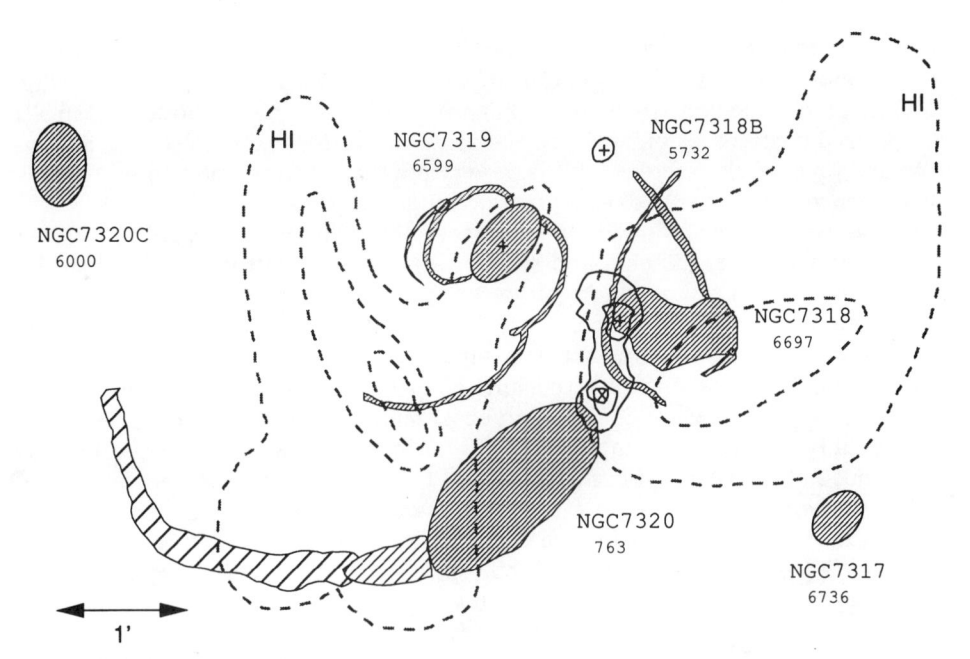

Figure 6. Schematic representation of Stephan's Quintet. High redshift HI and radio continuum (20 cm) contours are superimposed. Redshifts are indicated under each galaxy.

Discordant Redshifts: Direct Evidence

Figure 5 shows an image of Stephan's Quintet and its neighborhood. The quintet consists of four galaxies, NGC 7317, 18, 18B and 19, with redshifts near 6000 km s^{-1} and one galaxy NGC 7320, with a redshift near 800 km s^{-1}. The discovery of the "discordant" redshift for NGC 7320 was announced by Burbidge and Burbidge (1961) well before the discovery of quasars. There are two possible interpretations for this puzzling result. NGC 7320 is either a member of the group or is a foreground projection. If the former is true then either the redshift of NGC 7320, or those of its companions, do not obey the standard redshift-distance relation. There are three approaches for deciding between the possible interpretations. 1) One can look for evidence that NGC 7320 is interacting with the other galaxies in the Quintet. 2) One can try to derive redshift independent distances for the galaxies in SQ or 3) One can estimate the probability that NGC 7320 is accidentally projected on the accordant redshift quartet.

Work before 1982 tended to focus on the two former approaches because compact groups were known to be so rare that any estimate of the latter kind provided a disturbing answer. The probability calculation also had the disadvantage of being *a posteriori*, since at first only one (the Burbidges' estimated 10^{-3} probability of chance projection for SQ), and then three examples (SQ, SS and VV) were known. Any such calculation will almost always yield a very small probability because it depends upon the (very small) surface area of the group. The Burbidges discussed the size and luminosity of the discordant galaxy implied both by its redshift and the mean redshift for the remainder of the group. They had earlier noted (before the redshift of NGC 7320 was measured: Burbidge and Burbidge 1959) that the velocity dispersion for this group was quite high (about 400 km s^{-1}). Using virial theorem arguments popular at that time, they concluded that SQ was unbound (in an early reference to a now very popular concept, they argued against the existence of sufficient "dark matter" to bind the system).

The two decades following the discovery of the discordant redshift in SQ saw a large number of arguments and counter-arguments about the relationship between the high and low redshift galaxies. This group has probably been observed with more telescopes and at more wavelengths than any other galaxy aggregate. The results are a lesson in how subjective the interpretation of data can be and how difficult it can be to estimate a redshift independent distance. They also reveal, unfortunately, how the credibility of an observation or interpretation depends less on its quality than upon its implications for the standard paradigm. We summarize many, but not all, of the arguments raised. Figure 6 presents a schematic of SQ which can be used as a reference to the cited work along with Figure 5. Redshifts are indicated under each of the galaxies in Figure 6. NGC 7320C, which may be related to SQ, is also shown. The bright Sb spiral NGC 7331 (with redshift similar to NGC 7320) is located one half degree to the upper left (NE).

1) Kalloglyan and Kalloglyan (1967) pointed out that "The only—though fairly strong—argument against NGC 7320 being a member of the quintet is its radial velocity." They presented three color photometric data which showed that NGC 7320 was brighter in the U band on the side towards the high redshift galaxies. They argued that the data was more consistent if NGC 7320 was at the higher redshift distance.

2) In a paper titled "Stephan's Quartet?" Allen (1970) presented the first neutral hydrogen measures for NGC 7320. As the title suggests, he found the HI measures consistent with the redshift distance of the galaxy.

3) Arp (1971) suggested that the four higher redshift members of the quintet might be at the same distance as NGC 7320. He proposed that the group was ejected from NGC 7331, located 30 arcmin distant.

4) Arp (1972) argued that there was an excess of strong radio sources in the region between SQ and NGC 7331. He suggested that they were related to SQ and the proposed ejection event.

5) Lynds (1972) published redshifts for a number of galaxies in the region of NGC 7331 and SQ. He argued that the results were consistent with two unrelated sheets of galaxies at distances consistent with the two redshift systems in SQ.

6) Allen and Hartsuiker (1972) published a high resolution radio continuum map for SQ. They found a source coincident with NGC 7319 and a peculiar "arc" of emission between NGC 7319 and 7318B. It was noted that the arc extended south to the interface with NGC 7320.

7) Van der Kruit (1973) employed new Westerbork observations to argue that there was "definitely" no excess of radio sources in the region.

8) Arp (1973) presented images and spectra of the HII regions in the high and low redshift members of SQ. He showed that the HII regions in NGC 7320 were concentrated on the side towards the higher redshift quintet members. Arp argued that the largest HII regions in NGC 7320 and higher redshift NGC 7318 showed similar apparent size. He argued that this was most consistent with the

known properties of the largest HII regions in galaxies if both were located at the redshift implied distance of NGC 7320.
Observations of a supernova in NGC 7319 were found to be inconsistent with either redshift distance, but more seriously with the lower value.

Finally, Arp presented new images that revealed a curved filament extending from the SE end of NGC 7320 opposite the center of SQ. He argued that the width and resolution of this feature suggested that it was an appendage of NGC 7320. He argued that this feature was strong evidence for interaction between NGC 7320 and the rest of SQ.

9) Balkowski et al. (1973) published HI observations for NGC 7319 and concluded that this high redshift member was most likely at a distance near NGC 7320.
10) Shostak (1974ab) published HI observations for NGC 7319 and concluded that this high redshift member was most likely at its higher redshift distance. The principal difference between 9 and 10 was the calibration sample used.
11) Kaftan-Kassim and Sulentic (1974) published new low frequency radio continuum observations of the SQ and NGC 7331 region. They found evidence both for a diffuse radio connection between these two objects and for an excess of (steep spectrum) radio sources in the field.
12) Kaftan-Kassim et al. (1975) published a high resolution radio continuum map of SQ. They resolved the radio "arc" discovered previously. The southern component of the "arc" falls near the interface between NGC 7318ab and 7320 (see X in Figure 6).

Stridency alert: The establishment papers begin to take on a much more strident tone at this point. The frequency of words like "normal", "not unusual" and "typical" show a dramatic increase.

13) Gillespie (1977) published deep radio continuum survey data for the NGC 7331/SQ area. He concluded that the region was "normal in all its radio properties except for the distribution of the brightest sources." He concluded that these were unrelated to SQ or NGC 7331.
14) von Kap-herr et al. (1977) published radio continuum maps of the NGC 7331/SQ region. They concluded that no extended emission is present between the objects. They confirmed an excess of steep spectrum sources in the area.
15) Gordon and Gordon (1979) report on a search for HI emission in the velocity range between the two redshift systems in SQ. They detect no emission and conclude that "the data are consistent with the standard cosmological interpretation of redshifts."
16) Allen and Sullivan (1980), Peterson and Shostak (1980) and Shostak et al. (1984) reported new high resolution HI measures at both the high and low redshifts associated with SQ. They found the high redshift HI was not coincident with the optical galaxies but, instead, is displaced in several large clouds (Figure 6). This result invalidated all previous attempts to assign distances to the galaxies using HI properties. They found that low redshift HI associated with NGC 7320 showed "no peculiarities which cannot reasonably be related to known instrumental deficiencies." The latter authors found HI derived distances for galaxies near the quintet (with redshifts similar to the higher system). Those HI properties were found to be consistent with the redshift implied distance.

17) Kent (1981) used velocity dispersion measures for the high redshift members of SQ and the HI profile width for NGC 7320 to conclude that "All objects are found to be at distances consistent with their redshifts."
18) Van der Hulst and Rots (1981) published VLA radio continuum maps of SQ. They confirmed the previous resolution of the radio "arc" into two sources. The lower source peaks at the interface between the high and low redshift systems (NGC 7318B and 7320).
19) Sulentic and Arp (1983) reported new HI measures for NGC 7320. They argued for the presence of low redshift HI significantly displaced from NGC 7320. They showed that NGC 7320 is strongly HI deficient for its type. They argued that these results indicate that NGC 7320 is interacting with its higher redshift neighbors.
20) Bahcall et al. (1984) reported the X-ray detection of SQ. A diffuse source was found essentially coincident with the radio continuum emission at the interface between the high and low redshift systems (X in Figure 6).

At this point we enter the "rejection phase" where it is considered inappropriate for respectable scientists to carry out further observations of discordant redshift systems. We find both sides of the controversy convinced that their interpretation is the correct one. What have the data told us? The optical observations are extremely difficult to interpret. Comparisons between control samples and the high redshift SQ members are unlikely to prove anything. These galaxies are unambiguously involved in a strongly interacting system. Thus any discussion of abnormality in their properties cannot prove anything except that they are perturbed by interaction. NGC 7320, from this point of view, is favored for study and it does show some peculiarities. The most striking optical peculiarities include the luminous tail as well as the U band and HII region asymmetry. Does this evidence point towards interaction with the high redshift galaxies in SQ or a long past encounter with NGC 7331? Are the same peculiarities sometimes observed in isolated galaxies?

HI data for the high redshift galaxies reinforces the previous point about the uselessness of inferences concerning strongly interacting galaxies. The peculiar state of the HI is consistent with the general HI deficit in compact groups. The HI is not missing in this case, but merely displaced. It certainly establishes the physical nature of this compact quartet (or, at least, triplet). There is general disagreement over the level of peculiarity in the HI properties for NGC 7320. The claimed deficiency in HI is surprising and is a general property of galaxies in compact groups. If NGC 7320 is peculiar, who is responsible?

The radio continuum and X-ray data provide perhaps the most striking puzzle. The presence of diffuse X-ray emission supports the idea of SQ as a physically dense (and dynamically evolved) system and provides a further source for some of the missing HI. The centroid of the X-ray emission and one of the radio sources (X in Figure 6) falls near the SQ–NGC 7320 interface. It is not clear why this should be if NGC 7320 is a foreground projection. The lack of a clearcut answer from any of the above observations leads us to look for a statistical resolution of the problem.

Discordant Redshifts: Statistical Arguments

Much of the work on SQ was stimulated by the discovery of discordant redshift galaxies in two more of the most famous compact groups: Seyfert's Sextet

(SS) and VV172 (VV) (Sargent 1968, 1970; Burbidge and Sargent 1970). Seyfert's Sextet is the most compact (highest surface brightness) and one of the most isolated compact groups. The probability of a chance interloper in this case is vanishingly small. The problem with all of the individual calculations of interloper liklihood is that they are *a posteriori*. Further it is difficult to frame the appropriate question in the first place, since one can obtain any desired value for the probability of chance occurence by phrasing the question suitably. For example, the probability that a bright background Sc spiral falls along the same line of sight as a bright galaxy quartet is very small. In spite of this, the existence of three such discordant associations by the early 70's was a general source of uneasiness. Anyone familiar at the time with galaxy statistics and distributional properties appreciated the general rarity of compact groups and the consequent improbability of so many chance alignments.

The return to a statistical analysis of the problem was motivated by these three compact groups. A definitive statistical result, in the minds of many, was published in 1977 (Rose 1977). First it was noted that SQ, SS and VV are all quintets with a single discordant component. This reduces the problem to estimating how many real quartets of galaxies should have an unrelated galaxy superimposed. In principle, the simpler and better defined the question, the more credible and tractable the answer. One can simply estimate the interloper fraction by counting the number of quartets on the sky. It then follows that $N_5 = N_4 R$, where $R = A\sigma$ (N_4 and N_5 are the number of quartets and false quintets, A is the average group surface area and σ is the surface density of field galaxies). Rose (1977) reported the results of a search for galaxy quartets on the Palomar Sky Survey. He reported 26 definite and seven probable quartets on 69 (6° × 6°) Sky Survey fields. He extrapolated from this result an estimate of 433–550 compact quartets on the sky. This value, coupled with an estimate for the field galaxy density, yielded an expectation of 1.5–2.0 discordant redshift compact groups—in "remarkable agreement" with observation. This result was widely cited as the definitive resolution of the redshift controversy.

I can remember the uncritical and unreserved relief with which the above study was greeted. I can also remember how clear it was that the result was incorrect. There are nothing like 400–500 compact quartets on the sky with properties remotely similar to SQ, SS and VV. This fact became obvious with the publication of the HCG (Hickson 1982). Contemporaneously, I carried out my own reanalysis of the Rose survey (Sulentic 1983). The reanalysis was carried out by myself and, independently, by two of my undergraduate students. The idea was to compare an experienced and potentially biased result (my own) with the results of two completely naive catalogers. They were given a set of finding charts with SQ, SS and VV depicted at the same scale as the Palomar Sky Survey. They were told to catalog all systems of four or more galaxies with similar brightness and isolation. Results were compared and a final list of compact groups was assembled and compared with the list of Rose (1977).

The results of the reanalysis revealed that Rose (1977) had overestimated the number of galaxy quartets by a factor of between 2 and 10. Some of the objects were not even quartets. Many of the quartets were much fainter in apparent magnitude and lower in surface brightness than SQ, SS and VV. We found 2–3 times fewer objects in the same fields using soft selection criteria. We found ten times fewer objects when we required group properties to be similar to the famous three quintets (hard criteria). Note that this is true even if we neglect the contribution of the discordant components. At the same time, Hickson (1982) catalogued 100 compact groups down to declination −33°. He found 60 quartets which agreed well with our results. The HCG selection criteria accepted many compact groups with properties much less striking than SQ, SS and VV. Perhaps the most remarkable result of the Rose analysis was that no more than 10–12 of the 400–500 claimed compact groups had measured redshifts in 1977. Yet *all* of the expected discordant quintets had been discovered! It was immediately clear that this was extremely unlikely. This was pointed out in our reanalysis, which was accepted for publication in the *Astrophysical Journal*. Sadly, the referee commented that while acceptable for publication, our results were "unlikely to have any effect on thought."

The Latest Statistics on Discordant Redshifts

As stated earlier, the HCG has opened a new chapter in the study of compact groups. The existence of a reasonably complete sample has stimulated much study and underscored the difficulties summarized in earlier sections of this paper. Recently a redshift survey was completed for the HCG. The data have been kindly made available by Paul Hickson. The result is that approximately *37 out of 100* compact groups have at least one discordant redshift component (a galaxy with velocity differing from the mean of the accordant members by more than 2000 km s^{-1}). With complete redshift information we are in a position to obtain a reliable statistical expectation for interloper contamination in compact groups. We proceed by asking the simplest possible question. How many interlopers are expected in the HCG given the local surface densities. The HCG provides measured diameters for all HCG groups. Three independent sets of galaxy surface densities have been measured in the HCG fields (Sulentic 1987, Rood and William 1989, Kindl 1990). My survey counted galaxies in two zones of radius 0.5 and 1.0 degree. Two zones were used because not all neighboring galaxies are expected to have discordant redshift. In a superclustered universe, one expects many galaxies near any CG to have redshifts similar to the CG members. This is especially true if these neighbors have apparent size and magnitude similar to the CG components. Our inner count was intended to estimate this accordant redshift contribution. The outer annulus from 0.5 to 1.0° was regarded as the best estimate of the discordant redshift field. One can argue about the details of the procedure, but most of the numbers are small no matter how you do it. We used the local surface density and compact group diameter to estimate the number of discordant interlopers expected in each case. We then summed the expectations for the entire catalog and found values of 6, 0.2 and 0.02 for the number of single, binary and triplet interlopers. About one third of these objects are estimated to show accordant redshift with their respective compact groups. This leaves an expectation of four discordant compact groups in the HCG. The HCG redshift survey shows 26 single, 3 binary and 2 triple galaxy discordant interlopers. In addition we find five groups where all of the galaxies are discordant. In summary, we find about nine times more discordant cases than expected.

There is another viewpoint. Hickson *et al.* (1988) published an independent estimate of the interloper fraction. They concluded that the observed large number of discordant compact groups agrees well with statistical expectation. How could there be such a disagreement over such a straightforward calculation? We derived

Figure 7. A comparison of the group diameters used in Sulentic (1987) and Hickson et al. (1988) for estimating the discordant redshift interloper fraction in Seyfert's Sextet.

an estimate for all of the compact groups using their local surface densities and observed diameters. Hickson et al. (1988) returned to the question of the number of discordant quintets expected. This difference is irrelevant. The two estimates use similar galaxy surface densities but they use very different estimates for the area subtended by the groups. This difference will dominate any estimate, be it for one or one hundred groups. Sulentic (1987) used the actual group diameters as tabulated in the HCG. Since the group diameters include the discordant components, this seems a reasonable approach. The other advantage was that the definition of group diameter was independent of the test. Our approach becomes a simple question of how many interlopers are expected in a 0.5 deg^2 patch of sky (the combined area of the HCG groups).

Hickson et al. (1988), on the other hand, argued that the correct group diameter should represent the maximum group size where an interloper would still pass the HCG selection criteria. Figure 7 illustrates this point for SS. This group has an HCG defined diameter of approximately one arcmin. However, if an appropriately bright interloper had fallen up to approximately 8 arcmin from SS it would have passed the selection criteria. This argument leads to an increase in the surface area of "hypothetical SS" by more than 250 times. Using this approach Hickson et al. (1988) concluded that the large interloper population was expected. They argued that the Sulentic (1987) approach was valid only for internal discordant members. We find 10 HCG with internally discordant components (15 if one includes the 5 groups that are totally discordant) which is still a large excess over expectation.

We disagree with the conclusions of Hickson *et al.* (1988) and argue that their approach does not allow for the physical uniqueness and rarity of compact groups. This refers us back to Figure 3 which shows that compactness, and not isolation, is the determining factor for most compact groups. It seems unreasonable to base a calculation upon what groups might have been, when we have a well defined catalog of real groups to work with.

It still might be argued that the discordant component excess is an effect of the incompleteness of the HCG. The lower surface brightness compact groups are expected to have a much higher fraction of interlopers, as shown by Prandoni *et al.* (1992). Perhaps the discordant components pushed many lower surface brightness groups above the selection threshold. This seems unlikely to remove the excess for two reasons: 1) Only four groups fall in the surface brightness range $\mu = 25.0–26.0$ (the HCG cutoff). In most cases the addition of a discordant galaxy will not produce a large enough enhancement to elevate a group from below $\mu = 26.0$ to above $\mu = 25.0$, and 2) There is also a large excess of discordant groups among the subsample thought to be most complete (surface brightness above $\mu = 24.0$). It is not obvious how a selection effect can explain the excess of discordant redshift groups.

It has become unpopular (if it ever was popular) to discuss discordant redshifts, and especially to argue that there are too many of them. The bottom line, however, is whether we evaluate results based upon their scientific correctness or by the degree with which they fit into our preconceived ideas. There are far too many compact groups with discordant redshift components. The reason for this is not yet clear. As suggested earlier (and in Sulentic 1987), this excess may be a part of the solution to the more conventional challenges raised by the groups.

References

Allen, R.J. 1970, *Astr. Astrophys.* 7:330.
Allen, R.J. and Hartsuiker, J. 1972, *Nature* 239:324.
Allen, R.J. and Sullivan, W.T. 1980, *Astr. Astrophys.* 84:181.
Arp, H. 1966, *Astrophys. J. Suppl.* 14, No. 123.
Arp, H. 1971, *Science* 174:1189.
Arp, H. 1972, *Astrophys. J.* 174:L111.
Arp, H. 1973, *Astrophys. J.* 183:411.
Bahcall, N., Harris, D. and Rood, H. 1984, *Astrophys. J.* 284:L29.
Balkowski, C., Bottinelli, L., Chamaraux, P., Gouguenheim, L. and Heidemann, J., 1973, *Astr. Astrophys.* 25:319.
Barnes, J.E. 1985, *Mont. Not. R. Astr. Soc.* 215:517.
Barnes, J.E. 1989, *Nature* 338:123.
Burbidge, G.R. and Burbidge, E.M. 1959, *Astrophys. J.* 130:15.
Burbidge, E.M. and Burbidge, G.R. 1961, *Astrophys. J.* 134:244.
Burbidge, E.M. and Sargent, W.L. 1970, in: *Nuclei of Galaxies*, ed. D. J. K. O'Connell, (Amsterdam: North Holland), p. 351.
Carnevali, P., Cavaliere, A. and Santangelo, P. 1981, *Astrophys. J.* 249:449.
Gillespie, A.R. 1977, *Mont. Not. R. Astr. Soc.* 181:149.
Gordon, K.J. and Gordon, C.P. 1979, *Astrophys. Lett.* 20:9.
Governato, F., Bhatia, R. and Chincarini, G. 1991, *Astrophys. J.* 371:L15.
Hickson, P. 1982, *Astrophys. J.* 255:382.
Hickson, P., Kindl, E. and Auman, J. 1989, *Astrophys. J. Suppl.* 70:687.
Hickson, P., Kindl, E. and Huchra, J. 1988, *Astrophys. J. Lett.* 329:L65.
Hickson, P. and Rood, H. 1988, *Astrophys. J. Lett.* 331:L69.

Iovino, A., *Prandoni, I., MacGillivray, H., Hickson, P. and Palumbo, G.*, 1993, in: Proceedings of *Observational Cosmology*, in press.
Ishizawa, T. et al. 1983, *Pub. Astr. Soc. Jpn.* 35:61.
Kaftan-Kassim, M.A. and Sulentic, J.W. 1974, *Astr. Astrophys.* 33:343.
Kaftan-Kassim, M.A., Sulentic, J. and Sistla, G., 1975, *Nature* 253:1.
Kent, S. 1981, *Pub. Astr. Soc. Pac.* 93:554.
Kalloglyan, A.T. and Kalloglyan, N.L. 1967, *Astrophysics* 3:99.
Kindl, E. 1990, Ph.D. Thesis, Univ. of British Columbia.
Lynds, C.R. 1972, in: *External Galaxies and Quasi-Stellar Objects*, eds. D. S. Evans, (Reidel: Dordrecht), p. 376.
Mamon, G.A. 1986, *Astrophys. J.* 307:426.
Mamon, G.A. 1989, *Astr. Astrophys.* 219:98.
Mamon, G.A. 1993, in: Proceedings of HST Workshop on *Groups of Galaxies*, in press.
Mendes de Oliviera, C. and Hickson, P. 1991, *Astrophys. J.* 380:30.
Peterson, S.D. and Shostak, G.S. 1980, *Astrophys. J.* 241:L1.
Prandoni, I., Iovino, A., Bhatia, R., MacGillivray, H., Hickson, P. and Palumbo, G., 1992, in: *Digitized Optical Sky Surveys*, eds. H. T. MacGillivray and E. B. Thompson, (Kluwer Academic Publishers), p. 361.
Rabaça, C. and Sulentic, J.W. 1993, *Astr. Astrophys.*, submitted.
Rood, H. and Williams, B. 1989, *Astrophys. J.* 339:772.
Rose, J. 1977, *Astrophys. J.* 211:311.
Rubin, V.C. et al. 1991, *Astrophys. J. Suppl.* 76:153.
Sargent, W.L. 1968, *Astrophys. J.* 153:L135.
Sargent, W.L. 1970, *Astrophys. J.* 160:405.
Shostak, G.S. 1974a, *Astrophys. J.* 187:19.
Shostak, G.S. 1974b, *Astrophys. J.* 189:L1.
Shostak, G.S., Sullivan, W.T., III, and Allen, R., 1984, *Astr. Astrophys.* 139:15.
Stephan, M. 1877, *Mont. Not. R. Astr. Soc.* 37:334.
Sulentic, J.W. 1980, *Astrophys. J.* 241:67.
Sulentic, J.W. 1983, *Astrophys. J.* 270:417.
Sulentic, J.W. 1987, *Astrophys. J.* 322:605.
Sulentic, J.W. and Arp, H. 1983, *Astron. J.* 88:267.
Sulentic, J.W. and de Mello Rabaça, D. 1993, *Astrophys. J.* June 20.
Sulentic, J.W. and Lorre, J.J. 1983, *Astr. Astrophys.* 120:36.
Sulentic, J.W. and Rabaça, C. 1993, in: Proceedings of HST Workshop on *Groups of Galaxies*, in press.
van der Hulst, J.M. and Rots, A.H. 1981, *Astron. J.* 86: 1775.
van der Kruit, P.C. 1973, *Astr. Astrophys.* 29:249.
von Kap-herr, A., Haslam, C.G. and Wielebinski, R., 1977, *Astr. Astrophys.* 57:337.
Vorontsov-Velyaminov, B. 1959, *Atlas and Catalog of Interacting Galaxies*, (Moscow).
Vorontsov-Velyaminov, B. 1977, *Astr. Astrophys.* 28:1.
Williams B. and Rood, H. 1987, *Astrophys. J. Suppl.* 63:265.
Zepf, S.E. and Whitmore, B.C. 1991, *Astrophys. J.* 383:542.

Is there Matter in Voids ?

Bogdan Wszolek

Jagiellonian University Astronomical Observatory
ul. Orla 171, 30-244 Kraków, Poland
Fax: (12) 37-80-53
E-mail: bogdan@oa.uj.edu.pl (internet)

The results of studies of the South Coma void are compared with searches for intergalactic dust in the same region. The presence of dust in voids is suggested.
Key words: intergalactic matter, voids, redshifts

1. Introduction

Measurements of redshifts of galaxies allow us to speculate about the 3-dimensional distribution of these objects in the Universe. The most widely accepted interpretation of the observed redshifts is a Doppler shift. If this is the case, measured redshifts combined with the Hubble-law give direct information about the distance to objects. Observations of the 3-dimensional distribution of galaxies reveal a bubble-like structure of the Universe. Groups and clusters of galaxies occupy about $1/10$ th of the available space. The remaining volume is taken up by voids, which contain almost no luminous matter. The spatial scale of these voids is (10–40)h^{-1} Mpc, where h is the Hubble constant in units of 100 km s^{-1} Mpc.

On the other hand, there also exists non-luminous intergalactic matter, which theoretically may occupy galaxy clusters as well as large volumes within voids. The interactions of intergalactic matter with galaxies, and with radiation produced by them, may affect the interpretation of the extragalactic observations. Unfortunately, our present knowledge of the content and the distribution of intergalactic matter is very limited.

Intergalactic clouds are believed to be common in the remote Universe (absorption features in quasar spectra), but nearby examples are exceedingly rare. Good examples of large intergalactic clouds that are sufficiently close for there to have been extensive multifrequency studies are in the M96 group of galaxies at a distance of about 10 Mpc (Schneider et al. 1983) and in the vicinity of NGC300 at a distance of about 3 Mpc (Mathewson et al. 1975). These clouds were originally discovered by means of their radio emission at 21 cm wavelength from atomic hydrogen (HI). A multifrequency survey of the cloud in M96 group (Schneider et al.

1989) indicates that intergalactic gas in this cloud is nearly devoid of stars. For this reason, other objects of this type will be difficult to find, even though they may constitute an important population in the Universe. A few other nearby intergalactic cloud candidates have also been discovered by means of the their resulting extinction (Hoffmeister 1962, Okroy 1965, Rudnicki & Baranowska 1966).

The first systematic search for non-luminous matter in cosmic voids was made by Krumm and Brosch (1984). The authors tried to detect isolated HI clouds in two nearby voids in the direction of the constellations of Perseus-Pisces and Hercules. They concluded that there are no signs of the existence of proto-galactic-size HI clouds in these voids.

Yet there are good arguments for the existence of intergalactic matter within voids. In the Bootes void, several faint emission-line galaxies have been detected (Moody et al. 1987, Weistrop 1989). Gondhalekar & Brosch (1986) have observed $Ly\alpha$, SiIV and CIV absorption lines in background quasars in the direction of the Bootes and Perseus-Pisces voids, and they have argued that these lines are produced within voids.

In the present paper the possible presence of dust in voids is suggested.

2. Observations in South Coma Void Region

In 1965, long before hearing about voids in the distribution of galaxies, Okroy noticed a large area (about 150 square degrees) with a visible lack of galaxies in the direction of the Coma/Virgo constellations. Murawski (1983) suggested that this deficit was due to an obscuring cloud of intergalactic nature rather than to a variation in the distribution of galaxies. He showed that the N(m)-curve of galaxies in the direction of the cloud is shifted towards fainter stellar magnitudes (Wolf-diagram) as compared to a control region. He also noticed that the galaxy colors in the cloud area are redder than in the surrounding parts.

The presence of a large void in redshift space in the region of the Okroy cloud (The South Coma void) has been proposed by Tifft and Gregory (1988). They showed that a region of about 110 square degrees at velocities between 5000 and 6000 km s^{-1} is quite empty of galaxies with $m \leq 15.7$.

It should be mentioned that the break in the Wolf diagram and the redder average colors can, in principle, be well explained by the missing galaxy population at intermediate z-values, as occurs in the presence of a void. It simply indicates that the more distant and on the average redder galaxies contribute relatively more to the total sample. The numerical results due to the two effects (obscuration and true void) may be quite different, and are strongly dependent on the distance and size of the void. In the most general case, the two effects occur together.

Wszolek et al. (1989) have carried out an analysis of the Okroy cloud using infrared data (IRAS). We have found infrared emission from the cold dust from the central part of this cloud.

3. Discussion

For the South Coma void there is a very good coincidence with the Okroy cloud (see also Wszolek 1992). The 100μm radiation has a clear maximum near the center of the void and could extend to the outer parts, where the presently available infrared signals are too weak to be differentiated from noise. The other known intergalactic clouds also seem to coincide with voids, but the overlap is not as clear as for the Okroy cloud. For these clouds no infrared emission was found in IRAS data.

While dust absorption as the sole cause for the absence of galaxies in the region of the Okroy cloud can be ruled out, the presence of dust in the direction of a well established void may be of great importance. The existence of non-luminous matter in voids would be very significant for cosmological theories because such theories consider mainly gravitational masses. The matter distributed on the line of sight to the observed voids interacts with electromagnetic radiation coming from galaxies to the observer. If scattering processes are not purely elastic, this matter can produce additional redshifts, independent of any Doppler shift. However, no physical process has so far been accepted to explain the origin of such additional redshifts. Unfortunately, to show that voids are occupied by a reasonable amount of matter, or that any hypothetical process produces redshifts, and perhaps voids, more data are needed.

References

Gondhalekar, P.M. and Brosch, N., 1986, *IAU Symp.* No.119:575.
Hoffmeister, C., 1962, *Zeitschr. f. Astrophys.* 55:46.
Krumm, N. and Brosch, N., 1984, *Astron. J.* 89:1461.
Mathewson, D.S., Cleary, M.N. and Murray, J.D., 1975, *Ap. J.* 195:L97.
Moody, J.W., Kirshner R.P., MacAlpine, G. and Gregory, S.A., 1987, *Ap. J.* 314:L33.
Murawski, W., 1983, *Acta Cosmol.* 12:7.
Okroy, R., 1965, *Astron. Cirk.* 320:4.
Rudnicki, K. and Baranowska, M., 1966, *Acta Astron.* 16:65.
Schneider, S.E., Helou, G., Salpeter, E.E. and Terzian, Y.,1983, *Ap. J.* 273:L1.
Schneider, S.E., Skrutskie, M.F., Hacking, P.B., Young, J.S., Dickman, R.L., Clausen, M.J., Salpeter, E.E., Houck, J.R., Terzian, Y., Lewis, B.M. and Shure, M.A., 1989, *Astron. J.* 97:666.
Tifft, W.G., Gregory, S.A., 1988, *Astron. J.* 95:651.
Weistrop, D., 1989, *Astron. J.* 97:357.
Wszolek, B., Rudnicki, K., Masi, S., deBernardis, P. and Salvi, A., 1989, *Astroph. Sp. Sci.* 152:29.
Wszolek, B., 1992, *Apeiron* 13:1.

Redshifts and Arp-like Configurations in the Local Group

M. Zabierowski

Wroclaw Technical University
Wyb. Wyspianskiego 27
50370 Wroclaw, Poland

It is shown that each of the geometrically defined "lines" (subgroups) of galaxies in the Local Group of galaxies (considered by Iwanowska) contains members of various redshifts. In each line, however, we can distinguish members of quasi-discrete "quantized" redshift states. The so-called "possible" members of LG are real members of LG, according to the criterion developed in present paper. The redshift state of our Galaxy is represented by $k = -1$.

I. Velocity Dispersion s_{n-1} in the Local Group

1. All Galaxies in Iwanowska Lines

We consider all galaxies classified by Iwanowska (1989) as members of Arp-like lines contained in the Local Group (LG). The mean residual velocity of all galaxies listed by Iwanowska is

$$v = 60 \pm 84(s_{n-1}), \quad n = 37, \quad (s \geq v) \tag{1}$$

It is evident that the dispersion $s_{LG, n-1}$ (all velocities are given in km s^{-1}) is higher than the mean velocity itself. Here, n is the total number of galaxies forming all the Iwanowska's line configurations (resembling Arp configurations).

Let the velocity dispersion s_{n-1} be denoted simply as s. The Local Group dispersion s is comparable with the dispersion of the Virgo ellipticals (E+SO) investigated, among others, by Sulentic (1977), although the Local Group and Virgo Cluster are not comparable with respect to number of members. Hence the following question arises: Is the high value of dispersion s_{LG} a subclustering effect (caused by the existence of galaxy subgroups)? Furthermore, will it be possible to reduce this high value of s_{LG} by dividing it into subgroups?

For this purpose, we consider a division of the Local Group into its two most natural parts: the M31 galaxy system and the system our Galaxy. Afterwards, we will also investigate the subgroups distinguished by Iwanowska. To ensure the uniformity of our material and to avoid the objection that we may have selected the material expressly for our purpose, we will use only Iwanowska's data. She distinguished the Arp lines of galaxies associated only with the Andromeda Galaxy (M31) and with our Galaxy.

2. The M31 Group

The mean residual velocity in the case of the M31 system of galaxies is

$$v = 71 \pm 86, \quad n = 15, \quad (s \geq v) \tag{2}$$

the dispersion value 86 km s^{-1} and all values of dispersion being hereafter denoted simply as s (standing for s_{n-1}). In the present case of the M31 Group, the value of dispersion s_{M31} is roughly equal to s_{LG}. With this division, the large value of s cannot be reduced.

3. Our Galaxy Group: the Next Iwanowska Class

The Milky Way system of Iwanowska lines yields

$$v = 53 \pm 83, \quad n = 22, \quad s \geq v \tag{3}$$

Again the dispersion of values is higher that the mean value itself. All the numbers given in (1), (2) and (3) are of the same order of value. Again, the division of LG into subgroups causes neither any reduction of redshift dispersion nor any substantial change of the mean redshift. Thus, the following question appears: can we improve the situation by dividing the groups considered above into Iwanowska lines.

4. Five Subgroups in the Local Group: All Iwanowska Lines

The next possible partition of the Local Group follows from the Iwanowska's interpretation of the spatial distribution of galaxies in LG. She obtained five, physically related, straight galaxy lines (called by Iwanowska "bipolar jets").

The three bipolar jets connected with Andromeda Galaxy are denoted by Iwanowska as A, B and C. For these three jets we have obtained respectively:

$$v = 5 \pm 71, \quad n = 6, \quad (\text{jet C}) \tag{4a}$$

$$v = 72 \pm 120, \quad n = 4, \quad (\text{jet B}) \tag{4b}$$

$$v = 88 \pm 83, \quad n = 7, \quad (\text{jet A}) \tag{4c}$$

again the value of s is close to the value of s_{LG}.

For the bipolar structure associated with our Galaxy, denoted by Iwanowska as "a" and "b", we have obtained respectively

$$v = 44 \pm 102, \quad n = 8, \quad (\text{jet a}) \tag{5a}$$

$$v = 51 \pm 78, \quad n = 15, \quad (\text{jet b}) \tag{5b}$$

and again, the dispersion values of single jets are similar to the values of s_{Gal}, s_{M31} and s_{LG}.

We come to a highly illuminating conclusion: it is impossible to reduce the velocity dispersion by dividing galaxies into separate geometrical groups. The same, very large s is preserved for all five subgroups (lines of galaxies); the value of s actually increases instead of decreasing. This means that the high value of the velocity dispersion is a basic regularity.

5. Some General Remarks

In this work, we have not used any *a posteriori* partitioning of the Local Group. We have used only the partition proposed by Iwanowska. We can see that the two kinds of space partition do not lead—in the case of the Local Group—to any substantial reduction of the velocity dispersion. For all galaxies in the Local Group, the dispersion s is about 80 (1). For smaller, geometrically distinguished narrow groups, we obtained values of the same order—(4a), (4c) and (5b)—or even as high as 100 km s^{-1} and more—(4b) and (5a). It would be quite unnatural for galaxies to form some peculiar configurations (several geometrically recognizable lines) in physical 3-space; nevertheless, the velocity dispersion s cannot be reduced by space subdivisions.

Although each bipolar line can be divided into two segments, no velocity improvement (reducing s) has been reached in this way. Our conclusion is that the redshift states were well mixed even in quite isolated segments of Iwanowska's geometrical 3-dimensional lines in the Local Group. Thus, taking into consideration present results, it becomes even more difficult to argue that stability of the chainlike configurations of galaxies when we try to retain the Doppler interpretation of redshifts.

6. Towards an Explanation

Instead of the spatial partition, we can examine the velocity (redshift) partition. Surprisingly, all the galaxies considered above may be grouped into five distinct redshift groups:

1:    Galaxy, Dra, M31, IC 10, 1613   (6a)

$$s_{n-1} = 8, \quad s_n = 7, \quad n = 5$$

2:    LMC, SMC, Agr, For, Leo A, Scl, Sgr, UMi, NGC 6822, IC 5152, WLM   (6b)

$$s_{n-1} = 17, \quad s_n = 16, \quad n = 11$$

3:    Peg, M32, M33, NGC 147, 185, 205, 1569, 6456, A92   (6c)

$$s_{n-1} = 16, \quad s_n = 15, \quad n = 9$$

4:    Sex A, Sex B, NGC 1560, 3109, 4236, DDO 187, GR 8, Maffei 1, 2   (6d)

$$s_{n-1} = 20, \quad s_n = 19, \quad n = 9$$

5:    Car, NGC 404, DDO 47   (6e)

$$s_{n-1} = 24, \quad s_n = 20, \quad n = 3$$

This grouping of galaxies minimizes s, which varies by about 10. The values of s_n are presented for comparison. The values of s are stable. The expected value of s (random distribution of redshifts) is 22. The value obtained is slightly larger only in one group (6e). The values of s increase systematically from class 1 to 5.

In the calculated mean velocities v for the respective groups we also find a permanent regularity resembling Tifft's redshift states among residual velocities v = 72 k km s^{-1}, k = 0,±1,.... Comparison of observational data with expected Tifft-values yields the following:

Group 1: $v = -58$ ($v_T = -72$, $k = -1$) (7a)

Group 2: $v = 1$ ($v_T = 0$, $k = 0$) (7b)

Group 3: $v = 65$ ($v_T = 72$, $k = 1$) (7c)

Group 4: $v = 141$ ($v_T = 144$, $k = 2$) (7d)

Group 5: $v = 219$ ($v_T = 216$, $k = 3$) (7e)

It is worthwhile to point out that not only is the observed periodicity consistent with the Tifft's 72 km s^{-1} periodicity, but also one of distinguished mean velocities coincides almost exactly with the expected k = 0 value, i.e. no phase shift was revealed. This fact cannot be explained easily in terms of Doppler effect.

The important question arises: How does the grouping of redshifts repeat in the case of the individual lines; or how are individual groups occupied?

II. Distribution of Redshifts in Iwanowska's Richest Line, the Milky Way "b"-Chain

1. k = 0 State

The zero velocity group is made up of the following Milky Way companions: Aqr, For, Leo A, Scl, Sgr, NGC 6822, IC 5152, WLM. The mean residual velocity is close to zero. Moreover, the values of all the residual velocities are close to zero, thus $v = 0$ for the whole group and v_o is nearly equal to zero $|v_o| < 27$; the index "o" stands for an individual velocity. The exact value is

$$v = -1 \pm 14(s_{n-1}), \quad n = 8, \quad (s_8 \ll s_1) \tag{8}$$

Indexes 1,2,3,... denote cases (1),(2),(3),... respectively. It is evident that $s_8 \ll s_1$ and the value of v_8 is quite different from v_1, v_2 and v_3 (Equation 3). The outstanding value of v (here v_8) confirms the hypothesis that it is a separate" state" of redshifts.

2. k = −1 State

The Companions of Our Galaxy. If one adds, to the galaxies mentioned above, all the other companions (excluding the Galaxy itself), then s rapidly increases from $s = 14$ to $s = 74$. This fact justifies the isolation of the zero-velocity group. For all the companions we have:

$$v = 59 \pm 74, \quad n = 14, \quad (s_9 \gg s_8) \qquad (9)$$

The Companions Plus Our Galaxy. Our Galaxy itself seems to form a separate z-state. The "apex" velocity of our Galaxy determined in a classical way from "radial velocities" of Local Group galaxies is accepted here as the "redshift" of our Galaxy. In this case, we have a highly negative state, i.e. $v_o = -60$ (here v_o means the individual velocity of our Galaxy). This negative state does not fit the general tendency given by $v = 59$ (Equation 9). When our Galaxy is included, the value n_9 increases, but s_9 also increases because the outstanding value of $v_o \neq v_9$ is not compensated by increasing n. For the whole "b"-jet (companions plus the Milky Way) we have

$$v = 50 \pm 78, \quad n = 15, \quad (s_{10} > s_9 \gg s_8) \qquad (10)$$

The value obtained for s_{10} supports the hypothesis advanced in section I.6 that there is a negative state ($k = -1$) of z such that its v is close to $v_o = -60$. If this hypothesis is correct, we should expect a low dispersion around 10 for the $k = -1$ state. This claim can be easily verified empirically in the future. Some arguments in favor of it will be given below.

3. High Velocity State: k = 2

There are two arguments that the five N-companions (the North galaxies of the "b"-jet), i.e. Sex A, B, NGC 3109, GR 8, and DDO 187, form a separate redshift class:

a. All these galaxies constitute the substantial part of the northern half of the b-jet ("substantial", i.e. without Leo A and DDO 47—these two galaxies fit a new redshift state). They form the distinguished, i.e. nonrandom concentration of galaxies in the "b"-jet and, moreover, there is no standard-explanation of such a high concentration of the same redshifts in a single position of the LG-space.
b. Comparison of v and s with the values given by Equations (8) and (9) is decisive, namely:

$$v = 128 \pm 8, \quad n = 5, \quad (v_{11} \gg v_9 \gg v_8) \qquad (11)$$

This result for v and s (large v-excess $v_{11} - v_8$) indicates that it is not the zero-velocity group; nor is it the all companions group (9), because v is too large ($v_{11} - v_9$, $v_{11} - v_{10}$) and s is small ($s_9 \gg s_{11}$ and $s_{10} \gg s_{11}$); evidently s_{11} has a value similar to s_8.

Hence, the above-mentioned group of five galaxies does not belong to the zero-velocity group (8), to the all companions group (9), to the galaxies of the "b"-jet [the jet includes our Galaxy, (10)], or to our Galaxy state, called the $k = -1$ state. The pair of values of v and s compared with the previous results (other groups in the "b"-chain) reveals that the (11)-group has to be treated as separate.

The pairs of v and s values obtained above show that the separate redshift states appear also in a single Iwanowska line. The next questions are: Do there exist unoccupied states? Can the so-called "possible" members of the Local Group fit the proposed redshift grouping scheme? Is there any small redshift dispersion group of the LG-galaxies not fitting the empirical "quantization" law $v = 72$ k km s^{-1}?

4. DDO 47: An Observational Outsider and Possible $k = 3$ State

This companion probably forms a separate redshift state with a proposed $v_{k=3}$ near the individual velocity v_o characterizing DDO 47

$$v_k \simeq v_o = 191, \quad s = 10, \quad (n = 1) \tag{12}$$

If the companion DDO 47 is classified as belonging to the $k = 2$ state, then the addition of this one single member makes the dispersion increase from $s = 8$ to $s = 27$. Thus, this high velocity outsider ("possible" companion) belongs instead to the higher state $k = 3$.

5. The Five Different States in the Case of the "b"-Chain

We have obtained the five different states:

$k = -1$, our Galaxy, $n = 1$, $s \simeq 10$? (hypothetical estimate) (13a)

$k = 0$, the zero velocity group, $v = -1$, $s = 14$, $n = 8$ (13b)

$k = 1$, the unoccupied state (and there are no galaxies in the "b"-chain between $v_o = 33$ and $v_o = 118$; moreover the dispersions s of the neighborhood states are very low and reliable due to large n) (13c)

$k = 2$, the high-velocity group, $v = 128$, $s = 8$, $n = 5$ (13d)

$k = 3$, DDO 47, $v \simeq v_o = 191$, $s \simeq 25$? (13e)
(estimate from the difference $216-191 = 25$)

The empirical status of s in (13a) and (13c) is different because the redshift of DDO 47 is established directly, and the redshift of our Galaxy is established from statistical investigation, as noted in section II.2.b.

III. The Milky Way: "a"-Straight Chain

The recognized companions of the "a"-line (the companions directly included in Iwanowska's diagram (Figure 6 in the referenced work by Iwanowska) yield

$k = 0$, $v = 6$, $s = 26$, LMC, SMC, UMi, $n = 3$
(tides are often considered as influencing their velocities) (14a)

$k = 1$ unoccupied (14b)

$k = 2$ unoccupied (14c)

$k = 3$, $v = 234$, $s \simeq 18$? Car, $n = 1$, (14d)
(estimated from the difference $234-216 = 18$)

and there is one negative state in the case of this Iwanowska line, represented by two members—Dra and Milky Way:

$$k = -1, v = -59, s = 1, \text{Dra, Milky Way}, n = 2. \tag{14e}$$

Iwanowska has found the geometry of the "a"-bipolar line of the companions of our Galaxy. Nevertheless, two galaxies, *i.e.* NGC 6456 and 4236, have been treated in two ways in her work, since, on the one hand, the unusual 3-space geometry forces us to include these companions in one bipolar jet, while on the other hand, some authors consider them only as "possible" (the term "possible" stands for the whole astronomical tradition; we can recognize the meaning of this word from the context of the astronomical works) member of the Local Group. Thus Iwanowska also treats them as uncertain because there was no extra criterion for their membership. Now, using the regularity discovered in the present paper, we are forced to fit them into the empty states presented in (14b) and (15c):

$$k = 1, v_o = 73, \text{NGC 6456}, s = \sim 10, 74\text{--}73 = 1, n = 1 \tag{14b'}$$

$$k = 2, v_o = 142, \text{NGC 4236}, s = \sim 10, 144\text{--}142 = 2, n = 1 \tag{14c'}$$

There is no question whether the two mentioned galaxies considered as "uncertain" belong to the former or the latter state. Of course, they fit the regularity of z obtained just from "certain" members. I follow Iwanowska's terminology, which is part of the whole internal problem (history and logic) situation in astronomy.

IV. The M31 Configuration of Galaxy Lines

1. All M31 Jets

The full set of M31 jets yields the following redshift states:

$$k = -1, v = -57, s = 11, \text{IC10, 1613, M 31}, n = 3, \tag{15a}$$

$$k = 0, \text{unoccupied (expected redshift state)}, \tag{15b}$$

$$k = 1, v = 68, s = 16, \text{NGC 147,205,1569, M32,33, Peg, A92} \; n = 7, \tag{15c}$$

$$k = 2, v = 156, s = 16, \text{NGC 1560, Maffei 2}, n = 2. \tag{15d}$$

We ignore here NGC 185, which is considered by Iwanowska as a "certain" member of the M31 lines. Its velocity is $v_o = 41$; thus, it could be classified as a $k = 1$ state. However, no included galaxy has such great difference $|72_\odot k - v_o|$, which is 31 in the case of NGC 185. If one includes NGC 185 as a member of the most abundant state ($k = 1$, $n = 7$) then the result for v and s remains almost unchanged. Nevertheless, we must point out this individual case.

In the astronomical literature, there are often questions as to whether NGC 404 is a member of the Local Group or not. Iwanowska's approach creates a new criterion, and we can argue that NGC 404 is a member of the M 31 line system. The approach presented here (similar to Tifft's) creates a second criterion: again NGC

404 is a LG member. The velocity of NGC 404 is $v_o = 232$. Thus it fits the k = 3(!) state. As a result, we have the following redshift state

$$k = 3, v = 205, s = 38, \text{NGC 404, Maffei 1}, n = 2. \tag{15e}$$

2. The Three Bipolar Jets

There are three bipolar jets in M31 system selected by Iwanowska. However, all three jets are poor in galaxies with measured z. We can observe that the general regularity discovered for the Local Group, the Milky Way and the entire M31 system is repeated in each individual case of these three chains. This explains why the sub-partition of the Local Group does not show the decreasing s. It is expected that in the individual cases, a Tifft-like picture of the galaxy redshifts could be only slightly influenced by the tides. The connection between the hypothesis of the bipolar jets, Arp's hypothesis, and the role of tides can be considered an open problem (cf. Grabinska and Zabierowski 1979).

3. Some Peculiarities of the M31 Jets

Assuming that there exist no Tifft (intrinsic) redshift states, i.e. that redshifts are explained in terms of velocities and these velocities v_o are dynamical, caused by gravitational attraction, we are unable to explain the observed highly unbalanced distribution of positive and negative values of v_o.

Iwanowska (1989) has noted that galaxies located on both ends of the A line of the M31 system show a negative v_o. M31 has itself the same negative value. However, the A-line galaxies located between the main galaxy and the end companions all have positive v_o. In the Doppler approach, this phenomenon is very strange. It appears exactly as if the intermediate galaxies of this line are not bound physically to the mother-galaxy, and do not participate in the common group movement. Nevertheless, they are located on the same geometrical line. In the Tifft approach, this phenomenon is at least completely possible and perhaps even points toward a new regularity.

Discussion

In the investigations presented here, all galaxies considered by Iwanowska were used. Only one galaxy (NGC 185) did not fit the "quantization" scheme very well, but even this one would not spoil the general picture in a statistical sense (the s criterion is fulfilled: see section IV.1). It is impossible to identify any group of redshifts outside Tifft's scheme. Thus, it is hard find any convincing argument that single galaxies or groups of them move one another.

The LG-velocity dispersion s was compared with the velocity dispersion of the ellipticals of the cluster in Virgo, which is $s_E = 70\text{-}80$ km s^{-1} (Sulentic 1977). The peculiar form of Iwanowska's configurations contradicts the high value of s_{LG}.

Iwanowska (1989) doubted that bipolar galaxy lines considered by her were formed by capture during collapse (isotropic or anisotropic, cf. Rudnicki et al. 1989). Arp's hypothesis is particularly relevant for the problem. But why did lines of galaxies survive? Iwanowska's application of Arp's hypothesis to the galaxy lines in the LG is valid if we assume that basic components of redshifts are due to quantization of an unknown physical nature and small relative motions of galaxies produce only low Doppler deviations from these main "quantum" states, causing the observed dispersions s.

Another result of our analysis is that the redshift state assigned to our Galaxy is not $k = 0$ but $k = -1$.

References

Grabinska, T. and Zabierowski, M., 1979, *Astrophys. Sp. Sci.* 66:503.
Iwanowska, W., 1989, in: *From Stars to Quasars*, S. Grudzinska and B. Krygier (eds.), Mikolaj Kopernik University, 159.
Rudnicki, K., Grabinska, T. and Zabierowski M., 1989, in: *From Stars to Quasars*, S. Grudzinska and B. Krygier (eds.), Mikolaj Kopernik University, 119, 125, and 137.
Sulentic, J., 1977, *Astrophys. J.* 211:L59.
Tifft, W., 1975, *Discrete States of Redshift and Galaxy Dynamics*, Steward Observatory Preprint No 44.

Are the Galaxies Really Receding?

Fred L. Walker

3881 S. Via del Trogon
Green Valley, Arizona 85614

Sixty years of competing speculation and theory have failed to establish conclusively whether the Big Bang expansion concept is valid. Three methods to resolve the continuing impasse are discussed. First, it can be shown that two initial assumptions of the theory directly contradict each other, producing inconsistent and unacceptable results. Secondly, based on well established facts and assumptions, an organized, disciplined proof shows that observed conditions in a hypothetical expansion would be contrary to those which are now actually observed. Finally, a new method is presented to directly determine whether galaxies are receding or not based on rotation velocity and independent of the redshift. If no recession is found, there can be no universal expansion.

Introduction

Although serious questions about the validity of the Big Bang expansion concept have been raised by recent astronomical evidence, most scientists still support the concept. Various alternative concepts have been presented by some scientists, but these have received little notice. Thus, the growing diversity of ideas continues, with no conclusive results in sight.

The chief difficulty with proving or disproving the universal expansion theory is that all arguments on both sides of the question are, to some extent, speculative and are, therefore, subject to misinterpretation and some degree of uncertainty.

Such uncertainty is unavoidable in many areas under investigation, but in the case of expansion theory, a body of empirical evidence has accumulated during the sixty years since the idea was launched, such that it may now be possible to resolve this issue on a less speculative and more factual basis. Two possible methods for doing this are discussed here: that of formal proof and that of direct empirical measurement.

The Formal Proof

One method to formally disprove a theory, on a purely logical basis, is to show that the initial argument behind the theory is inconsistent with the rules of logic. For example, the expansion theory derives from a number of assumptions, and it can be shown that two of these assumptions directly contradict each other, resulting in mathematical inconsistencies (Walker 1989). Thus, the expansion theory assumes that the present time recession velocity of galaxies in an expansion is proportional to the distance light has traveled from each galaxy to the observer and also to the actual present time distance of each galaxy.[1] Yet, it can be demonstrated that these two distances are different, that this fractional difference varies with galactic distance, so that both assumptions cannot be true.

Another method to formally disprove a theory such as the Big Bang would be to show that an observer's view of such a theoretically expanding universe would be contrary to what astronomers actually observe in the real universe. To do this, there must first be available sufficient empirical evidence or well established assumptions to determine clearly how the universe *does* appear to astronomers and how it *would* appear in a hypothetical expansion.

On this basis, using presently established facts and generally agreed upon assumptions, it can be logically demonstrated that the observed density of galaxies along any line of sight would increase systematically with distance in an expanding universe. This means that galactic densities at 10^{10} light years would be at least several times greater than the density nearby. This is contrary to the actual observed results from recent galaxy mapping projects, and the logical conclusion is that the universe is not expanding (Walker 1991).

Although physicists are skeptical about this type of approach based primarily on logic, the method may offer considerable advantage in specific problem areas where direct conclusions cannot otherwise be reached.

Direct Measurement

Another more direct method might be to directly measure the recession or non-recession of the galaxies independently of the cosmological redshift. Such a measurement, if successful, would provide conclusive and final evidence that space is or is not expanding.

Edwin Hubble and Richard Tolman attempted such a direct measurement approach in 1935, based on decreased surface brightness as a function of redshift. Their test, together with recent applications, was described by Jaakkola (1988) as a powerful indication that the galaxies are not receding.

However, neither this test nor other proposed tests of the expansion hypothesis have been generally accepted as conclusive.

---

1 Ed. Note: The measured spectral shift is believed by most astronomers to refer to the time at which the light left the galaxy, and thus the spectral shift should be proportional to the distance of the galaxy. At the present time, both are greater, and hence there is no contradiction with a theory of an expanding universe. See discussion in *Apeiron* No. 6, 1990. H.C.A.

At the same time, it is well known that a number of scientists have proposed non-Doppler mechanisms to explain the redshift without any withdrawal of the galaxies (Jaakkola 1978, Vigier 1990, Arp 1991, LaViolette 1986, Marmet 1991). Consequently, the redshift itself is not finally conclusive as an indication of expansion.

For a fully conclusive test, it would be necessary to observe some regularly periodic event (a Cepheid variable, for example) which is intrinsic to all galaxies. In an expanding universe, the rate of any regular periodicity should appear to slow down as galactic distances and recession velocities increase.

The fact that such a slow-down does occur as any recurring event moves away from an observer was confirmed by observations made in 1676 by Swedish astronomer Olaf Roemer, who discovered that the regular orbital rate (eclipse rate) of Jupiter's moons appears slower when the distance between the earth and Jupiter is increasing and faster when the distance is decreasing (Singer 1990).

The reason why such regularly periodic events appear to slow down when they are receding is easily seen in Figure 1, which gives a simple illustration of the mechanics. A rapidly rotating disc is at a distance d from an observer. If the disc rotates at regular time intervals, each equal to Δt, an observed point P on the rim at a time t will be seen to return to that same location after each time interval Δt if the disc remains at the same distance d from O. Thus, if the observed starting location of P at time t is at the end of a radius, CP, which is perpendicular to the line of sight OC, P will complete one revolution and return to that location at time $(t+\Delta t)$.

However, if the disc is receding from the observer at velocity, V_r, while it completes one revolution, the disc will have moved away by a distance $V_r \Delta t$ during that time, and its center C will have arrived at O'. The observed point P would now be located at P' at time $(t+\Delta t)$.

Now, the light from P' must travel the additional distance $V_r \Delta t$ to reach the observer, so that its arrival at O will be delayed by an additional time interval $V_r \Delta t / c$, where c is the speed of light.

Consequently, when the disc is withdrawing, the observed time for the disc to complete one revolution is $\Delta t(1+V_r/c)$, and its period of rotation has increased by a fraction V_r/c. Since the observed time required for a point on the periphery to complete one revolution is greater, its observed orbital velocity V_o is lower.

Here, however, it is important to note that, although the observed orbital velocity V_o, relative to O, is slower, the actual orbital velocity V relative to the center of the disc remains always the same. V_o is observed to be slower only

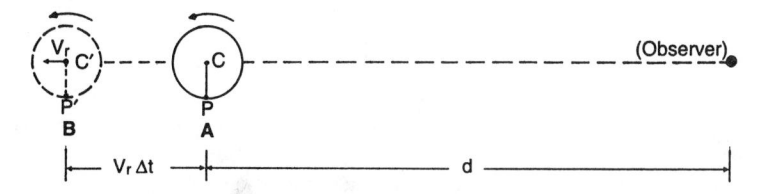

Figure 1. Rotation period of a receding disc.
    A. Stationary disc completes one observed rotation during each time period Δt
    B. Receding disc completes one observed rotation during a longer time period $\Delta t + (V_r \Delta t / c)$

because of the increasing time lag required for light from point P to reach O as the disc recedes.

With this in mind, the relation between the actual rotation velocity V of point P and its observed rotation velocity V_o is readily seen. Since P travels the same orbital distance $V\Delta t$ in an observed time period which has increased to $\Delta t + (V_r/c)\Delta t$, the *observed* orbital velocity of a point on the rim is now:

$$\frac{V\Delta t}{\Delta t\left(1+\frac{V_r}{c}\right)} = V_o$$

$$V = V_o\left(1+\frac{V_r}{c}\right) \qquad (1)$$

Because of this relationship, Roemer was able to compute, with remarkable accuracy for that time, the speed of light which was indicated by the slower eclipse rate (and reduced rotational velocity) of Jupiter's moons when the planet was receding, and *vice versa*.

Similarly, it can be shown that the observed time period for any regularly periodic event will increase as it recedes from an observer.

Periodicity of Galactic Rotations

The only regularly periodic event other than the frequency of light which is now observable in a distant galaxy is its orbital period, which is a function of its orbital velocity.

Rotational velocity can be determined either at radio wavelengths, by measuring the global HI profile width (Tully and Fisher 1976), or optically from a spectral profile of redshifts along a galactic diameter perpendicular to the line of sight (Rubin 1983). In the latter case of a spectral profile, after deducting the central, cosmological redshift, the remaining "Proper motion" redshifts indicate tangential velocities of visible material in the galaxy at each radial distance from the galactic center.

In this way, an astronomer can directly measure the observed orbital velocity of the material objects (gas and dust clouds, stars, *etc.*) which exist at any point along a galactic radius. The entire assemblage of such materials at the observed point corresponds, on a far vaster scale, to the dot, P, on the disc in Figure 1, as previously discussed.

A most interesting fact which results from such spectral examinations is that spiral galaxies of the same morphological type (Sa, Sb, Sc) and absolute magnitude have the same rotation velocities *independently of their distance from the observer*.

First indications of this were reported by Tully and Fisher (1975) who proposed that there is "a good correlation between the global neutral hydrogen line profile width, a distance-independent observable, and absolute magnitude". Here, the profile width also gives maximum rotation velocity.

This Tully-Fisher relation clearly establishes that, within distances of about 375 million light years (115 Mps), all observed galaxies of the same absolute magnitude have approximately the same maximum rotation velocity (V_{max}), independent of distance. Yet, these same galaxies are thought to be receding from the Milky Way at velocities which vary from less than 110 km s^{-1} for the closest galaxies, at about 2.5 million light years (.75 Mps), to more than 6,000 km s^{-1} for the farthest galaxies at about 375 million light years (115 Mps).

Accordingly, as previously discussed, these *observed* galactic rotation rates should not be the same in an expansion. Instead, the observed rotation rate should decrease as galactic distance and recession velocities increase. Since, according to the TF relation, they do not decrease, this would indicate that the galaxies are not receding.

Optical Measurements of Rotation

These radio observations are confirmed by an optical survey of galactic rotations made by scientists at the Carnegie Institution (Rubin *et al.* 1980 and 1982). Their analysis confirms that, for galaxies of the same S type (Sa, Sb, or Sc), there is a good correlation between rotation velocity (V_{max}) at the isophotal radius and absolute blue magnitude. Their farthest reading, for galaxy U12810, was at a distance of 165 Mps with magnitude –22.6, and a redshift recession velocity of 8124 km s^{-1}. (Using an H value of 50 km s^{-1} Mps^{-1}, and $z = 0.02708$).

Accordingly, if U12810 is, in fact, receding, its observed rotation velocity V_o should be less than its normal rotation velocity, as previously discussed.

Since the *observed* orbital velocity V_o is 235 km s^{-1}, the relation given in equation (1) may be used to determine what the *actual* orbital velocity V would be if the redshift, $z = 0.02708$, represents a velocity of recession. At this relatively short cosmological distance, $z = V_r/c$, so that $V = 241.36$ km s^{-1}.

The actual rotation velocity should be greater than the observed rotation velocity by 6.36 km s^{-1} or 2.6%—if the galaxy is receding.

A consistent variation of this magnitude between normal and observed rotation rates could probably be detected if there were sufficient observational evidence.

Testing the Galactic Recession Hypothesis

Now, considering all the factors discussed so far, the procedures necessary to determine whether the galaxies are actually receding becomes apparent.

First, we would need to determine the rotational rate for nearby galaxies of a particular spiral type and magnitude at distances within several Mps. Here, distances and magnitudes are well established from the distance scale. Observed information on a number of galaxies could be averaged to obtain their most accurate rotational velocity. At this short distance, (small z), this observed rotation velocity should be approximately equal to the normal rotation velocity, V, for galaxies of that type and magnitude.

Unfortunately, the nearby galaxy samples used by Tully and Fisher were of various types and magnitudes which are not suitable for our present purpose.

Next, rotational information should be obtained, at the farthest observable distances (largest redshifts), for a reasonable number of galaxies of the same spiral type and magnitude used for the nearby galaxy sample. From this, the observed rotation velocity at that distance would be approximated.

Here again, the most distant observations made by Rubin et al., are not suitable for our purpose (Table 1). Instead it would be desireable to have a larger galaxy sample with less variation in luminosity reaching to a greater and distance.

However, if suitable data were obtained, the nearby (normal) rotation velocity of a particular galaxy sample could be compared with its observed rotation velocity at a distance in order to discover any difference.

Obviously, such a comparison cannot be made from existing data. Extensive additional research would be required for meaningful results.

Table 1. Optical samples for distant galaxies. Data is for the 5 most distant Sb galaxies discussed by Rubin et al. in their examination of 23 Sb galaxies for optically observed rotation properties. V_r is galactic recession velocity in km s^{-1} indicated by the redshift. V_o is observed rotation velocity (V_{max}) in km s^{-1}. (Source: Rubin et al., 1982)

NGC	M_B	Distance (Mps)	V_r	V_o	z
U11810	-21.2	98.3	4709	197	.0157
2590	-21.9	95.8	4985	256	.0166
1417	-22.3	81.5	4114	330	.0137
1085	-22.4	136.0	6784	310	.0226
U12810	-22.6	165.0	8124	235	.0271
Total	-110.4	576.6	28716	1328	.0957
Average Values	-22.08	115.3	5743	265.6	.0191

Conclusion

Even so, the stakes are high, and the results should be worth the extensive time and resources needed. Such a direct measure of the withdrawal or non-withdrawal of the galaxies could finally resolve any questions about the validity of the expansion idea and Big Bang once and for all.

If the galaxies are receding in a general expansion of universal space, their observed rotation rates *must* decrease with distance. However, if the universe is not expanding, no such decrease would be found.[2]

If an adequate research effort confirms that there is no such decrease, then the unavoidable conclusion would be that the Big Bang never happened.

In that case, a new cosmological concept would be needed to replace the Big Bang. It would have to provide some plausible alternative mechanism for the creation of matter ranging in scale from the subatomic size of the smallest particles to the vast magnitude of the largest galaxies. Since a continuous creation of matter

---

[2] Ed. Note: If clocks run slower on younger galaxies, more distant galaxies, because of look-back time, would show slower rotation in a non-expanding universe. H.C.A.

could not be accommodated indefinitely in a static space, provisions would be needed for a balanced cycle of events in which the creation of matter in parts of the cycle would be balanced by the conversion of matter back into energy in other parts.

One possible concept might incorporate a static universal space, occupied by a dynamic ether in which both electromagnetic wave (photon) energy and atomic particles (rotational energy) such as electrons and protons would exist. Energy radiation from stars and galaxies would power the synthesis of basic particles (rotational energy) in interstellar regions of the ether while large scale currents and eddies in the ether would then impel these particles inertially into the centers of stars and galaxies to provide fuel for the fusion (and other) processes going on there. Such a universe would be quasi-static, stable, infinite, and would remain always in a state of equilibrium. Such a concept is no more speculative than the Big Bang, but, as I have shown elsewhere (Walker 1992), it might explain a number of mysteries which are now unresolved.

References

Arp, H., 1991, How non-velocity redshifts in galaxies depend on epoch of creation, *Apeiron* 9-10:18.
Encyclopedia Britannica, 15th Edition, Vol. 10, p. 164.
Jaakkola, T., 1988, Four applications of the surface brightness test of the cosmological expansion hypothesis, in: *Proceedings, 6th Soviet-Finnish Astronomical Meeting*, eds. V. Hanni and J. Tuominen.
Jaakkola, T., 1978, The redshift phenomenon in systems of different scales, *Acta Cosmologica* 7:17.
LaViolette, P., 1986, Is the Universe really expanding? *Ap.J.* 301:544.
Marmet, P., 1991, A new mechanism to explain observations incompatible with the Big Bang, *Apeiron* 9-10:45.
Rubin, V.C., Burstein, D. and Thonnard N., 1980, A new relation for estimating the intrinsic luminosities of spiral galaxies, *Ap.J.* 242:L149.
Rubin, V.C., Ford W.K. Jr. and Thonnard, N., 1982, Rotational properties of 23 Sb galaxies, *Ap.J.* 261:439.
Rubin, V.C., 1983, Dark matter in spiral galaxies, *Scientific American* 248:6:96.
Singer, C., 1990, *A History of Scientific Ideas*, Dorset Press.
Tully, R.B. and Fisher, J.R., 1976, A new method of determining distances to galaxies, *Astron. Astrophys.* 54:661.
Vigier, J.P., 1990, Evidence for nonzero mass photons, *IEEE Transactions on Plasma Science* 9:1.
Walker, F., 1989, A contradiction in the theory of universal expansion, *Apeiron* 5:1.
Walker, F., 1991, The stationary universe theorem, Unpublished.
Walker, F., 1992, Where is the cosmological alternative?, *Physics Essays* 5:3:340.

The Case Against the Big Bang

Eric J. Lerner

Lawrenceville Plasma Physics
20 Pine Knoll Drive
Lawrenceville, New Jersey 08648

Despite its widespread acceptance, the Big Bang theory is presently without any observational support. All of its quantitative predictions are contradicted by observation, and none are supported by the data. Its predictions of light element abundances are inconsistent with the latest data. It is impossible to produce a Big Bang "age of the universe" which is old enough to allow the development of the observed large scale structures, or even the evolution of the Milky Way galaxy. The theory does not predict an isotropic cosmic microwave background without several additional *ad hoc* assumptions which are themselves clearly contradicted by observation. By contrast, plasma cosmology theories have provided explanations of the light element abundances, the origin of large scale structure and the cosmic microwave background that accord with observation. It is time to abandon the Big Bang and seek other explanations of the Hubble relationship.

I. Introduction

While the Big Bang is widely accepted as a scientific explanation of the Hubble relation, it rests on very few quantitative predictions. The most definite such quantitative predictions are the abundances of the light nuclides He^4, D, and Li^7. Well-known results, (Wagner, Fowler and Hoyle 1966) based on nuclear physics and the assumptions of the Big Bang, show that the pre-galactic abundances of these three nuclides are a function only of the photon-proton ratio, which, by the Big Bang theory, is an invariant in the universe. Since the number of photons, (which is dominated by the number of photons in the cosmic background radiation) is known accurately, this ratio is in effect a function of the baryon density in the present day universe. In practice, this density is not known very accurately, and it has been treated as a free variable. The Big Bang predictions, therefore, reduce to a prediction of the abundances of two of the light nuclides, given the abundance of one. This one abundance is used to derive the "true" photon-baryon ratio and thus the other two abundances. The apparent validity of this prediction, based on data available in the late 1960's, was one of the main reasons for the general acceptance of the Big Bang.

The second and less specific prediction is that no object in the universe is older than the Hubble age, the age of the universe estimated from the inverse of the

Hubble ratio relating the redshifts and distances of galaxies. While this is qualitatively a very firm prediction of the theory, it is quantitatively vague for two reasons. One, the Hubble constant itself is not accurately determined, and second, the deceleration parameter, which indicates how rapidly the assumed expansion of the universe proceeded in the past, is also not known, and is dependent on the actual density of all matter in the universe. However, the range of these values compatible with observations does, as we shall see, set very real limits on the age of objects in the universe, if the theory is valid.

The third and final quantitative prediction of the theory, in its current form, is the existence of an isotropic Planckian background cosmic radiation. The temperature of this radiation is not predicted by theory, but, in the inflationary form currently popular, its isotropy and blackbody spectrum are. As we shall note below, these are not valid predictions of the Big Bang in its most general form.

These three predictions and their claimed correspondence with observation are the entirety of the evidence cited in favor of the hypothesis that the universe originated in an instant in an intensely hot, extremely dense state. What is striking is that at the present time, not one of these predictions can be validly cited as evidence for the Big Bang, which, therefore, is entirely unsupported by observations. The first two predictions are flatly contradicted by observation, while the third does not actually constitute evidence as to the primordial state of matter, and involves additional predictions which are themselves contradicted by observation.

II. Light Element Abundances

The key problem for Big Bang nucleosynthesis (BBN) predictions of the light element abundances lies in the discrepancy between the predictions for deuterium and He^4. The predicted abundance of He^4 decreases with decreasing baryon-photon ratio, η, while the predicted abundance of D decreases with increasing η. Since the observed upper limits on the pre-galactic abundances of both elements have been declining, we now have a situation where there is no value of the density parameter which yields predictions agreeing simultaneously with both observed He^4 and D abundances.

We begin with deuterium: Hubble Space Telescope observations by Linsky fix the current abundance of D at $1.65 \pm 0.1 \times 10^{-5}$ by number relative to H (Linsky 1992). Thus the 3 sigma upper limit is 2×10^{-5}. Of course, much of the pre-galactic deuterium could have been destroyed in stars. However, many authors have placed strict limits on how much of the primordial D could have been destroyed. Yang *et al.* (1984), for example, show that while D is easily burned to He^3, He^3 is destroyed only where temperatures are high enough to burn H to He^4. Considerations of the amount of He^3 produced in the galaxy, combined with calculations as to the production of He^3 in stars, leads to the conclusion that the current sum of abundances of D and He^3 is at least half the pre-galactic value. This leads to an upper estimate of the pregalactic D abundances of about 8×10^{-5}. Delbourg-Salvador *et al.* (1987) conclude that a destruction of more than $2/3$ of the pre-galactic D by astration would lead to great variations in current D, depending on the exact history of a given region. Since such variations are not observed, they calculate that primordial D abundance is less than 3 times present, or, based on the Linsky observations, less than 6×10^{-5}.

Physically, both of these arguments are related to the fact that if the whole of the galactic material is processed on average once through stars, about e^{-1} of the deuterium will not have been so processed. Deuterium destruction much greater than this requires two such processings. But this rate of nuclear processing would produce about twice as much energy, He^4 and heavier elements, such as CNO, as are observed.

By the BBN formula, upper limits on D abundance set lower limits on the density parameter η. D abundance of 6×10^{-5} implies $\eta > 3.4 \times 10^{-10}$ while D abundance of 8×10^{-5} implies a limit of 3.0×10^{-10}. These limits in turn imply lower limits on the predicted primordial abundance of He^4 of 23.9% by weight and 23.6% respectively. These estimates are based on an assumed neutron lifetime of 882 sec, the current two sigma lower limit. A "best" value of 888 sec would increase these limits to about 24.1% and 23.8% respectively.

The abundance of He^4 is therefore a crucial test of BBN. Since He^4 is produced by stars, observers have long sought to focus on those galaxies with the least stellar production, which in turn is indicated by the abundance of heavier elements, C, N and O. It is assumed, based on theories of stellar evolution, that He^4 abundance should increase as a function of heavy element abundance, so that the pre-galactic

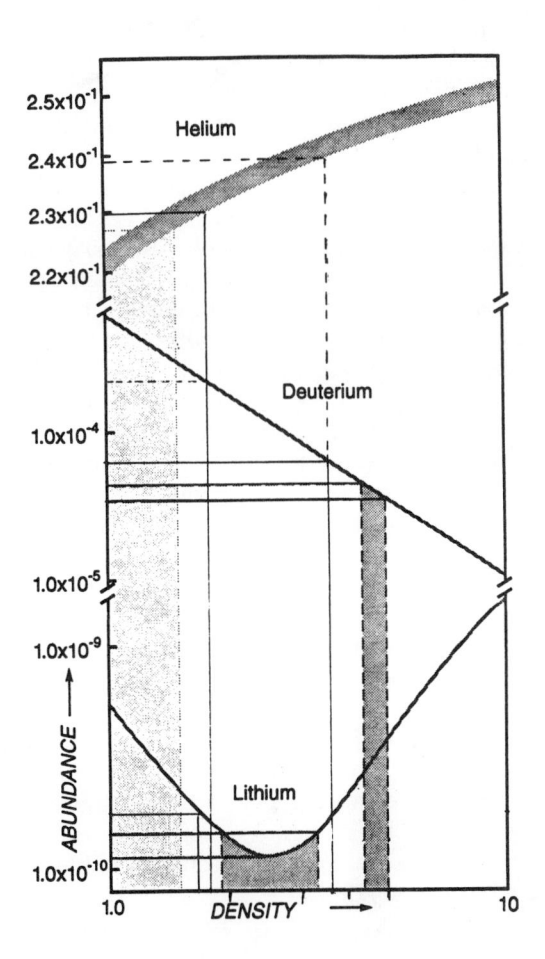

Figure 1. Abundances of elements predicted by Big Bang nucleosynthesis vs. observed abundances (horizontal lines). Predictions from Olive (1990) observed values from Olive (1990) and Fuller (1991). There is no value of η (ratio of protons to photons $\times$ 10^{10}) that gives accurate abundances for all elements.

abundance should be determinable by extrapolating He^4 abundance to the zero-heavy-element condition. There are three ways of doing this—linear extrapolation, nonlinear extrapolation, and simply averaging the galaxies with the lowest He^4 or CNO abundances.

All of these approaches show that the observed He^4 pre-galactic abundance is less than the lower limit predicted by BBN. Melnick *et al.* (1992) using a linear correlation, obtain 21.6±0.6%, which is 3.3–4.4 standard deviations below the BBN value, depending on which BBN limit is used, 23.6% or 24.1%. Using the six galaxies with O abundance below 5×10^{-5} (the next highest galaxy in the sample has an O abundance above 8×10^{-5}) and averaging the He^4 abundances, we obtain 22.4±0.3% which again rules out BBN at the 4 sigma level.

Using a different sample, and a nonlinear regression of He^4 on N and O, Mathews *et al.* (1990) obtained a primordial He^4 abundance of 22.3%±0.2 which puts the lowest BBN limit 6.5 sigma too high.

Olive *et al.* (1990), averaging the ten lowest He^4 galaxies, obtained a 2-sigma upper limit of 23.7%. Even this higher value put the lowest BBN limit at 2-sigma, implying only a 5% chance of agreement with observation. However, this lowest BBN limit already assumes a neutron lifetime at the 2-sigma lower limit. In addition, the sample used did not include the new measurement of SBS 0335-052 by Melnick (1992) of 21.1+1.9% nor did it include UM461 with a value of 21.9±0.8%. The inclusion of these two galaxies would push the BBN lower limit back up to 3 sigma above observations, in agreement with the other results.

Finally, Pagel *et al.* (1992) use linear extrapolation to derive an abundance of 22.7±0.5%, or 2–3 standard deviations below the Big Bang predictions.

If we use the He^4 observations to estimate η by BBN predictions, we conclude that, at the 2 sigma level, He^4 abundance is at most 23.0% and $\eta < 1.8 \times 10^{10}$. The BBN prediction for deuterium abundance would then be $> 1.74 \times 10^{-4}$. This is 8 times the upper limit for current D abundance in the ISM and twice the highest plausible estimate for a pre-galactic abundance compatible with current observations. To achieve this 8-fold decrease in D abundance would require that the average parcel of the ISM has been processed twice through stars. As Yang *et al.* (1984) point out, this would lead to overproduction of He^4, C, N and O relative to observations. The complete processing of matter through a generation of stars with the present luminosity function of the galaxy enriches the ISM by approximately 12% He^4. Two such processings would lead to an enrichment by about 24%, giving a total abundance of 46%, far above current observations. In addition, the production of this much He would yield 1.7×10^{18} ergs per gram of galactic mass. This would produce a luminosity/mass ratio 1.7 times that of the sun or about 8 times that observed.

Finally, as Delbourg-Salvador *et al.* note, such an extensive destruction of D would lead to great variations in current D abundance, which have certainly not been observed.

Thus, if D observations are used to predict He^4 abundance, the lowest possible BBN observations are ruled out at beyond the 3 sigma level, while if He^4 observations are used to predict D abundance the gap between theory and observation is at least a factor of two, grossly more than the observational uncertainty. *We must conclude that standard BBN is excluded by the observations.* This conclusion has also been reached by others, such as Riley and Irvine (1991).

Since the recognition by many researchers that the He^4 and D observation together are incompatible with standard BBN, a number of efforts have been made to propose "fixes" to the theory. The two possibilities discussed are inhomogeneous BBN and a very massive tau-neutrino.

Inhomogeneous BBN, as elaborated by, for example, Mathews *et al.* (1990), hypothesizes small-scale density fluctuations in the early universe that would affect BBN. It should be emphasized that this hypothesis is entirely *ad hoc*, not emerging from any physical theory. While BBN has relatively unambiguous predictions, inhomogeneous BBN introduce 4 more free variables—the radius, density contrast, spacing and shape of the inhomogeneities—which produce an enormously larger range of possible predictions. Despite this large number of free variables, the range of possible predictions does not include points that simultaneously have He$^4 < 23.0$ and $D < 8 \times 10^{-5}$, the requirements of current observations. Indeed, the situation is somewhat worse than in standard BBN, since with inhomogeneous BBN, lithium production is enhanced. For conditions that produce He^4 abundance of 23.0%, D is about the same as in standard BBN, but Li^7 is predicted at around 2×10^{-9}. This is in gross contradiction with an observed upper limit of 2.3×10^{-10} for Li^7 abundance in PopII stars. Arguments that hypothesize lithium depletion to produce exactly the same depleted abundance in a range of PopII stars are too implausible on their face to merit discussion. So inhomogeneous BBN also fails to resolve the conflict between theory and observation.

A second fix is to assume a very massive tau-neutrino, which would lower the effective number of neutrinos in BBN calculations below three and thus lower He^4 abundance predictions. A massive tau-neutrino would begin to pair-annihilate during BBN, thus reducing the number of neutrinos. However, Kolb and Scherrer (1982) have shown that this effect occurs only with a tau-neutrino more massive than 15 Mev, relatively close to the observational upper limit of 35 Mev from particle physics experiments. Such a massive neutrino would have to conveniently decay rapidly to avoid a rapid gravitational collapse of the universe. Alternatively, one could hypothesize the non-existence of the tau-neutrino which has never been directly observed. This would, however, contradict existing particle physics theories.

Thus from any standpoint, the observational evidence is incompatible with Big Bang nucleosynthesis, assuming only that there are tau-neutrinos, as particle theory predicts.

III. Large Scale Structure and the "Age of the Universe"

The second basic prediction of the Big Bang theory is that no object in the universe can be older than the time since the Big Bang—the "age of the universe". The clearest contradiction of this prediction is in the data on the large scale structure. The large scale structure discovered by R. Brent Tully (1986) and confirmed by the IRAS survey of W. Saunders (1991) is widely interpreted as contradicting Cold Dark Matter theories of structure formation. However, an analysis of Saunders's data, combined with data on the observed average velocities of galaxies and other observational constraints, leads to contradictions with the entire Big Bang hypothesis, not just some versions of it. As I show here, it is impossible to generate the observed structures in less than three times the "age of the universe" determined by the Big Bang theory. In addition, even if we ignore this fundamental problem, such structures would generate distortions of the CBR spectrum that are incompatible with COBE observations, if, as is assumed in the Big Bang, cosmic density increases as $(1+z)^3$ and the CBR was thermalized prior to $z = 20$.

Saunders et al.'s survey of IRAS galaxies provides a high-quality picture of the current distribution of galaxies within a cube $560h_{50}^{-1}$ Mpc on a side, centered on the Milky Way. The survey shows that most of the galaxies are concentrated into two elongated concentrations whose centers are separated by about $360h_{50}^{-1}$ Mpc. In examining the formation of such large structures, the first question to be answered is: how far must matter travel to form such structures from an initial even distribution?

Table 1. Large Scale Structure

Density/Average Density (1)	Vol/Density (2)	Mass/Density (3)
2.5	17	42.5
2.2	33	72.6
1.93	56	108.1
1.63	89	145.1
1.48	130	192.4
1.40	141	197.4
1.29	170	219.3
1.14	199	226.9

Saunders provides data in the cited paper showing the number of bins (each $80h_{50}^{-1}$ Mpc on a side) having a given density of galaxies. This data, taken from Figure 3 (Saunders 1991) is shown in Table 1. Column (1) is the density of a set of grid points divided by average density, column (2) is the number of grid points per density interval (arbitrary units) and column (3) is the product of the first two cols. We model the data as two identical cylinders extending the full height of the survey volume. That is, the highest density region is assumed to be divided into two cylinders of radius $a_f(1)$ and the regions of progressively lower density are assumed to be annular cylinders with outer radii $a_f(n)$. To determine the average distance traveled, we expand each cylinder's volume by d, its density divided by the average density of the entire survey volume, obtaining an initial radius $a_i(1)$. It is then trivial to determine the average distance traveled by the matter in each annular cylinder and to thus obtain the mass-weighted average distance, $D = 38$ Mpc. The calculation is summarized in Table 2. Column (1) is the outer radius (Mpc) of each annular cylinder with the density defined in Table 1. Column (2) is the initial outer radius when matter had been spread out to average density. Column (3) is the average distance moved by matter in each annular cylinder.

To determine the apparent age of these structures, independent of any assumed cosmological theory of the early universe, we can simply divide D by V, the average velocity of matter on these very large scales. Various surveys of average peculiar velocities of galaxies are consistent with $V = 600$ km s^{-1}, the peculiar velocity of the Milky Way relative to the CBR. Thus, $D/V = 63 h_{50}^{-1}$ Gyr or over 3 Hubble times. The apparent age of the structures is thus triple the oldest possible age of the Universe in the Big Bang model—a serious contradiction to the theory.

Table 2. Calculation of Average Distance Traveled in Formation of Large Scale Structure

R_f (1)	R_i (2)	D (3)
41.9	66.2	25.6
53.4	82.6	26.8
87.4	126.7	33.7
110.6	153.3	41.1
119.9	163.5	43.2
129.5	173.4	43.8
150.2	193.7	43.7
171.4	212.7	42.5

Hypothesizing higher velocities in the past cannot overcome this problem without violating other observational constraints. Two deceleration mechanisms are possible: thermal pressure from a hot intergalactic medium or magnetic pressure from intergalactic magnetic fields. For a plasma cloud traveling through a hot plasma of temperature $T(eV)$ deceleration is:

$$\frac{dV}{dt} = 1.65 \times 10^{-6} nT^{-3/2} V \text{ cm sec}^{-2} \tag{1}$$

The X-ray background indicates $T = 40$ keV. Taking plasma particle density $n = 1.6 \times 10^{-7}$ cm^{-3}, the deceleration over 10 Gyr would be only 5.6 km s^{-1}. If a cooler temperature of the IGM is hypothesized higher decelerations are obtained: at $T = 9$ keV, $\Delta V = 51$ km s^{-1}. Further decreases in T cannot produce any higher decelerations, since below 9 keV total pressure would limit deceleration. Here

$$\frac{dV}{dt} = 1.6 \times 10^{-12} \frac{nT}{nRm_p} \tag{2}$$

where R(cm) is the present radius of the large scale structures, around 170 Mpc, and m_p(gm) is the proton mass.

For magnetic braking, Faraday rotation measurements limit $nBr < 2 \times 10^{13}$ g cm^{-2} (Valle 1975) where r is the current radius of the LSS and B(gauss) is average magnetic field strength. This yields a limit of .24 microgauss for the intergalactic field strength at these scales, maximum decelerating pressure of 2.4×10^{-15} dyne cm^{-2} and a maximum deceleration in 10 Gyr of 54 km s^{-1}. Thus the maximum average velocity over the past 20 Gyr is 708 km s^{-1} for an apparent age of 54 Gyr, still far longer than the Hubble time.

These age estimates are only lower limits since they assume, unphysically, that all the matter is accelerated instantaneously to the maximum velocity in the right direction.

If we try to estimate the length of time needed to form these structures assuming a Big Bang—that is assuming the cosmos was once in an extremely dense state—we still have a contradiction.

Gravitational attraction is clearly inadequate, as the many unsuccessful attempts with dark matter models have shown (Saunders 1991). This leaves some form of explosive mechanism. Here, an instantaneous velocity is imported to some matter, but the energy involved is only spread as the "bubble" formed in the explosion expands. Such explosive mechanisms are hypothesized in the currently popular "texture" models (Cen et al. 1991). It should be emphasized that such models are entirely *ad hoc*, since the underlying physical processes hypothesized are without any experimental or observational basis.

Even such an *ad hoc* explosive mechanism, however, fails to resolve the problem of the age of the structures.

For a dark matter, $\Omega = 1$ universe, the radius of the bubble formed is (Levin et al. 1991)

$$R = 3.4 \left(\frac{E}{10^{61}}\right)^{.2} \left(\frac{t}{10^{10}}\right)^{.8} \text{Mpc} = 6.4 \times 10^4 E^{.2} t^{.8} \text{ cm} \tag{3}$$

where E is explosive energy in ergs and t is time of formation in years. Now,

$$E = \frac{1}{2} m_p n V^2 R^3 \tag{4}$$

So,

$$t = 1.03 n^{-¼} V^{-½} R^{½} \tag{5}$$

taking $n = 2.9 \times 10^{-6}$ cm^{-3} ($\Omega = 1$, $H_o = 50$ km s^{-1} Mpc^{-1}), $V = 600$ km s^{-1}, $R = 38$ Mpc, we get $t = 34$ Gyr. For an $\Omega = 1$ Big Bang, the current age of the universe is $\frac{2}{3}$ the Hubble time = 13 Gyr, so the time of formation is 2.6 times the "age of the universe". For $H_o = 100$ km s^{-1} Mpc^{-1}, the situation is worse by a factor of $2^{½}$.

In fact, this is an underestimate, since a bubble 38 Mpc in radius will move the average particle only 9.5 Mpc. If we assume a cylindrically symmetric explosion we still need a bubble $38 \times 3 = 114$ Mpc in radius. In this case, $t = 59$ Gyr or 4.4 times the age of the universe.

The age of the universe can be pushed back by introducing a cosmological constant, another *ad hoc* hypothesis. For an $\Omega_{TOT} = 1$ universe, the age of the universe is

$$t_U = \frac{2}{3} H_o \frac{\ln\left(1 + \Omega_\Lambda^{½} / \Omega_B^{½}\right)}{\Omega_\Lambda^{½}}$$

For $\Omega_B = .04$, $t = 31$ Gyr for $H_o = 50$ km s^{-1} Mpc^{-1}, which is clearly inadequate. To achieve an age of 60 Gyr, Ω_B must be 4×10^{-4}, which clearly is not allowed by Big Bang nucleosynthesis predictions. By abandoning $\Omega_{TOT} = 1$, the inflationary model, any arbitrary age can be obtained. An age of 60 Gyr requires a careful tuning of Ω_Λ to 1.18, with $\Omega_B = .04$. Care is required because a slightly higher cosmological constant of 1.195 would eliminate a Big Bang altogether, with a minimum radius of the cosmos being about 2 Gpc. Aside from the implausibility and arbitrariness of such fine tuning, which eliminates any Big Bang prediction on the age of the universe, such a "dachshund universe" would spend most of its lifetime at a radius corresponding to a z of about 5.5. This would lead to a massive overabundance of objects as this z is approached and, would also, among other things, lead to an early formation of galaxies. Clearly the current universe does not have an average age of galaxies of 60 Gy, so the epoch of galaxy formation must have occurred considerably later than that of the formation of very large scale structures, again in contradiction to any Big Bang scenarios.

Thus there is no way to create the structures observed by Saunders *et al.* without either taking far longer than the time since the hypothetical Big Bang or without generating final velocities of matter far in excess those observed.

Even if we ignore this fundamental problem and allow arbitrary final velocities, we still have a contradiction with observation. Levin *et al.* (1991) have calculated the distortion of the CBR spectrum (y parameter) due to explosions of various magnitudes starting at a given z, assuming only that density increases as $(1+z)^3$ and that the CBR is not thermalized at low z. For a $\Omega = 1 > \Omega_B$ dark matter cosmos with explosions at $z = 9$, which they calculate to be the latest possible time,

$$y = 5 \times 10^{-5} h_{100}^3 \Omega_B f_o R^2 \qquad (6)$$

where R is in Mpc and f_o is the volume fraction of the cosmos in the bubbles. With $R = 114$ Mpc and $f_o = .25$ and $\Omega_B h_{100}^2 > 10^{-2}$ as required by Big Bang nucleosynthesis calculations (Olive *et al.* 1990) we have

$$y = 1.7 \times 10^{-3} h_{100} \qquad (7)$$

For $H_o = 50$ km s^{-1} Mpc^{-1}, $y = 8.5 \times 10^{-4}$, in clear contradiction with COBE upper limits on y of 3×10^{-4}. For a non-dark matter universe, y, according to Levin would be doubled, and early creation times or a higher value of h would only make things worse.

It should be noted that this problem cannot at all be solved by the introduction of "bias", an *ad hoc* assumption that the dark matter in the universe does not have the same spatial distribution as the galaxies. One still must explain how the baryonic matter came together in such large inhomogeneities. Only a bias between the bright, observable baryonic matter and non-observed baryonic matter would have any impact on the problem. Since the amount of matter observed in clusters is already close to the limits allowed by any Big Bang nucleosynthesis calculations, there is simply not enough baryonic matter left over to make a large difference.

The problems with large scale structure are not the only age-of-the-universe problems that contradict basic Big Bang predictions. Many researchers (for example Hatzidimitrou 1991) have shown that the oldest globular clusters in the Milky Way must be in excess of 15 Gyr old, making a minimum age for the start of the contraction of the protogalaxy that was to form the Milky Way of about 16 Gyr. At the same time, many measurements of the Hubble constant are converging on a value of about 85 km s^{-1} Mpc^{-1}, with a two sigma lower bound of 65 km s^{-1} Mpc^{-1}. Without a cosmological constant these figures imply an age of the universe of 11.8 Gyr and 15.4 Gyr respectively for an empty universe and 7.8 Gyr and 10.25 Gyr for an inflationary, flat universe. An inflationary universe is clearly ruled out, since it implies a universe at least 6 billion years younger than our own galaxy. A low-density universe is also at least marginally ruled out at the two sigma level.

Again, of course, the *ad hoc* introduction of a cosmological constant can relieve this contradiction, at the cost of eliminating any Big Bang prediction as to the age of the universe.

IV. Cosmic Background Radiation

The third key prediction of the Big Bang theory is the isotropy and Planck spectrum of the cosmic background radiation. Here again, there is a large gap between theory and prediction. As has been known for more than a decade, the original version of the theory does not predict even rough isotropy of the CBR, since portions of the sky separated by more than about 10 degrees would not have had time to be in contact with each other and so could not have reached thermal equilibrium with each other. Neither isotropy nor a Planck spectrum would thus be expected on large angular scales, in gross contradiction to observation. (This is the so-called horizon problem.) The widely accepted solution to this problem is the additional hypothesis of an inflationary stage that expands a small portion of the universe that has reached thermal equilibrium. This is another wholly *ad hoc* hypothesis, since it is derived from grand unified theories that are themselves without any experimental justification. Indeed, the only unequivocal predictions of these theories, the instability of the proton, has been contradicted by experiment (Becker-Szend 1990).

If inflation is accepted, it predicts an Ω of unity for the density of the universe. Since the directly observed density leads to an Ω of only .02 at most, this is another violent contradiction with observation. Indeed, an Ω of unity with ordinary matter is inconsistent with Big Bang nucleosynthesis predictions which allow at most a density of .04 (Olive *et al.* 1990). To overcome this inconsistency and contradiction with observation, a third *ad hoc* hypothesis has been widely proposed that a new and wholly unobserved type of matter, dark matter, constitutes 98% of the universe. Again, the existence of such dark matter is utterly without experimental or observational support.

It is usually argued that, although there is no evidence for an Ω of 1, there is indirect evidence for some dark matter. This evidence is in the flat rotation curves of galaxies and the velocities of galaxies in groups and clusters as well as in gravitational bending of light by clusters, all of which is used to argue that dark matter accounting for an Ω of 0.1 exists. All of this evidence, however, has major failings. Davies (1990) has pointed out convincing evidence that galaxies are optically thick. This both increases the amount of luminous matter and invalidates conclusions drawn from optical rotation curves. Observations of rotation curves by radio are observations of the motion of the plasma in the galaxy, not the stars. As Peratt (1986), and later Battaner *et al.* (1992) convincingly showed, such flat rotation curves are to be expected when the role of galactic currents are taken into account, and thus do not necessitate dark matter. Valtonen and Byrd (1990) have pointed out that background and foreground galaxies as well as non-viralized galaxies that are expelled from clusters would produce apparent masses about four times the actual masses of clusters. Finally, estimates of dark matter based on observations of gravitational bending of light are limited to the core regions of clusters and do not support amounts of dark matter greatly in excess of the amount of visible matter.

Perhaps the greatest contradiction between the inflationary-dark matter model and observation is in the large scale structure. As many authors have pointed out, if the true Ω is unity then overdensities on the scale observed in large scale structures would produce typical peculiar velocities among galaxies of 1000–2000 km s^{-1}, two to four times the value observed (Weinberg and Cole 1992). Yet another *ad hoc* hypothesis, biasing, is generally appealed to to overcome this contradiction. Here it is simply assumed that the non-baryonic dark matter has a different, smoother spatial distribution, presumably because of a higher effective temperature. The existence of this smooth, background of massive particles would of course have virtually no observable consequences, since the dark matter is *ex hypothesi* not observable except by its gravitational effects, and a smooth background would have no gravitational effect. Like any statement without observational consequences, the existence of such a background must be taken as a statement of faith, not a scientific hypothesis!

Thus the conventional explanation for the microwave background requires the acceptance of three purely *ad hoc* assumptions—inflation, dark matter and bias—and result in a final model that includes elements—the smooth dark matter background—that have no observational effects.

The only alternative to this scenario is the use of another *ad hoc* hypothesis, the cosmological constant, as a substitute for dark matter. In this model, matter is all baryonic, amounting to an $\Omega = .04$ while the remaining energy density, and $\Omega = .96$, is tied up in the cosmological force field. This variant, as we pointed out, makes the universe older, thus relieving in part the age of the universe problem, although not fully resolving it. It is not a very plausible approach, having the universe start off with a wholly negligible cosmological field, with an energy density 10^{-102} times smaller than the matter density, and just crossing over to a cosmological force-dominated universe at a z of around 2.

In addition, a high cosmological constant would result in a high density of objects at redshifts of around 2, yielding among other things a high number of gravitational lenses. Recent surveys of quasars contradict those predictions, assuming that quasars are at their cosmological distances (Maoz et al. 1992).

Despite these serious difficulties, the recent discovery by the COBE satellite of apparent anisotropies in the CBR has been hailed as definitive proof of the Big Bang theory. The rather tortured logic involved is that these fluctuations show that inhomogeneities in the early universe did indeed exist and that they led to the formation of large scale structure through gravitational instability. However, the existence of such fluctuations in fact does not help to resolve any of the previously cited contradictions between Big Bang predictions and observation—the light elements and age of the universe problems remain. Nor were the magnitude of the fluctuations accurately predicted by Big Bang theory. Without dark matter theorists predicted that fluctuations on the order of one part in 10^3 would be observed, a prediction falsified in the 1970's. With the addition of dark matter, the theories became sufficiently flexible that predictions in the year prior to the COBE announcement ranged from 2×10^{-5} to 10^{-7}. The actual claimed detection at 10^{-5} is within this very broad range, but the range is so broad as to negate any predictive value of the theory.

In fact, there remain additional contradictions between the predictions of Big Bang theory and the observed level of isotropy of the CBR. Observations of an upper limit on anisotropy of 1.4×10^{-5} at scales of one degree (Gorskii 1992) imply limits to peculiar velocities in the present day universe on length scales of 50–$100h^{-1}$Mpc. These limits of about 90 km s^{-1} are far below observations in the range of 300–500 km s^{-1}. Even with inflation and dark matter, it is not possible for such velocities to be produced gravitationally without violating isotropy constraints, assuming the CBR is indeed primordial.

V. Other Contradictions

Thus, not a single one of the Big Bang's quantitative predictions is in accord with observation. Even the very existence of substantial quantities of matter in the cosmos is in contradiction with the basic assumptions of the Big Bang. One of the most exactly confirmed laws of particle physics is baryon number conservation—that is, in all experiments when protons are generated from energy, anti-protons are generated in exactly equal numbers. In the hypothetical early phase of the Big Bang, at a cosmic age of about 2.2×10^{-6} seconds, protons and antiprotons would be formed by pair production when the temperature dropped below $10^{13}°K$. Mutual annihilation of newly formed proton-antiproton pairs will prevent particle density from rising above n_c.

$$n_c = (tc\sigma)^{-1} \tag{8}$$

where t is the cosmic age, 2.2×10^{-6} sec, and σ is the annihilation cross section of protons of Gev energy. Since $\sigma = 5 \times 10^{-26}$ cm^2, $n_c = 2.8 \times 10^{20}$ cm^{-3}. Since this density would exist at a z of 3.6×10^{12}, current density would be anticipated to be $n_c/z^3 = 5.7 \times 10^{-18}$ some 10^{11} times less than observed matter density. The only way around this problem is, again, an appeal to *ad hoc* hypotheses of baryon number non conservation at high temperature, hypotheses, again without any empirical support.

Two final contradictions can be added to this substantial list. While it is rarely cited as evidence for the Big Bang, the theory makes striking predictions about the relationship between the redshift and angular diameter of objects with the same linear dimension. While in Euclidian space, angular diameters decline with distance and thus with z, in the Big Bang universe this is not the case. In an empty universe, the angular diameter of an object asymptomatically approaches a lower limit as z goes to infinity, while in an inflationary universe, the angular diameter reaches a minimum at $z = 1.25$ and then increases. Observations of the angular diameter of clusters, large central galaxies in clusters and of radio galaxies have all been used to test this prediction (Kapahi 1987). Without exception, all have found that the data fit a simple $\theta = 1/z$ relationship, in violent contradiction to Big Bang predictions.

As in other cases of conflict between theory and observation, Big Bang proponents have simply introduced additional *ad hoc* hypotheses, in this case speculating that size evolution is just enough to compensate for the predicted angular diameter enlargement. Since this size evolution is supposed to affect such different phenomenon as galaxies, clusters and radio galaxies equally, it is extremely implausible.

Lastly, the Big Bang hypothesis demands that the Hubble constant be in fact a constant everywhere and that the assumed expansion is isotropic. Yet Ichikawa and Fukugita (1992) have shown that, using identical methodology, there is an approximately 20% difference in the value of H_o in the Perseus-Pisces direction as compared with the Coma direction. This discrepancy extends over a region of more than 100 Mpc and cannot be explained by any gravitationally driven peculiar motions.

We can thus see that the statement "the Big Bang is a well-verified theory" is closely akin to the statement "the Emperor's new clothes are beautiful" and is as little supported by observation.

VI. The Plasma Alternative

While Big Bang cosmology cannot be reconciled with the existence of large scale structure, such structures were successfully predicted by plasma models (Lerner 1986). In these models, without a Big Bang or high density phase of cosmic history, large scale vortex current filaments develop as instabilities within an initially more uniform cosmic plasma. Such filaments grow until they become gravitationally unstable and then begin to gravitationally contract, breaking apart into a hierarchy of smaller filaments. Plasma instability theory limits the characteristic gravitational orbital velocity within fully contracted filaments and the objects, such as galaxies and clusters of galaxies that condense from them, to between $(m_e/m_p)^{1/2}c$ and $(m_e/m_p)^{3/4}c$ or 160–1070 km s^{-1}, where m_e and m_p are the electron and proton masses respectively. Such vortices must be collisional to contract gravitationally, which, for a characteristic maximum velocity of $(m_e/m_p)^{3/4}c$, implies $n'D = 1.0 \times 10^{19}$ cm^{-2}, where D is the average distance between structures and n' is the density structures have

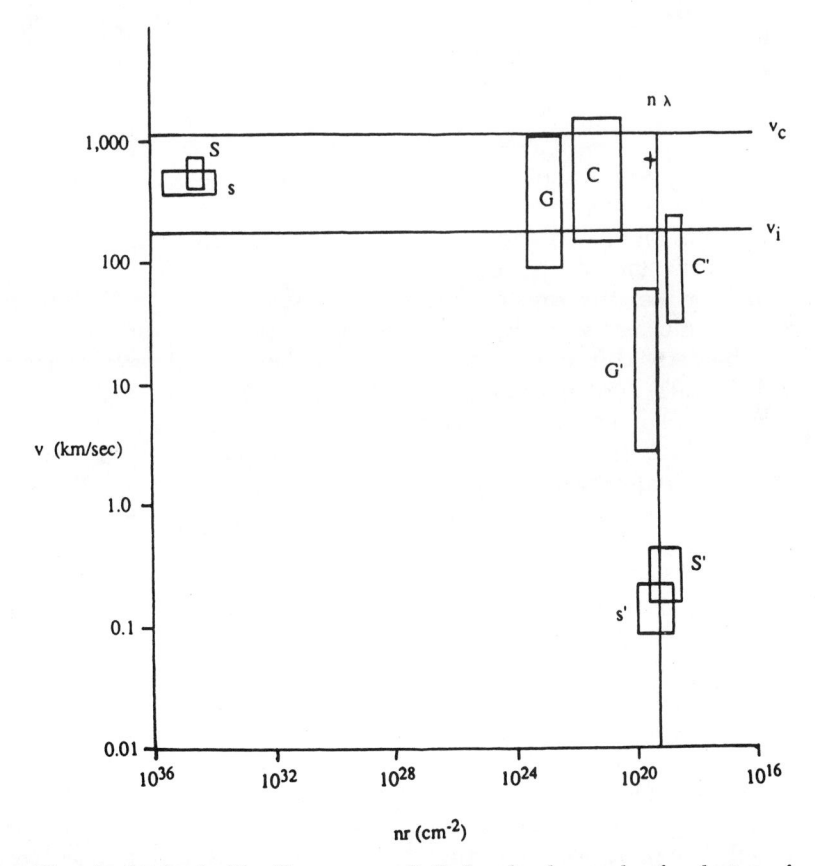

Figure 2. Plot of orbital velocities V versus nr. C, G, S and s show value for clusters of galaxies, galaxies stars with M > 1.8 M$_\odot$ and stars with M < 1.8 M$_\odot$. C', G', S' and s' are values for uncondensed clusters, galaxies, and stars. The vertical and horizontal lines come from plasma instability theory. The cross indicates the dimensions of the large scale structures mapped by Saunders (1991).

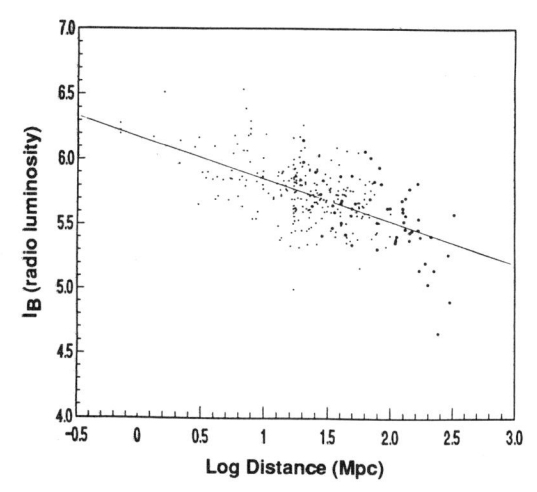

Figure 3. Index of relative radio luminosity I_B is plotted against Log distance (Mpc).(H_o = 75 km s^{-1}). Small dots are spirals, large dots interactives and starburst galaxies. Solid line is best fit, $I_B \sim D^{-.32}$. Since I_B is logarithmic (arbitrary units), the regression indicates a factor of 10 absorption over the range of distances represented.

before condensing (for example, for a galaxy this would be the average density of all matter in a cluster of galaxies).

As shown in Figure 2, these two simple relations predict the entire hierarchy of structures, from stars to large scale structures.

To see how the structures confirmed by Saunders et al. (1991) fit with these predictions, we must estimate the density of the structures. An upper limit on this density can be derived from estimates of observable matter density which is consistent with $\Omega = .02$ (Byrd and Valtonen 1990). For $H_o = 50$ km s^{-1} Mpc^{-1}, this yields $n = 6 \times 10^{-8}$ cm^{-3}. For a cylinder of radius 170 Mpc, orbital velocity is 1,000 km s^{-1}. Alternatively, we can take the observed average velocity of 600 km s^{-1} and assume that this is the orbital velocity. This yields a lower $n = 2 \times 10^{-8}$ cm^{-3}. For a spacing between cylinders of about 360 Mpc, with $n = 1-3 \times 10^{-8}$ cm^{-3}, $nr = 1-3 \times 10^{19}$ cm^{-2}. As can be seen from Figure 1, the structures are in good agreement with earlier plasma-based predictions.

In addition, theories based on the interactions of plasma filaments have explained in detail the formation of galaxies as shown in analysis and simulation (Peratt 1986). Such simulations show the development of the form of spiral galaxies and explain the flat velocity profiles without any recourse to dark matter. Extensions of these theories, using the above relationships of $n'D$, which imply a relation between density and mass of condensed objects, show that young galaxies are dominated by intermediate mass stars of 4–10 $M_\odot$ (Lerner 1988). Such stars produce the observed .22 proportion of He^4 by thermonuclear reactions and produce the observed abundance of deuterium and Li^7 by cosmic ray induced reactions (Lerner 1989).

The energy from these intermediate mass stars, thermalized by galactic and intercluster duets, provide the energy for the microwave background (Lerner 1988). The CBR is, in turn, isotropized by an absorbing and emitting thicket of dense, magnetically confined filaments of current in the intergalactic medium (Lerner 1988, 1990, 1992).

Such an absorptive and emitting IGM could account for the isotropy and spectrum of the CBR even with LSS. For example, for an approximate model of IGM density which declines as D^{-1} to 40 Mpc, remains constant to 600 Mpc and then remains constant at .125 lower density, the y parameter for an emitting and absorbing medium would be 1.8×10^{-4}. This results from the summation of radiation from elements of the IGM at higher and lower z. If a LSS is modeled as a 1.5 enhancement in IGM density extending from 300–500 Mpc, the resulting anisotropy from such LSS covering $\frac{1}{4}$ of the sky is 1.9×10^{-5}. More realistic inhomogeneous models would probably reduce these figures considerably.

Such an IGM would not eliminate radio emission from distant radio sources: average absorption at $z = 1$ would be about a factor of 6,000.

Previous evidence has supported the idea that the IGM is not transparent to radio and microwave radiation, but scatters and isotropizes it. The principal evidence for this is in the decrease with distance of radio luminosity of galaxies for a constant IR luminosity (Lerner 1990). More recent evidence (Lerner 1993) extends this relationship to 300 Mpc (Figure 3) and yields a statistical significance of 8 sigma. Such a relationship can only be explained by an absorptive medium, which by Kirchoff's law must be responsible for the blackbody CBR emission.

Thus, previous work has shown that a coherent, non-Big Bang approach, which assumes an evolving cosmos without a beginning in time and takes into account plasma phenomena, can successfully explain large scale structure, galaxy formation, the abundance of light elements and the characteristics of the CBR as well as predict new phenomena, such as the absorption of radio radiation by the IGM. In contrast, the Big Bang approach is incompatible with all data. To continue to cling to this theory and introduce an unlimited number of *ad hoc* hypotheses to bridge the gap with observation is to abandon the scientific method. It is clearly time to relinquish this theory and seek other explanations of the Hubble relation, such as, for example, the Alfvén-Klein ambiplasma model (Alfvén 1979, 1983; Alfvén and Klein 1962) or other alternatives that do not postulate a dense, hot epoch of cosmic evolution or a beginning in time.

References

Alfvén, H., 1979, *Astrophys. and Space Sci.* 66:23.
Alfvén, H., 1983, *Astrophys. and Space Sci.* 89:313.
Alfvén, H. and Klein, O., 1962, *Arkiv fur Fysik* 23:187.
Battaner, E., et al., 1992, *Nature* 360:652.
Becker-Szend, R., et al., 1990, *Phys. Rev. D.* 42:2974.
Campbell, A., 1992, *ApJ* in press.
Cen, R.Y. et al., 1991, *ApJ* 383:1.
Davies, J.I., 1990, *MNRAS* 245:250.
Delbourg-Salvador, P., Audouze, J., and Vidal-Madjar, A., 1987, *Astron. Astrophys* 174:365.
Fuller, G., et al., 1991, *ApJ* 371:111.
Gorski, K.M., 1992, *ApJ* 398:L5.
Hatzidimitriou, D., 1991, *MNRAS* 251:545.
Ichikawa, T. and Fukugita, M., 1992, *ApJ* 394:61.
Kapahi, V.K., 1987, in: *Observational Cosmology*, A. Hewitt et al. (eds.), IAU, p. 251.
Kolb,E.W. and Sherrer., R.J., 1982, M Phys. Rev. D 25:1481.

Lerner, E.J., 1986, *IEEE Trans. Plasma Sci.* 14:609.
Lerner, E.J., 1988, *Laser and Particle Beams* 6:457.
Lerner, E.J., 1989, *IEEE Trans. Plasma Sci.* 17:259.
Lerner, E.J., 1990, *ApJ* 361, 63.
Lerner, E.J., 1992, *IEEE Trans. Plasma Sci.* 20:935.
Lerner, E.J., 1993, *Astrophys. and Space Sci.*, in press.
Levin, J.J. et al., *ApJ* 389:464.
Linsky, J., 1992, *Space Telescope Science Institute Newsletter*, March 1992, p. 1.
Maoz, D., et al., 1992, *ApJ*, 394:51.
Mathews, G.J., et al., 1990, *ApJ*, 358:36.
Mathews, G.J., Boyd, R.N., and Fuller, G.M., 1991, *Bull AAS* 23:1394.
Melnick, J., Heydari-Malayeri, M., and Leisy, P., 1992, *Astron. Astrophys.* 253:16.
Olive, K. et al., 1990, *Phys. Lett B.*, 236:454.
Olive, K.A., Steigman, G., and Walker, T.P., 1991, *ApJ*, 380:L1.
Pagel, B.E.J., et al., 1992, *MNRAS* 255:325.
Peratt, A., 1986, *IEEE Trans. Plasma Sci.* 14:763.
Riley, S.P. and Irvine, J.M.,1991, *J Phys G.* 17:35.
Saunders et al., 1991, *Nature* 349:32.
Tully, R.B., 1986, *ApJ* 303:25.
Valtonen, M.J. and Byrd, G.G., 1990, *IEEE Trans. Plasma Sci.* 18:38.
Wagoner, R.V., Fowler, W.A., and Hoyle, F., 1966, *ApJ* 148:3.
Weinberg, D., and Cole, S., 1993, *MNRAS* 259:652.
Yang, J. et al., 1984, *ApJ* 281:493.

De Sitter Redshift: The Old and the New

John B. Miller

Advanced Technology Materials, Inc.
7 Commerce Drive
Danbury, Connecticut 06810
USA

Thomas E. Miller

Columbia University
One Atwell Road
Cooperstown, New York 13326
USA

The de Sitter redshift, a more or less moth-balled relativistic redshift, is re-examined in light of present-day observations and compared to the Hubble redshift. The redshift-magnitude relation given by the Hubble law is compared to that given by the de Sitter law. The intrinsic brightness of quasars, in the context of a de Sitter redshift, is not at all extraordinary. For high-redshift objects, such as quasars, redshift is an indicator, but a weak indicator, of distance. Could the "ancient" de Sitter redshift still play a role in modern astronomy?

Introduction

We shall consider three aspects of the de Sitter solution: first, the history of the de Sitter redshift; second, an analysis of recent observations in the context of this redshift; and finally, a brief discussion of the expected black-body radiation from a de Sitter event horizon.

Historical Overview of the de Sitter Redshift

A friend and colleague of Ehrenfest, Lorentz, and Einstein, the Dutch astronomer Willem de Sitter discovered the first-known cosmological redshift, the so-called de Sitter redshift, in 1917. Since 1992 is the 75th anniversary of the de Sitter solution, it seems appropriate to highlight the history of this venerable theory.

In the last of de Sitter's three papers comprising the first English language version of Einstein's general theory, de Sitter (1917) notes,

> Consequently the frequency of light-vibrations diminishes with increasing distance from the origin of co-ordinates. The lines in the spectra of very distant stars or nebulae must therefore be systematically displaced towards the red, giving rise to a spurious positive radial velocity.

This prediction of a redshift was apparently made in ignorance of the contemporary observational work by Slipher demonstrating such a redshift. "Slipher's list of 13 velocities, although published in 1914, had not reached de Sitter, probably as a result of the disruption of communications during the war" (Hubble, 1936).

There are three possibilities for a static universe: the flat spacetime of Minkowski, Einstein's "cylindrical" world in which space is curved but time is flat, and de Sitter's solution in which both space and time are curved. Ehrenfest actually had the basic idea behind the de Sitter solution (Kerszberg, 1989). In essence, Ehrenfest asked of de Sitter: if space is curved, why shouldn't time be curved as well? It is the time curvature which leads to the de Sitter redshift.

Although de Sitter theory has now slipped largely into obscurity, it was a dominant theory throughout the 1920's, sharing the spotlight with Einstein's "cylindrical" world. (Note that the de Sitter theory is distinct from the Einstein-de Sitter theory.) Edwin Hubble (1929), in his original paper demonstrating a correlation between redshift and distance, writes,

> The outstanding feature, however, is the possibility that the velocity-distance relation may represent the de Sitter effect.... In the de Sitter cosmology, displacements of the spectra arise from two sources, an apparent slowing down of atomic vibrations and a general tendency of material particles to scatter.

It is interesting to note that two types of de Sitter redshift were being discussed in the 1920's: metrical and kinematic. The kinematic redshift helped give rise to the idea of an expanding universe. However, we shall consider only the original, metrical, de Sitter redshift.

By the mid-1930's, the de Sitter solution seems to have vanquished its static rival, the Einstein "cylindrical" universe, only to succumb to the time-dependent solutions of Friedmann, Lemaître and others. The reasons for this decline in popularity are not entirely clear, but a general unwillingness to consider the possibility of negative pressure may have played a role. An equation arises in the de Sitter solution, $p + \rho = 0$, where p is pressure and ρ is density. Thus, assuming density is not negative, one must have either negative pressure or an empty universe in which $p = \rho = 0$. Apparently, neither of these alternatives seemed palatable at the time, and the de Sitter solution fell into disuse. Furthermore, World War II intervened, and the best and brightest were at work on more practical problems.

By the early 1960's, only a few authors were advocating the de Sitter solution. In one noteworthy study in *Nature*, Browne (1962) wrote, "In many respects the most satisfactory of the cosmological models based on solutions of Einstein's gravitational field equations is the steady-state de Sitter model." Earlier, Hawkins (1960) had advocated the de Sitter redshift in the context of the expanding steady-state cosmology of Bondi, Gold, and Hoyle. Ironically, Hawkins later (1969) abandoned de Sitter theory with the discovery of quasars:

> Actually both the steady state theory and the static universe were dealt a severe blow by the discovery of quasars. Neither theory can explain a concentration of these objects at the boundary of the universe, existing at an epoch of 3 billion years in the past.

It will be shown below that, despite Hawkins's capitulation, the de Sitter redshift provides a reasonable explanation for quasars.

Comparison of de Sitter Redshift with Current Observations

A metrical redshift, as its name implies, arises from the relativistic spacetime metric,

$$ds^2 = -\gamma^{-1}dr^2 - r^2 d\theta^2 - r^2 \sin^2\theta d\phi^2 + \gamma dt^2 \qquad (1)$$

(We choose units such that $G = c = 1$.) Redshift z is proportional to the percentage increase in wavelength and is defined as

$$z = \frac{\lambda_r}{\lambda_o} - 1 = \gamma^{-\frac{1}{2}} - 1 \qquad (2)$$

where λ_o is the wavelength of the unshifted photon and λ_r is the wavelength of the photon at a distance r.

There are many different formulations of the de Sitter solution depending on which coordinate transformations are employed. Our interest is in converting apparent brightness to intrinsic brightness. Since we wish to use the inverse-square law, we must choose so-called "pseudo-Euclidean" coordinates which preserve the Euclidean formula for surface area: $A = 4\pi r^2$. Other coordinate transformations would entail unfamiliar surface area formulae and will not obey inverse-square law dimming. For the de Sitter solution, assuming "pseudo-Euclidean" coordinates,

$$\gamma = 1 - \left(\frac{r}{R}\right)^2 \qquad (3)$$

where

$$R^2 = \frac{3}{8\pi\rho} \qquad (4)$$

r represents the distance to the de Sitter event horizon; and ρ is the mean mass density (*e.g.*, Eddington, 1923). In a de Sitter world, combining equations (2) and (3), distance r can be given in terms of redshift z as

$$\frac{r}{R} = \left[1 - (z+1)^{-2}\right]^{\frac{1}{2}}. \qquad (5)$$

This relationship is plotted in Figure 1.

Figure 1. The de Sitter metrical redshift is plotted versus distance as given by equation (5).

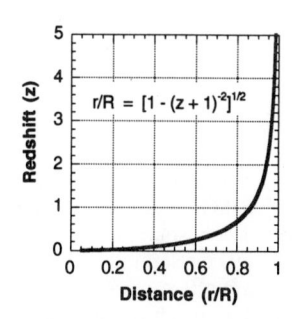

Assuming inverse-square law dimming, absolute magnitude M and apparent magnitude m are related to distance r by

$$M + C = m - 5\log_{10}(r) \qquad (6)$$

where C is a constant determined by observation. In a Hubble world, redshift z is directly proportional to distance, thus

$$M_H + C_H = m - 5\log_{10}(z) \qquad (7)$$

where M_H is absolute magnitude and C_H is a constant. Combining equation (5) and (6), the magnitude-redshift relation in a de Sitter world is given by

$$M_{DS} + C_{DS} = m - 2.5\log_{10}\left[1 - (z+1)^{-2}\right] \qquad (8)$$

where M_{DS} is absolute magnitude and C_{DS} is again a constant.

From combined catalogs (Huchra and Clemens, 1991; Huchra et al., 1992) of more than 20,000 objects (including galaxies, active galaxies, and quasars) with published measurements of both redshift and apparent magnitude, we transformed apparent magnitude to absolute magnitude assuming a Hubble law, equation (7), and a de Sitter law, equation (8). The scatter plots of Figure 2 show magnitude plotted versus redshift: first the raw data, second assuming the Hubble law, and third assuming the de Sitter law. A least-squares fit line is drawn for each. The vast majority of objects are buried within the dense, central portion of the graph. Almost all objects fall within a band spanning about ten magnitudes vertically. This band is clearly down-sloping assuming the Hubble law but is roughly horizontal assuming the de Sitter law. We find that absolute magnitudes are practically uncorrelated with redshifts given the de Sitter redshift-distance relation. This may have interesting implications for high-redshift objects such as quasars.

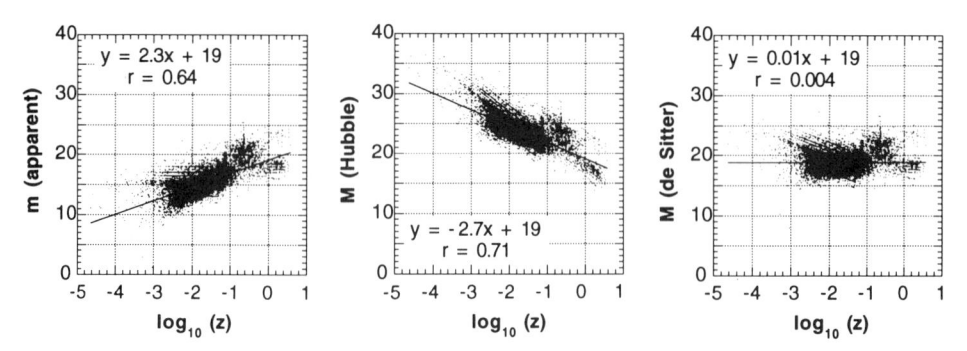

Figure 2. ZCAT and ZBIG catalogs combined (n = 20272)
    (left)   The apparent magnitudes of objects from The CfA Redshift Catalogue and the ZBIG catalog are plotted vs. redshift.
    (center) The absolute magnitudes obtained by assuming the linear Hubble redshift-distance relation and applying equation (7) to the data set on the left. (The values are offset, since the constant C_H has been arbitrarily set equal to zero.)
    (right) The absolute magnitudes obtained by assuming the de Sitter redshift-distance relation and applying equation (8) to the data set on the left. (The values are also offset, since the constant C_{DS} has been arbitrarily set equal to zero.)

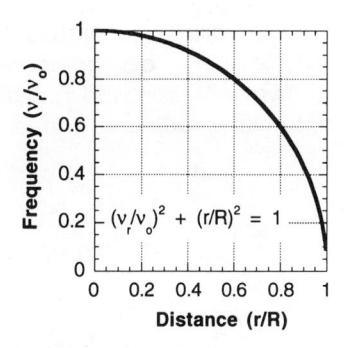

Figure 3. Fractional frequency is plotted versus distance as given by Eq. (9).

Assuming redshift increases with distance, for incomplete catalogs such as we are using, some increase in intrinsic brightness may be expected as a result of Malmquist bias: *i.e.* the more distant objects will tend to be brighter because we will only be able to see distant objects if they are intrinsically brighter than closer objects. Assuming a linear Hubble redshift-distance law, high-redshift objects, such as quasars, must be extraordinarily powerful since they are assumed to be much farther away than objects of intermediate redshift, and yet appear nearly as bright. This is difficult to explain as a result only of Malmquist bias. Of course, quasar luminosities may be understood as an evolutionary effect: as we are looking out into space, we are looking back into time when objects existed that were intrinsically much brighter than anything currently existing.

Because of the asymptotic character of the de Sitter redshift, quasars with high redshifts may not be much more distant than intermediate-redshift objects. Quasars may be considered in the context of a nonlinear de Sitter redshift-distance law (equation 5) without requiring an evolutionary effect. Absolute magnitudes are roughly the same at all redshifts, assuming a de Sitter redshift.

Although normally defined in terms of wavelength, the simplest way to think of the de Sitter redshift is in terms of frequency. The de Sitter frequency-distance law is given by

$$\left(\frac{v_r}{v_o}\right)^2 + \left(\frac{r}{R}\right)^2 = 1 \qquad (9)$$

where v_o is the frequency of the unshifted photon and v_r is the frequency of the photon at a distance r. This curve forms an easily visualized quarter circle as shown in Figure 3. However astronomers do not report their results in terms of fractional frequency but in terms of redshift z (or, more often, in terms of the *inferred* quantity "velocity of recession.")

Black-Body Radiation from a de Sitter Event Horizon

The 2.7° K black-body radiation is often cited in support of the Big Bang theory since it was predicted by this theory in advance of the observation. But note that Eddington (1926) predicted a 3° K temperature before he adopted Lemaître's Big Bang model.

The de Sitter event horizon, like a black hole event horizon, emits black-body radiation. It can also absorb radiation. Emitting and absorbing over a long time, the horizon should come to equilibrium with the energy contained within it. This is consistent with the well-known observation that the energy density of starlight is roughly equal to the energy density of the microwave background radiation. The observed 2.7° K radiation can be thought of as Hawking radiation from a de Sitter event horizon in a quasi-static universe.

References

Browne, P.F., 1962, The case for an exponential red shift law, *Nature* 193:1019.
de Sitter, W., 1917, On Einstein's theory of gravitation, and its astronomical consequences, Third Paper, *MNRAS* 78:3.
Eddington, A.S., 1923, *The Mathematical Theory of Relativity*, Cambridge University Press.
Eddington, A.S., 1926, *The Internal Constitution of the Stars*, Cambridge University Press.
Hawkins, G.S., 1960, The redshift, *Astronomical Journal* 65:52.
Hawkins, G.S., 1969, *Splendor in the Sky*, Revised Edition, Harper and Row, New York, p. 282.
Hubble, E., 1929, A relation between distance and radial velocity among extra-galactic nebulae, *PNAS* 15:168.
Hubble, E., 1936, *The Realm of the Nebulae*, Yale University Press, p. 109.
Huchra, J. and Clemens, C., 1991, *ZBIG Catalog*, Personal Communication.
Huchra, J., Geller, M., Clemens, C., Tokarz, S. and Michel, A., 1992, *The CfA Redshift Catalogue*, Springer-Verlag.
Kerszberg, P., 1989, *The Invented Universe: The Einstein-De Sitter Controversy (1916-17) and The Rise of Relativistic Cosmology*, Clarendon Press, p. 178.

Equilibrium Cosmology

Toivo Jaakkola

Tuorla Observatory
University of Turku
SF-21500 Piikkiö, Finland

In equilibrium cosmology (EC), the Universe neither expands nor changes globally in other respects, *i.e.*, it is in equilibrium. Theoretically, EC is based on the (empirically supported) strong ("perfect") cosmological principle (CP) and—as a consequence of the CP—electrogravitational coupling (EGC). The essence of and arguments for the EGC are summarized. EC divides into three fields: radiation and gravitation, connected via EGC, and research into equilibrium evolutionary processes (EEP), which in the CP-universe, ensure the properties of matter are unchanging, even though all individual systems evolve. The foundations and construction of EC are simple and coherent, and the CP-property makes it extremely sensitive to the empirical tests.

In radiation cosmology, the EGC explains the redshift effect with all its observed properties, including distance-dependent and distance-independent z's; in the former, varying steepness of the (r,z) relation depending on density is predicted and also observed. Cosmological redshift is a special case of the universal z-effect; EC predicts a unique (m,z)-relation, which is in form different but numerically coincides with the Hubble relation, which has depicted the observations well for six decades. The value of the Hubble constant is derived theoretically. The quantization of both the (cosmological) redshift of galaxies and intrinsic z of QSO's also follows from the theory.

The cosmic background radiation is a re-emission of the electromagnetic energy originating in galaxies and absorbed by EGC in the cosmological redshift. Its energy density, temperature, Planckian spectrum, photon-baryon number ratio, similarity with the local starlight energy density, and general isotropy are all derived from the EC-model. The predicted dipole anisotropy vector fits the observations well. In particular, it is emphasized that the reported extremely exact blackbody form of the spectrum and the still extreme isotropy of the CBR are firm fingerprints of a Universe in equilibrium, while these are implausible in the other models.

In gravitation cosmology, an explicit formulation of Mach's principle is first given. Following from a finite limiting value of the "Machian force" of cosmic masses, the solution of the Seeliger-Neumann gravity paradox is obtained. It appears that the Machian force ($a_c = G_c\rho/\alpha_c H \approx 10^{-8}$ cm s^{-2}) determines the structure in the Universe: uniform cosmological distribution, the transition to hierarchical local structure, and mass-to-size structure of supergalaxies, clusters, groups and individual galaxies. The gross internal structure and its evolution are derived, and the flat rotation curves of spiral galaxies are explained without hypothetical dark matter. The field equation defining EGC and equation of state joining the two long-range forces to large-scale structure are given.

Equilibrium evolutionary processes (EEP) are not examined in detail. The problem is to find, for galaxies (and stars) and their systems, and for the intergalactic background, those processes which establish physical parameters as invariant on the large scale, even though these are known to change locally. Important parameters are, *e.g.*, density distribution, rotation, morphological type, stellar populations, galaxy age and mass distributions, element abundances, and photon/baryon number ratio. Some preliminary discussion is presented. In all cases, processes in the diffuse cosmological background, those in the dense galactic nuclei, and, of course, stellar astrophysics, have important roles. It should be noted that, while Olbers' paradox is solved easily by the finite value of the integral of the redshifted radiation of the galaxies, this works only on the first level; the final solution will probably parallel the solution of the photon/baryon EEP problem. The whole problem of the EEP forms a new and extremely rich field for physical science, and will probably be the focus of research in the next century.

1. Introduction

For over two decades, the author has been involved in a project to test the cosmological expansion hypothesis in a systematic manner (Section 2). The results have been quite consistent: the Universe does not expand. Nor does it evolve globally in other respects. Consequently, the Universe must be in equilibrium, and this necessitates a corresponding theory.

Of course, this result destroys the standard big-bang theory of cosmology (ST). Nor does it it offer support to the classical steady-state cosmology (SST), which has been the main challenger of the ST for almost half a century. The distinguishing feature of the SST is the perfect cosmological principle, which is confirmed by the results in Section 2. Consequently, the new theory described here will be steady-state, though it will not be a static version of the classic SST. There is no expansion, while the theoretical basis is different. The cornerstone of the equilibrium cosmology (EC) will be the strong cosmological principle (CP, Section 3.i.); its physical foundation is not general relativity, as in ST and SST, but a hypothesis of electrogravitational coupling (EGC, Section 3.ii). Consequently, in most respects (*e.g.* as regards the redshift, microwave background, origin of baryonic matter, galaxies and light elements, solutions in the background paradoxes, *etc.*), the theory differs not only from the ST but also from the SST.

The general structure of the EC will be presented in Section 4. Its three components—radiation cosmology, gravitation cosmology and the search for equilibrium evolutionary processes (EEP)—will be outlined in the subsequent sections 5 through 7. The most obvious testing possibilities are pointed out in Section 8. The discussion in Section 9 deals with the status of ST, SST and the new theory.

All the sections below confront the reader with empirical or theoretical arguments which are quite new or controversial in the prevailing model. The only way to solve the dilemmas posed is by logical reasoning combined with careful analysis of observations specific to the problems. In particular due to the strong form of the CP, but also due to its theoretical conceptions, the theory presented appears to be an exceptionally testable theory. Perhaps the most intriguing—and most difficult—problem is that of the EEPs. This field of research is established by the most characteristic feature of the Universe which leads to a third cosmological paradox, analogous to the earlier paradoxes of Olbers-de Cheseaux and Seeliger-Neumann. It states: everything evolves, yet the whole does not evolve. Together with its associated parameters, this paradox manifests itself in cosmic nature in a wide variety of forms. Its full solution will be a matter for science in the next century.

2. Observations in Favour of an Equilibrium Universe

An equilibrium Universe means essentially that, in the first instance, the hypothesis of cosmological expansion is incorrect, and second, there exist no other global evolutionary effects; *i.e.*, the galaxies as a class, their systems as a class, and properties of intergalactic matter on the large scale remain unchanged. The author is well aware that both of these assertions contradict most of what is presented in the current cosmological and astrophysical literature. Even so, I consider them well justified.

There is a widespread belief that theories of the Universe are only weakly testable. This is not at all true. There are at least four groups of tests of the cosmological expansion hypothesis, each containing tens of separate tests along with the classes of objects and the parameters that can be examined. The results in the four test groups are reviewed below and references to more detailed studies are presented. Three more theoretically controversial items are then discussed.

i. Properties of the Redshift Effect in Systems of Different Scales

As the first test, an examination of the properties of the redshift effect in systems of different scales will inevitably reveal whether it is a Doppler effect or something else (Jaakkola 1978a). Taking into account the effect of gravitation, the z-effect should be, in the Doppler case, fainter within the systems than between the systems. In the case of interaction redshifts, the situation would be the reverse. There is some evidence for the strength of redshift (α) being higher within the Local supergalaxy (LSG) than in the homogenous metagalactic distribution. In clusters, groups and pairs of galaxies, z appears to depend on, *e.g.*, the type (Jaakkola 1971, Moles and Jaakkola 1976), compactness and status of the galaxies, high z-values usually being connected with features pointing to youth of the galaxy (Jaakkola 1973). There are indications that redshift is also a function of position in the systems, an argument in favour of strong intergalactic z-fields. These results remove the missing mass problem in systems of galaxies (Jaakkola 1976, 1993 a).

In individual galaxies, there are redshift gradients from the near to the far sides (Jaakkola et al. 1975a). A redshift field is found also in the plane of the Milky Way, with α ten times larger than the cosmological value $\alpha_c = H/c$. This field has distorted the structural maps of our galaxy derived from the kinematical maps (Jaakkola et al. 1978, 1984). The lines originating in the nucleus of the Galaxy are redshifted by 40–75 km s^{-1} (Jaakkola 1978b). The O–B type stars in the solar neighborhood, in star clusters and in the Small and Large Magellanic clouds show excess z of 5–15 km s^{-1} (Jaakkola 1978a, Arp 1992). The solar limb redshift is larger than Einstein's prediction by $2\times10^{-7} - 10^{-6}$. Significant redshifts of the lines emitted by Taurus A and spacecraft Pioneer 6 are found symmetrically before and after eclipse by the Sun (Pecker 1976). These effects, like the related excess light deflections observed for tens of stars, cannot be explained by relativity theory.

Bringing the results together, the strength of redshift appears to depend on the density according to $\alpha \propto \sqrt{\rho}$; the value of $h = c\alpha$ ranges from the metagalactic value

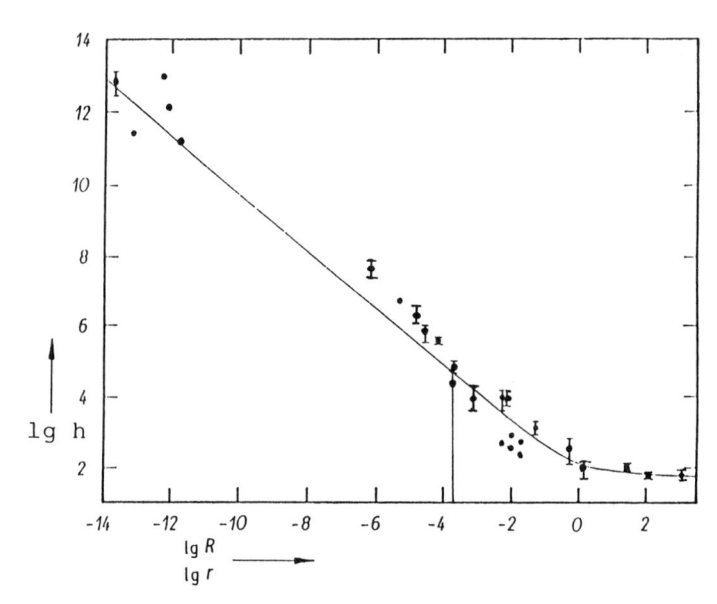

Figure 1. Strength of redshift, $h = c\alpha$, given in units of km s^{-1} Mpc^{-1}, in systems of different scales. R is the radius of a system, r the distance from the centre, or (as for three values on the right which refer to z_c) distance from the observer (in Mpc). In the middle, galaxies and their systems, on the left, solar observations.

$H \approx 60$ km s^{-1} Mpc^{-1} to 10^{13} km s^{-1} Mpc^{-1} on the surface of the Sun (Jaakkola 1978, Figure 1 here). The results indicate unambiguously the test alternative where α is stronger within the systems than between them, proving that redshift is an interaction effect and not due to expansion. The problem of the QSO redshift is omitted here; I would only point out that the large intrinsic redshifts in a sub-class of QSO's (Jaakkola et al. 1975b, Jaakkola 1982, 1984) are due to one and the same interaction effect as the cosmological redshifts.

ii. Consistency of the Cosmological Test Results

The second test of expansion involves a systematic review of the global and the local cosmological tests for both the expanding and the static theories (Jaakkola et al. 1979). Within the frame of the standard theory, the empirical value of the deceleration parameter q_0 is dependent on the method: the Hubble diagrams give it the mean value +0.93 ± 0.19, the local tests +0.03 ± 0.08, the (Θ, z)-relation for galaxies +0.3 ± 0.2, for clusters −0.9 ± 0.2, and for radio sources symbolically < -1; the latter results are inconsistent with all existing relativistic models. The empirical inconsistency of the standard cosmology appears to be of a systematic and stable character (see Figure 2).

In the static Universe interpretation, the (m,z)-relation is practically identical with Hubble's linear relation and with that for $q_0 = +1$. The value $q_0 \approx +1$ found above indicates that the data fit the static model. The (Θ,z)-diagrams for clusters, and for radio sources (Nilsson et al. 1993) also fit the static prediction (Figure 3). The observed convergence of the test results, opposing the trend in the standard cosmology, provides strong evidence of the static character of the Universe.

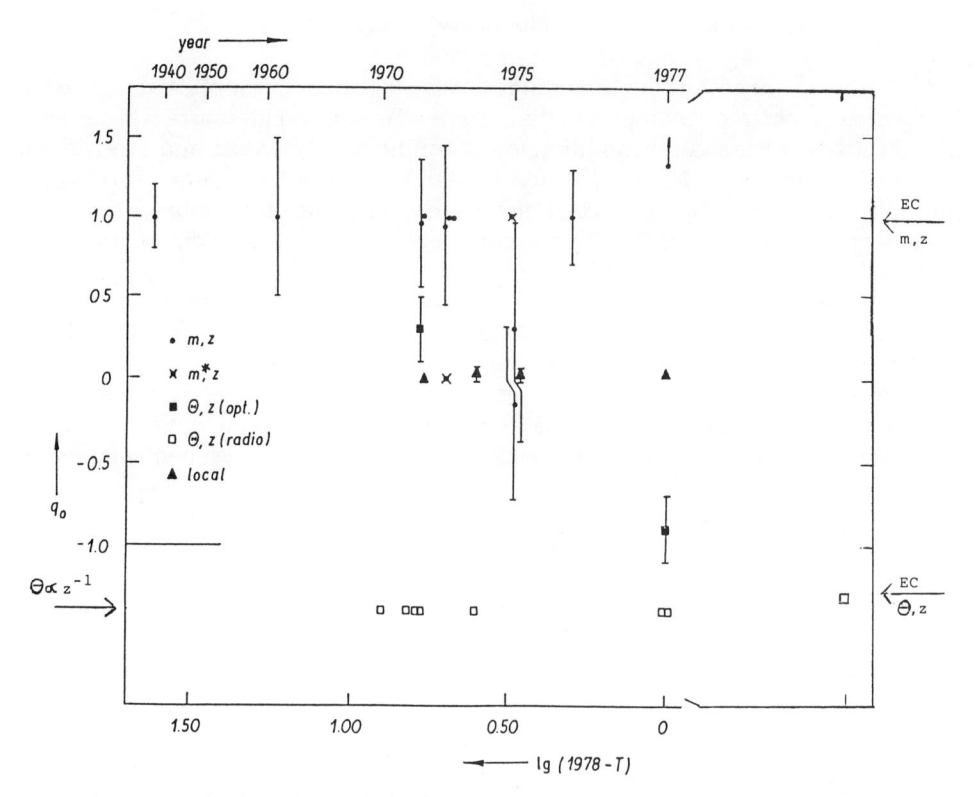

Figure 2. Systematization of cosmological test results in the framework of the ST (q_o -values), and EC (arrows on the right), as a function of the year of publication. Note systematic and constant spread in the former model. Angular diameters (θ, squares) fall below any relativistic prediction.

iii. The Hubble-Tolman Test

The surface brightness test (SB $\propto (1+z)^{-a}$; $a = 4$ for expansion, $a = 1$ for the static model), first suggested by Hubble and Tolman (1935), forms the third test group. It has been applied to four different kinds of data (Jaakkola 1986). Galaxies in six clusters have been argued to support expansion (Crane and Hoffman 1976). However, agreement (Hoffman and Crane 1977) is due to other reasons: procedure of absorption correction, a K-related selection effect and circular reasoning involved in usage of the metric SB. After proper corrections, $a \approx 1$. Even without such corrections, the 1st, 3rd and 10th brightest galaxies give $a \leq 1$. A similar criticism applies to the analysis of Sandage and Perelmuter (1991). The SB-profiles of six cD clusters, measured at 0.5 and 1.0 Mpc from the centres, favour the static solution. So do the QSO host galaxy SB-profiles. The contrary conclusion by Gehren (1985) results from inhomogeneity of the data and use of the $q_o = +1$ metric. The lowest SB-values in the subsequent z-intervals for QSO double radio sources, based on component sizes and LAS, follow the static model prediction. Hence, the Hubble-Tolman test also argues consistently against the expansion hypothesis.

iv. Non-Existence of Cosmological Evolutionary Effects

The theme of cosmic evolution in respects other than the supposed expansion is one of the most frequent topics in the current astronomical literature. The absence of such effects is the second condition for an equilibrium universe, and I consider it the fourth test group of the expansion hypothesis. If the Metagalaxy expands, it must also change in other respects; if not, its age is infinite and it must not change. There are several kinds of data that exclude evolution (Zwicky 1957). Other data, such as the counts of bright QSO's (Green and Schmidt 1978), have been claimed to support evolution. It has been shown (Jaakkola 1982) that this result is due to a morphological selection effect, and if we add the type 1 Seyfert galaxies which are physically identical to the QSO's, the counts fit the static unevolving model closely. The argued spectral evolution of galaxies results from a selection effect arising from the K-term (Laurikainen and Jaakkola 1985a, b). There are several ways to interpret the radio counts without evolution (Jaakkola *et al.* 1979). The experimental failure of

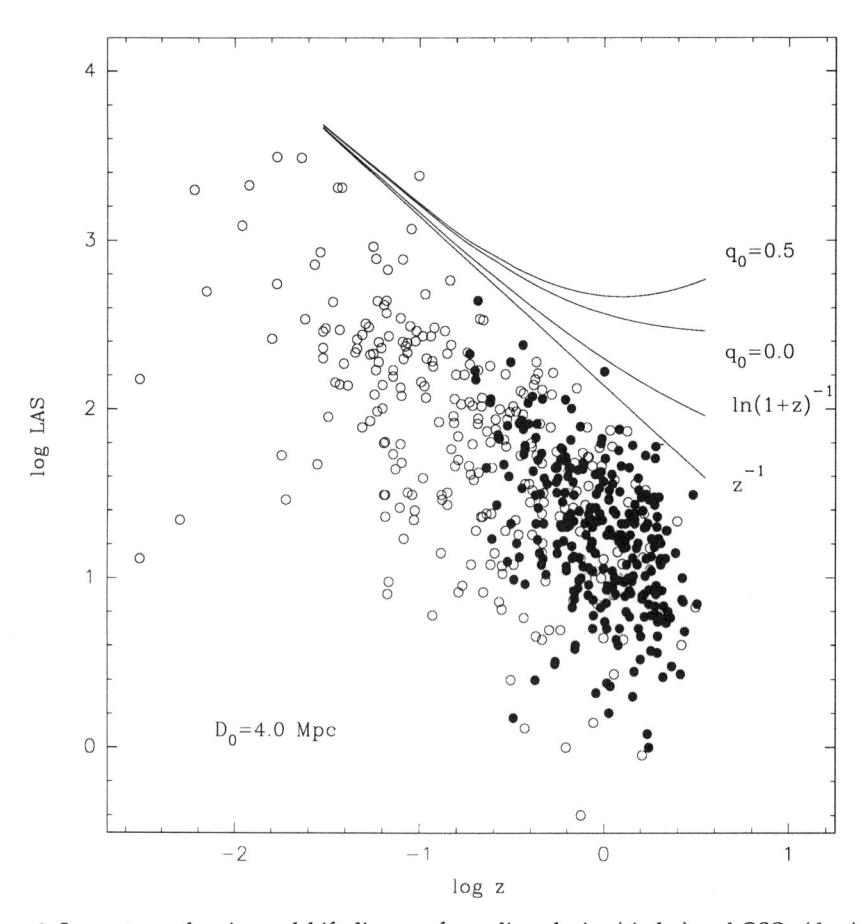

Figure 3. Largest angular size–redshift diagram for radio galaxies (circles) and QSOs (dots), from Nilsson *et al.* (1993). Two ST predictions, the (unphysical) linear relation and the curve for EC (tired light) are shown.

the most strongly urged claims of cosmic evolution demonstrates that such effects do not exist, and hence contradicts the expansion hypothesis.

v. Homogeneity and Isotropy

Statistical non-evolution and homogeneity of a cosmic distribution are probably complementary, becoming true at the same scale. Subregions of a stationary homogeneous distribution must be in mutual equilibrium: the contrary would be hard to imagine. Hence, a homogeneous, isotropic cosmological distribution argues for a state of equilibrium; the inference is straightforward, and it may be countered that homogeneity and isotropy are basic features of the ST models as well. Anything, even a homogeneous explosion cloud, can be ordered mathematically in a specific model, but this is not a physical statement, while the inference of equilibrium from isotropy definitely is.

Empirically, the CBR is isotropic, past the local dipole deviation, to the level of $< 10^{-5}$. In EC, one half of the CBR originates within $z = 1$. The X-ray background is isotropic at < 1.3 percent level; the effective distance of the source is not known well, but it could be the scale of galaxies and quasars. As for the galaxies and their systems, there is no reliable evidence of structures larger than superclusters, which have sizes of some tens of Mpc. For the time being, the reported structures and voids of hundreds of Mpc should be treated with caution: this result may simply be due to the redshift quantization, as suggested by the concentric form of many such structures. Galactic absorption may be at play in some data. From isotropy, and from counts of galaxies and X-ray sources, we conclude tentatively that a volume containing a number of superclusters, say with radius 100 Mpc, forms a homogeneous region where most of the equilibrium processes can be tested with a degree of confidence.

vi. Planck Spectrum of the CBR

Quite surprisingly, nobody has yet noticed that the exact black-body form of the spectrum of the CBR can be regarded as firm evidence of a Universe in equilibrium. Such a Universe is an ideal cavity (blackbody) radiator, in which emission and absorption are in equilibrium. In cosmic nature there probably does not exist any other such radiator. The exploding cloud of the ST-model cannot be a black-body radiator with emission and absorption in equilibrium.

vii. Similarity of Cosmological and Local Energy Densities

Tremendous significance should be attached to observation of a cosmological radiation background at a temperature $T \approx 3°$ K, since this is also the temperature of interstellar space. Accordingly, the radiation energy densities U_c and U_l for CBR and local starlight are similar. Moreover, it is known that the (local) energy densities of the magnetic field, cosmic rays and motions in gas clouds are close to U_c (Peebles 1971). The fact that $U_c \approx U_l$ indicates that the galaxies are in equilibrium with their cosmological environment, *i.e.* the Universe is in equilibrium. In EC, many threads lead from the fact that $U_c \approx U_l$ to the dynamics of the interconnection between local and global structure. In ST, where $U_c \propto (1+z)^4$, this fact can only be due to a very improbable chance coincidence.

viii. Discussion and Implications of Empirical Data

Therefore, several different kinds of evidence point consistently to the conclusion that the Universe is in equilibrium. This means there is no expansion. In the author's estimation, there are already arguments enough for this statement. Without doubt, these are not enough to convince the majority of the researchers working in the field. However, a consensus is not necessary; obviously the tests must be continued, both those described above and those following from a new theory. Ideally, the tests should be performed by independent groups using the most advanced data. What we wish to stress here is that there are already arguments enough in favour of developing a theory of a non-expanding universe and deriving from it further tests of the real state of the Universe.

Homogeneity and non-evolution mean that, on the scale of the observed Metagalaxy, we possess a representative sample of the Universe. This finding opens the door to a genuine cosmology. It is worth noting that Hubble (1934) reached that conclusion sixty years ago on the basis of galaxy counts from $12^m 7$ to $16^m 0$.

The non-existence of expansion and other cosmic evolution means that there is no time characteristic of the cosmological scale. In this regard, EC differs radically from the ST and also from the SST, since the latter involves a cosmic time in the sense of recession of the existing galaxies. On the other hand, EC, unlike its rivals, involves a large number of definite cosmological time-scales determined by certain important evolutionary periods.

The most important implication of the above results is the requirement that a solution be found to the following paradox: all individual systems are known to evolve, yet the Universe does not evolve. In other words, galaxies evolve, but galaxies do not evolve as a class. In addition to the paradoxes of the dark night sky and the finite background gravitation, this is the third major problem confronting cosmological research. All three suggest a characteristic property of the cosmological scale, namely, that it is different from the other scales not only quantitatively but also in a qualitative respect.

3. Theoretical Foundations of Equilibrium Cosmology

The cosmological problem can also be approached from the direction of theoretical conceptions. This affords a consistent logical structure of the theory, with observations remaining as testing possibilities. Naturally, in the theoretical approach some pre-existing knowledge about the subject matter—the Universe—and about the laws of nature is necessary. In equilibrium cosmology, a theoretical assertion about the Universe is given in the form of the (strong) cosmological principle (CP). The conception of physics here is radically different from the relativistic physics; it is contained in the hypothesis of electrogravitational coupling (EGC). A new conception of gravitation will also be presented.

i. The Cosmological Principle

Any non-empirical idea about the Universe is, of course, philosophical in character. Throughout the history of natural philosophy since antiquity, one of the main motifs in both idealist and materialist modes of thought has been a belief that behind the visible, changing and vanishing local world there exists an invisible eternal and unchangeable level of reality. A cosmological model—like the ST—which contradicts this great line of philosophical thinking can hardly be correct. The cosmological principle (CP) in its strong ("perfect") form conforms to the old philosophical conviction and presents it in a scientific language. It says, essentially, that the Universe is seen as similar from any place at any time. This definition already extends beyond any local inhomogeneities.

We shall call this statement about the Universe simply the "cosmological principle," and refer to it as CP. It may be pointed out that the "strong" CP is not completely identical to the "perfect" CP in SST. In the latter there is, *e.g.*, a continuous increase in distance between any two galaxies or systems of galaxies, and this signifies the existence of a cosmic time, which is absent in EC. Hoyle (1980) also suggests an increase of information content in his model. Other philosophical and physical problems are numerous in SST, in particular with the topic of "continuous creation" versus the conservation principles; these difficulties are not encountered in EC.

That the CP should be a foundation stone of the new theory is also suggested by the history of cosmology, which can be seen as "variations on a theme of one". There is the well-known progression: geocentric–heliocentric–galactocentric world pictures. The presently prevailing model of an expanding universe is obviously only a continuation of the outdated ideas which have the common conception about the Universe as a single structured and (in the modern view) evolving system. While the spatial aspect of CP has been generally accepted, the Copernican revolution in science remains still unfinished in its temporal content. In the structure of natural sciences cosmology is the science which extends beyond any local, structured, evolving system and aspires to the universal, permanent properties of matter. Accordingly, cosmology can be defined logically as the study of an entity which fulfills CP, the strong cosmological principle.

It is an empirical, not a philosophical, question whether the observed part of the Universe, the Metagalaxy, fulfills the CP. In Section 2, it was shown that it probably does. The tests must be continued, since an answer to this primary question fixes the character of the Universe better than anything else.

If we adopt the CP as the foundation, the theory becomes extremely testable. It covers all the parameters which characterize the properties of matter: their number is practically endless.

The position of the CP is so central in EC because, in the first instance, it defines the position of cosmology among the sciences; second, it defines the present model in an exact and clear manner; third, it defines its position with respect to the other theories, mainly ST and SST; fourth, it makes the theory testable; and fifth, the other fundamental hypothesis behind the theory, the ECC, can be reduced to it.

Other discussions of the cosmological principle can be found, *e.g.*, in Bondi (1960), Hoyle (1980), Rudnicki (1989) and Jaakkola (1989).

ii. Electrogravitational Coupling

A physical theory is necessary in order to deal with the various aspects of the equilibrium Universe in an explicit manner. Equilibrium in the Universe means on one level a large-scale equilibrium between baryons (b), photons (γ), and gravitons (g). Radiation (γ) is produced chiefly as described by the current theories. Gravitation is the countervailing equilibrium effect which maintains the energy in the systems of baryonic matter unchanged on the large scale. The γ's and the g's also interact and are in large-scale equilibrium. This is called electrogravitational coupling (EGC). The EGC is the physical theoretical basis of the EC, while the CP is its cosmological basis.

The hypothesis of EGC was found necessary when searching for a physical mechanism of the redshift effect. Three main arguments suggest EGC; these are deeply rooted in the history of physics. Both for radiation and for gravitation, there exist the well-known parallel paradoxes of an infinite background field, known as the Olbers-de Cheseaux and Seeliger-Neumann paradoxes. EGC provides a simultaneous solution for both. Second, there is a long history, originating from Faraday (1849) and recently advanced by Woodward (1983), of laboratory experiments designed to produce electricity by accelerated masses. Third, the solar tests of GR, combined with obviously related data concerning redshifts of sources occulted by the Sun, speak for a connection between gravitation and redshift, *i.e.*, for the EGC. Moreover, at all macroscopic scales of nature there are "anomalies" which can possibly be explained by EGC. The latter features and the three historical deductions have been discussed in more length in Jaakkola (1991), and are listed in Table 1. The table should be considered only as indicative: it may include phenomena that are actually unrelated to EGC. The rationale for Table 1 is to call attention to the fascinating possibility that such a wide range of phenomena might really have a single physical basis.

The hypothesis of EGC is applied in cosmology by studying how radiation and gravitation respond to the effect of coupling (Sections 5 and 6). Section 6 includes a discussion of the nature of gravitation.

4. Structure of Equilibrium Cosmology

According to its theoretical foundations, EC can reasonably be divided into three sections. Radiation cosmology (Section 5) deals with such effects as redshift, cosmic background radiation and Olbers paradox in the framework of an equilibrium Universe. Gravitation cosmology (Section 6) attempts to solve the problems of the large-scale homogeneity of matter distribution, the design of the hierarchic structure on the smaller scale, the Seeliger-Neumann gravity paradox, and the missing mass problem, to mention a few historical and currently discussed open problems. The EGC hypothesis connects the two sections, which can be seen as studies of the responses of the two long-range forces to their mutual coupling.

The third branch of cosmology deals with a question: how can the Universe fulfilling the CP remain unevolving, even though every one of its constituent systems evolves? The problem is approached by considering equilibrium evolutionary processes (EEP), *i.e.* chains of processes in galaxies and intergalactic space that ensure the properties of matter remain unchanged on the large scale, while these properties are changing locally within any system. The number of cosmologically interesting parameters is very large. A theoretical understanding and observational check of the constancy of such parameters, such as density distribution on various scales, morphological distribution of galaxies, angular momentum, element abundances and photon/baryon/graviton number ratios - to mention only a few - forms the contents of the third section of EC (Section 7). The EEPs are actually the core of EC. These will also bring the strong and the weak nuclear forces into the purview of

Table 1. Empirical Arguments in Favour of Electrogravitational Coupling (EGC)

No	System	Scale (cm)	Effect or Problem	Possible Link to EGC
1.	Universe		Cosmological redshift z_c non-Doppler	EGC is redshift mechanism
2.	-"-	∞	Finite CBR (Olbers paradox)	z_c (EGC) part of solution
3.	-"-		Finite CBG (gravity paradox)	Due to EGC
4.	-"-		Origin of CBR	Cosmic EG induction
5.	-"-		Homogeneity and isotropy	Finite CBG
6.	-"-		Local structure (galaxies, clusters)	Finite CBG
7.	-"-		Dynamical stability	Finite CBG + item 11 instability
8.	-"-		Steady state	Involves EG and other forces
9.	Hierarchic structure	10^{25} -10^{11}	Strength of z $h \propto \sqrt{C}$	Redshift coupled with mass
10.	Outer parts of gal.	10^{23} -10^{22}	Missing mass problem	From EGC $F \propto 1/r$ dynamics
11.	Nuclei of galaxies	10^{21} -10^{18}	Missing energy problem	Light weakens gravity
12.	Quasars, compact gal.	10^{19} -10^{17}	Intrinsic redshift z_i	Light coupled with matter
13.	Solar vicinity	10^{12} -10^{11}	Light deflection $\delta > \delta(GR)$	-"-
14.	Solar vicinity	10^{12} -10^{11}	Passing light redshifted	-"-
15.	Solar surface	10^{11}	Center-to-limb redshift	As other (r,z) gradients
16.	Earth surface	10^{5}	"Fifth force"	As other deviations from $1/r^2$ gravity
17.	Laboratory experiment	10^{2}	Electrograv. induction	Self-explanatory
18.	Sun-Moon-Earth	10^{13}		Gravity anomalies in solar eclipses
19.	Sun-Mercury	10^{13}	Mercury orbit acceleration	
20.	Planet-satellite	10^{10} -10^{9}	Anomalous orbit accelerations	Gravitation different from usual conception
21.	Laboratory experiment	10^{2}	Grav. depends on composition?	

cosmology. In previous cosmological theories, there are no elements comparable to the study of the EEPs. Consequently quite little is known about them. The EEPs will possibly be a major field of research in cosmology and astrophysics, as well as in theoretical and laboratory physics in the coming decades.

Looking at the foundations and the structure of EC, both are compact and intrinsically consistent: EGC can be derived from the CP, and the latter is in agreement with the essence of philosophical tradition, with the history of cosmology, and with the observations. The structure of the theory follows logically from the two theoretical principles. The foundation provided by the CP guarantees that EC is a highly testable scientific theory.

5. Radiation Cosmology

In radiation cosmology the main task is to derive from the fundamental theory, *i.e.*, from the CP and EGC, the existence and properties of the redshift effect and of the cosmic background radiation, and also to deal with the Olbers-de Cheseaux paradox of the finite background light.

i. The Redshift Effect

Let us summarize the properties of the redshift effect that need to be explained by a cosmological theory. The following represents a personal view of the effect.

F1. Spectral shift is always toward the longer wavelengths (with a few local exceptions).
F2. The redshift $z = \Delta\lambda/\lambda_o$ is independent of λ_o, the rest wavelength of a line.
F3. In a homogenous distribution, for small distances r, z is approximately proportional to r: $z \propto H$. For any distance, z fits well Hubble's linear (m, $\log z$)-relation

$$m = 5\log z + K(z) + C_1 \tag{1}$$

where the K-correction $K(z)$ is observational in character.
F4. The parameter α depends on matter density ρ, with $\alpha \propto \sqrt{\rho}$ found for the main classes of systems in various levels of hierarchy (Jaakkola 1978a).
F5. On the cosmological scale $\alpha = \alpha_c = H/c \approx 6.33 \times 10^{-29}$ cm^{-1} ($H \approx 60$ km s^{-1} Mpc^{-1} is the Hubble constant and c the velocity of light).
F6. For most QSOs, z is independent of distance and it originates in the source: $z \approx z_i$. Equation (1) does not apply. The brightest QSOs in each z-interval follow a relation (Jaakkola 1984)

$$m = 2.5\log(1+z) + C_2 \tag{2}$$

There appears to be a cut-off of at $z_i \approx 2.5$, with a strong drop-off at higher values.
F7. Both r-dependent and r-independent redshifts obtain quantized, preferred values according to the equation

$$\Delta\ln(1+z) = C_3 \tag{3}$$

For cosmological redshift, $C_3 = 2.4 \times 10^{-4}$, corresponding to $\Delta(cz) = 72$ km s^{-1}, as observed for nearby galaxies by Tifft (1976, 1988), Napier (1991, 1993) and Guthrie and Napier (1990). For the QSOs, $C_3 = 0.206$, as discovered by Karlsson (1971) and confirmed by Arp et al. (1990).

F8. Intrinsic redshift is time-variable, with young objects observed at excess z_i, which decreases over cosmic periods (Jaakkola 1973, Arp 1987).

F9. z in general may be slightly time-variable by a few km s^{-1} of (cz) on a time-scale of years; presently z's are decreasing. This seemingly incredible finding by Tifft (1988) should be taken seriously.

We now attempt to derive properties 1 to 9 from the fundamental theory (CP and EGC). The existence of $z > 0$ (point F1) is required by CP: otherwise the total energy in the form of radiation would grow continuously. The physical mechanism is the EGC. If $\Delta\lambda \propto \lambda_o$ (F2) would not be valid, the spectrum of the background radiation would change, contrary to the CP; moreover, gravitation as the z-mechanism is not a selective effect.

The Hubble relation (F3) is deduced as follows. The effect of a homogenous graviton bath (or other tired light medium) on a photon moving in it is a linear function of distance:

$$\frac{dJ_x}{J_x} = \frac{dE}{E} = \frac{dv}{v} = -\alpha\, dr \qquad (4)$$

i.e. the photon loses part of its forward momentum J_x and energy $E = hv$ into the graviton bath through which it moves. Integration from the source, where $E_o = hv_o$, to the observer at distance r, and using $z = E_o/E - 1$ gives

$$\int_{E_o}^{E}\frac{dE}{E} = \int_{v_o}^{v}\frac{dv}{v} = -\alpha\int_0^r dr \qquad (5)$$

$$E = E_o e^{-\alpha r} \qquad (6)$$

$$\ln(1+z) = \alpha r \qquad (7)$$

Equation (7) gives the redshift-distance relation in the EGC model.

Before deriving the coefficient α, we will present the redshift law in observable terms of magnitude m,

$$m = -2.5\log f = -2.5\log\frac{f_o}{r^2(1+z)}$$

where f_o and f are the emitted flux and redshifted flux at distance r. Then we have

$$m = 5\log\ln(1+z) + 2.5\log(1+z) + K(z) + C_1 \qquad (8)$$

where $K(z)$ is the K-correction (effect of z on m through the form of the spectrum and broadening of the waveband), and constant C_1 contains the absolute magnitude and the distance scale α. The first term on the right comes from the geometrical $1/r^2$ effect of distance via equation (7), and the second term is the energy effect of redshift diminishing the flux by a factor $1/(1+z)$. Now equation (8) is a very close approximation of the original Hubble relation [equation (1)]. At $z = 0.1$ the difference between equations (1) and (8) is $\Delta m = 0.0008$ mag, at $z = 1$, $\Delta m = 0.043$ mag, and at as high as $z = 10$, $\Delta m = 0.50$ mag. Hubble's linear $(m, \log z)$ relation is commonly adopted intuitively without any theoretical justification. However, it has been observed repeatedly in the $(m, \log z)$ diagrams of galaxies from the 1930s to the present day (Section 2.ii., Jaakkola et al. 1979). The embarrassing situation, where the data actually fit the linear $(m, \log z)$ relation deduced through erroneous steps from an erroneous linear redshift-distance law $[z = \alpha r$, instead of equation (7)], becomes clear through the practical identity of equations (8) and (1).

By the same token, the EC prediction for the global z-angular diameter relation is from equation (7)

$$\Theta = d \frac{\alpha_c}{\ln(1+z)} \tag{9}$$

where d is linear diameter.

The gravitational field depends on density ρ, and, hence, in the EGC scheme, α depends on ρ (F4). From equations (26) and (29), to be derived in Section 6, we obtain

$$\alpha(r) = \sqrt{\frac{A}{a}} \cdot \sqrt{\rho(r)} = 6.34 \times 10^{-14} \sqrt{\rho(r)} \left(\frac{cm}{g}\right)^{\frac{1}{2}} \tag{10}$$

With regard to F5, a value of $\alpha_c = H/c$ can be predicted theoretically from EGC, without expansion (Jaakkola 1991).

$$\alpha_c = 1.04 \frac{\sqrt{G_c}}{c} \tag{11}$$

Adopting $\rho_c = 10^{-30}$ g cm^{-3} and $G_c = 10 G_o$ (Section 6), we obtain $\alpha_c = 2.83 \times 10^{-29}$ cm^{-1}. This yields a reasonably good fit with the observed value $\alpha_c(obs) = 6.33 \times 10^{-29}$ cm ($H = 60$ km s^{-1} Mpc^{-1}).

The large intrinsic redshift, z_i, of a sub-class of the QSOs (F6), like that of the compact and late-type galaxies and early type stars, is due to one and the same physical effect—EGC—as z_c. The length-scales are here small, and values of ρ and α high. The particular form of the bright edge of the QSO Hubble diagram, fitting equation (2), follows neither from the distance effect nor from the energy effect, since the graviton atmosphere absorbing the factor $z/(1+z)$ of the incident energy must reradiate it. More obviously, the form of equation (3) is due to the broadening of the rest-frame spectral band by the redshift, usually included in the K-term; the K-effect due to the form of the spectrum is close to zero for this class of QSOs.

The quantized fine-structure of the z-effect (F7) is expected in view of the quantum character of the photons and of the gravitons with which they interact. In the case of z_c, energy is lost in constant fractions ΔE over constant distance intervals Δr, i.e. $\Delta E/E = -\alpha_c \Delta r = -C_3$. Equation (3) follows from equations (4) to (7). Hence periodicity is not exactly in z but in $\ln(1+z)$; for $z < 0.05$ the two are practically the same. This brings the periodicies in z_c and z_i to the same figure. In z_i, the energy losses correspond to non-constant distance intervals, mean free path $L(r)$, and varying $\alpha(r)$, obeying $L(r) = C_3/\alpha(r)$.

The temporal decrease of z_i and in some classes of objects (F8) probably follows from decreasing density evolution. High z_i in young objects indicates their origin as dense splinters thrown out of the nuclei of the parent systems; the formation of a gravitational field *ab novo* is also possible in the pressure-induced gravitation scheme. This resembles the Hoyle-Narlikar-Arp interpretation of excess redshift (Arp 1987).

A small change in accurate z-values over a few years (F9), as argued by Tifft (1988)—which at first sight sounds impossible for a sample of astronomical objects—could possibly, if true, be explained in the EGC scheme. The redshift energy, *i.e.* the fraction $z_i/(1+z_i)$ of the incident energy, caught by the gravitational atmosphere of QSOs, must be reradiated, and this could explain the rapid light variation and its z-dependence (a broad maximum around $z/(1+z) = \frac{1}{2}$, Jaakkola 1984). The light variations might be accompanied by fluctuations of the gravitation field. The same is valid with smaller amplitudes for normal galaxies, including the Milky Way, where the "galactic aurora" and its gravitational counterpart are predicted by the EGC. Tifft's time-dependent redshift could be due to the latter, a local aurora-like variation of the g-field of the Galaxy. In due course, the sign of the change of z is expected to change.

Therefore, EC offers an effective theoretical apparatus to explain the redshift phenomenon, which itself is much richer in features than usually imagined. Obviously, in this respect, EC is superior to the ST, which actually contains no theory of redshift. Its fundamental theory, the GR theory of gravitation, actually predicts contraction and a cosmological blueshift effect. Whereas the Doppler mechanism is a theory of symmetric shifts toward the red and blue, galactic redshift is a quite different, asymmetric effect. To explain it, ST resorts to an *ad hoc* assumption of a metaphysical (beyond-physics) big bang event. In EC, redshift is an equilibrium effect predicted by the fundamental theory.

ii. Cosmic Background Radiation

In EC, while redshift is an equilibrium process providing for the energy balance between electromagnetic radiation and the other substances in the Universe, the cosmic background radiation (CBR) is an equilibrium effect maintaining energy balance between the gravitational background and radiation. The CBR is reradiation of the redshift energy (RE) absorbed into the graviton bath in the redshift effect.

To broaden the context further: just as z_c is physically the same effect as z_i in QSOs, and the same as the z-gradient over the disk of the Milky Way, so the CBR is physically the same effect as reradiation of RE absorbed in z_i by the gravitational atmospheres of QSOs, or their more mundane analogues in galaxies. The latter obviously provides an interpretation of the light variations in QSOs and related objects, as well as the variable background radiation related to our Galaxy, as predicted by the EGC hypothesis. It is appropriate to refer to this new radiation mechanism and to call it "redshift radiation" (RR). Since most of the radiation energy density in the Universe is contained in the CBR, RR actually forms the principal mechanism of observed electromagnetic radiation.

The CP implies that the cosmic vacuum is in EC an ideal black body (cavity) radiator, *i.e.*, absorber and emitter of radiation. In fact, it may be the only black body expected to exist in nature. Therefore the reported extremely exact black-body form of the CBR spectrum is not a surprise in EC: it should be a surprise in ST, which asserts that the Universe is an exploding cloud. The observation of the Planckian spectrum is a strong evidence for an equilibrium state of the Universe, and an argument in favor of the present theory.

According to the CP, the Universe is structureless on the large scale and the observed extreme isotropy of the CBR is therefore expected, following from the isotropy of the galaxy distribution, cosmic background gravitation (CBG) and the redshift effect. Via the z-effect, the CBG thermalizes and homogenizes the radiation from discrete sources.

Let us attempt to derive the properties of the CBR quantitatively. The original CBR energy is in the emission processes of discrete sources. Therefore the energy density of CBR should equal the total redshifted light from galaxies. The intensity of radiation due to the galaxies from within distance r is the integral

$$I_r = \frac{nL}{4\pi} \int_0^r \frac{dr}{1+z} \tag{12}$$

where n is number density and L is the luminosity of an average galaxy. Using equation (7), we obtain

$$I_r = \frac{nL}{4\pi\alpha_c}(1-e^{-\alpha r}) \tag{13}$$

$$I_z = \frac{nL}{4\pi\alpha_c}\left(\frac{z}{1+z}\right) \tag{14}$$

which approaches the value

$$I_c = \frac{nL}{4\pi\alpha_c} \tag{15}$$

when $z \to \infty$. The function I_z and the contributions of the background intensity from the different z intervals are shown in Figure 4. It can be seen that half of the total background light comes from within $z=1$ and nine percent from $z>10$. The distribution in Figure 4 of the origin of CBR energy (and CBR flux) is quite specific to the present theory, and also applies to the Machian cosmic background gravitation (Section 6). If one adopts as parameters the values $n = 3 \times 10^{-2}$ Mpc^{-3}, $L = 3.9 \times 10^{44}$ erg s^{-1} (10^{11} stars of solar luminosity), and $H = 60$ km s^{-1} Mpc^{-1}, we obtain for the integrated radiation energy density $U_c = 4\pi I_c/c = nL/c\alpha_c$ the value $U_c = 2 \times 10^{-1}$ erg cm^{-3} = 0.13 eV cm^{-3}. Within the range of uncertainties, this is the same as the observed energy density of the microwave background, $U_c(\text{obs}) = 0.25$ eV cm^{-3}.

For the temperature of blackbody radiation, the Stefan-Boltzmann law can be applied:

$$R_c = \delta T_c^4 \tag{16}$$

where radiance $R_c = 4\pi I_c = cU_c = nL/\alpha_c$. With the above value of U_c, and $\delta = 5.67 \times 10^{-3}$ erg s^{-1} cm^{-2} K^{-4}, the prediction for the cosmic temperature is $T_c = 3.2°$ K. Agreement with the observed value 2.73° K is quite satisfactory, in particular as the accuracy of the ST prediction is much lower, from 1 to 30 Kelvins (Smirnov 1964,

Peebles, 1971). A more strict intercomparison of the values of nL, α_c, U_c and T_c is a matter for future study.

Though the way in which T_c and the helium abundance have been related is an admirable chapter of ST, as a test of the correctness of the theory it is rather loose, both theoretically and observationally. Derivation of U_c and T_c from the luminosity density of discrete radiation, and the Planckian spectrum and isotropy as implications of equilibrium—all straightforward from the fundamental theory—make the EC approach tighter than the ST approach, with the added virtue that EC works with terms which are acting in nature at the present time, not in the remote past.

The standard cosmology does not offer any explanation of the photon/baryon number ratio $n_\gamma/n_b \approx 10^8$ which is a fundamental cosmological parameter. Here the interpretation is immediate:

$$\frac{n_g}{n_b} = \frac{U_c}{(hc/\lambda)/\rho_c/m_p} = 0.51 \frac{m_p L}{chHMT_c} \tag{17}$$

where Wien's law, equation (15) and $U_c = 4\pi I_c/c$ are used in the latter step, h is Planck's constant and m_p is the proton mass.

Therefore, the problem reduces to the astrophysical problems of emissivity (L/M), on the one hand, and redshift (H), on the other hand. Taking $T_c = 3.2$ K and M/L = 10 in solar units, we obtain $n_\gamma/n_b = 3 \times 10^8$. This fits the observations.

The similarity of the cosmological radiation energy density U_c and the local starlight energy density U_l should be regarded as an extremely important cosmological observation. This may even be the key observation for attempts to understand the connection between local physics and the cosmological environment. Also the local energy density of the magnetic field, cosmic rays and motions in gas clouds have values close to U_c.

Standard cosmology does not provide any interpretation of these similarities, which appear as pure chance coincidences. At $z = 0.5$ the cosmic value of $U_c \propto (1+z)^4$ would change by a factor of five, and in the big-bang cosmology this is an insignif-

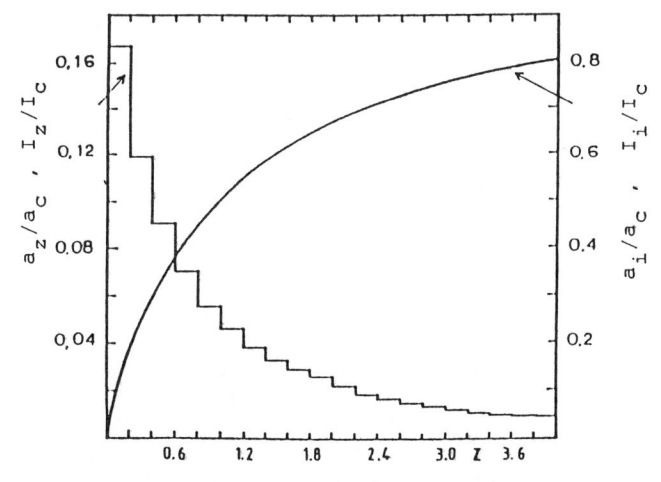

Figure 4. Contributions from within z (scale on the right) and from different z-intervals (left), to the CBR and CBG.

icant period as compared, e.g., to the time of the last scattering thought to have happened at $z \approx 1500$. The similarity of U_l and U_c follows from the adopted interpretation of the background radiation and from Machian control of galaxy formation, to be discussed in Section 6. Evidently

$$\frac{U_l}{U_c} \approx \frac{n_l L_* R_l / c}{n_c L_* / H} \approx \frac{n_l R_l}{n_c R_c} \tag{18}$$

Here L_* is the stellar luminosity, n the number density and R the radius; indices l and c refer to the local and the cosmological values, respectively. From $a_c \approx a_i$ (Section 6), $n_l/n_c \approx R_c/R_l$, which implies $U_c \approx U_l$, hence the observed similarity.

The dipole anisotropy and recently reported smaller scale fluctuations should then be addressed in the framework of EC. The observed dipole anisotropy, $\Delta T/T \approx 10^{-3}$, is not a problem of magnitude, but it is in terms of direction, which is almost opposite to the solar apex, due to galactic rotation ($V_\odot = 232.3$ km s^{-1}, $l = 87.°8$, $b = +1.°7$: de Vaucouleurs et al. 1991). The dipole measurement is, in terms of velocity, $V_B = 332$ km s^{-1}, $l = 287°$, $b = +61°$ (Cheng et al. 1979). The scope of the difficulties following from the Doppler ST interpretation of the dipole anisotropy, including a bulk motion of the whole LSG at 600 km s^{-1}, and a curious supermassive but invisible cluster, the "Great Attractor" (GA) causing these motions, should raise doubts about its correctness.

In EC we have the following prediction for V_B. The known local motions are: the Sun in the Galaxy, $V_\odot$, as above; the Galaxy in the Local group, $V_G = 61$ km s^{-1}, $l = 121.°2$, $b = -21.°6$ (as observed by the blueshift of M31); and the Local group due to rotation of the LSG, $V_{LG} \approx 300$ km s^{-1}, $l \approx 319°$, $b \approx -14°$. De Vaucouleurs (1958) has repeatedly presented evidence of supergalactic rotation, and it is also seen in the data presented in the GA context (e.g. Staveley-Smith and Davies 1989). After reduction of the non-velocity redshift effect which is obvious in the data, the local z-data reveals a clear differential rotation pattern, analogous to the kinematical data of the Milky Way. The three local velocity vectors summarize into the Doppler component of the CBR dipole:

$$\overline{V}_D = \overline{V}_\odot + \overline{V}_G + \overline{V}_{LG} = 236 \text{ km s}^{-1}, l = 24°, b = -22° \tag{19}$$

A further, non-Doppler component is predicted in EC due to the excess field in LSG, $\Delta\alpha/\alpha = (h_{LSG} - H)/H \approx (80-60)/60 = 0.33$. This excess strength of z in the LSG (Jaakkola 1978) is still present in a very distinct manner in the GA data, e.g. in Staveley-Smith and Davies. (1989). Because the CBR is emission of redshift energy, the excess $\Delta\alpha/\alpha \approx 0.33$ causes an excess temperature, ΔT, for the CBR. From equation (14), we expect for the non-Doppler component of the dipole anisotropy

$$\frac{\Delta T_{nD}}{T_b} = \frac{\Delta\alpha}{\alpha_c} \frac{z_{LSG}}{1+z_{LSG}} = 1.66 \times 10^{-3} \tag{20}$$

for $z_{LSG} \approx 0.005$, with coordinates $l = 284°$, $b = +74°.5$ (Virgo cluster). In terms of symbolic velocity, this is $V_{nD} = c\Delta T_{nD}/T = 497.5$ km s^{-1}. The total predicted anisotropy is

$$\overline{V}_T = \overline{V}_D + \overline{V}_{nD} = 456 \text{ km s}^{-1}, l = 347°, b = +51° \tag{21}$$

In view of the uncertainties, the fit of V_T with V_B is quite satisfactory: the angle difference is 30 degrees. Closer scrutiny will be needed to check whether the details of the nearby distributions of density, α and ΔT fit the interpretation.

Using the recently reported smaller scale and smaller amplitude fluctuations of the CBR (Smoot et al. 1991) as an ultimate verification of the big bang doctrine represents a very strange approach indeed. While the details of $\Delta T/T < 6 \times 10^{-6}$ are typically about 7 degrees wide, regions 3 degrees apart would have been causally unconnected at the time of last scattering (here neglecting a discussion of the inflation theory). In EC, there is some interest in comparing the position, size and scale of the faint fluctuations with those of the second-order clustering at small and moderate distances. Like the Sunyaev-Zeldovich effect, these fluctuations can be used as a test of the origin of the CBR.

In the author's estimation, the EC-approach has already achieved much more than ST in understanding the CBR: EC explains all the properties of CBR directly from the fundamental theory without *ad hoc* hypotheses, just as it did above for the properties of the redshift effect.

iii. The Olbers-de Cheseaux Paradox

We need not discuss the meaning and history of the Olbers' paradox here. I would only note that with the collapse of the big-bang hypothesis, this centuries-old riddle still remains open. The finiteness of I in equation (15) gives one part of the answer: the tired-light redshift effect makes the integrated incident radiation from galaxies finite. But as an absorption effect (of energy), any tired-light mechanism is open to the same criticism as was Olbers' interpretation in terms of interstellar absorption. It will be seen in Section 7.i., that EC is able to overcome this difficulty, though the details of the full solution still remain unclear.

6. Gravitational Cosmology

The discovery that the Universe is in equilibrium also requires a gravitational treatment. A century ago, Newtonian analysis came up against the Seeliger-Neumann gravity paradox. The big-bang cosmology solved this in its own peculiar manner; this is not applicable in EC, and hence we are still faced with the paradox, and with it, the stability problem. Other features which must be confronted by any cosmological theory are the observed homogeneity and isotropy, *i.e.* structurelessness on the large scale, and the observed hierarchical structure on smaller scales. Moreover, the problem of the existence and amount of invisible matter has grown during the last decades into an important cosmological topic.

Starting from the theoretical principles discussed in Section 3, the general task here is to examine, on the cosmological scale, the response of gravitation to coupling with the electromagnetic interaction, as asserted in the EGC hypothesis. We first consider the general nature of the effect of gravitation.

i. Nature of Gravitation

General relativity (GR), usually adopted as the fundamental theory in cosmology, is, in our view, unsatisfactory for both conceptual and empirical reasons (for a discussion, see Jaakkola 1991, 1993; Monti 1991). Moreover, the equilibrium theory is, in essence, contradictory to GR. It regards baryonic matter, its gravitational effect, and electromagnetic radiation as being in mutual equilibrium on the large scale. It is not plausible to assume that gravitation should differ from the other substances, working through space geometry instead of particle/quantum interactions like the other forces. This dilemma has been a general concern in physics, which has operated during several decades simultaneously with quantum mechanics and the geometrical tools of GR. It is straightforward to assume that gravitation works by gravitational quanta, gravitons (g), which reside and move in space like the CBR photons (γ_b), and interact with γ_b and incident photons from galaxies (γ_g). This interaction is just the EGC. On the cosmological scale, the g's form the background field which can be called cosmic background gravitation (CBG), analogous to CBR.

Gravitation appears to be a pressure effect of gravitons flowing from the background space. The flow is proportional to the mass of the body, and it is conducted by the gravitational field of the body. The field is formed and maintained by interactions of the gravitons streaming from the background space with those already existing in the field, with the radiation field, and with the particles of the body and its atmosphere. The background field, which is the source of the graviton inflow, is associated with the higher-order system—for the Earth the solar system and the Galaxy which form their own gravitation fields—and hierarchically thereafter, up to the homogeneous cosmological background field CBG. The strength of gravitation is a variable, $G(r)$, the locally measured value of which is Newton's constant G_o.

The apparent two-body attraction does not mean a direct gravitational interaction between the bodies. It is due to a mutual screening of the inflows towards the bodies. Newton's law of gravitation can be deduced in a simple manner (Jaakkola 1991, 1993), the inverse square distance factor resulting from the geometrical contraction of the solid angles and $1/r^2$ dependence of the surface density of the graviton inflows. At the scale of the solar system, the strength of the galactic background gravitation field, $G(r)$, stays nearly constant at G_o. The EGC does not have a measurable effect at that scale.

Gravitation as a pressure effect is not an attractive but rather a repulsive force. Like the other physical forces, gravitation is a local interaction. In a sense, the historical dilemma of local action versus action-at-a-distance vanishes in the two-body problem when it is treated as a screening effect: the screening body acts at a distance without mutual signals of interaction, but gravitation still remains a local action of gravitational quanta.

The idea of pressure-induced gravitation (PIG) can be traced into the nineteenth and eighteenth centuries, when most of the famous physicists of the time attempted intensely to solve the riddle of gravitation (North 1965). Most adopted a material field concept, as is the case here. Closest to the present approach is perhaps that of le Sage (1784). GR has been a milestone in this history, but because it was often mistakenly thought of as the ultimate answer, it has caused a stoppage in the centuries-long effort towards an understanding of gravity. More recent work by Broberg (1982) resembles the model of gravitation described above. A more detailed historical account, a more specific critical treatment of GR and a more extensive study of gravitation, including discussion of laboratory experiments, must be omitted. The properties of gravitons are still unknown, and a treatment of EGC on the

ii. Mach's Force and the Gravity Paradox

Opinions are divided as to whether the inertia of a body is already implied in Newton's mechanism or whether it requires a further explanation along the lines first suggested by Mach, *i.e.*, in terms of an action by distant masses (Mach, 1883; Bondi 1960). I am inclined toward the former explanation (with gravitation understood as above). However, the gravitational effect of the distant regions of the Universe, which we will call the Machian force, belongs to the same picture. Interconnection of local and global action is a characteristic feature of EC, which appears to be capable of giving an explicit expression for the Machian force.

According to the lines of the PIG-theory, there is no direct gravitational interaction between the mass systems over cosmological distances; their screening effects are hidden by more important local effects. However, the distant masses are in equilibrium with their gravitational background space and thereby affect our local gravitation through that background field. This kind of focusing of the distant action is very satisfying, since as far as I know, Newtonian (or relativistic) formulations of the Machian interaction have never worked well.

Consider a homogeneous cosmological distribution with mean density, ρ_c, and a graviton g with energy E_0 in a background field at distance r (or z) moving towards the Earth. E_0 is connected to a particular graviton spectrum, of which we have today no knowledge. In the presence of EGC, exactly as for the photons in equation (6), the graviton is observed on the Earth at $E = E_0 r^{-\alpha_c r}$, with the same value of $\alpha_c = H/c$ as for the redshift effect. This provides the theoretical justification for the Seeliger-Neuman-type potential function, deduced a century ago quite *ad hoc*, but here deduced directly from the fundamental theory.

Gravitational acceleration per steradian induced by the masses and their background field within radius r and the corresponding z is (cf. Jaakkola 1983, 1987)

$$a(r,z) = \int_0^z G_c \rho_c e^{-\alpha_c r} dr = \frac{G_c \rho_c}{\alpha_c}\left(1 - e^{-\alpha_c r}\right) = \frac{G_c \rho_c}{\alpha_c}\frac{z}{1+z} \tag{22}$$

Here G_c, α_c and ρ_c are the cosmological values of the gravity parameter (analog of Newton's constant), strength of redshift, and density, respectively. Equation (21) can be understood as an explicit formulation of Mach's force and "Mach's principle" (with a meaning slightly different from the original one; see above). Figure 4 shows α_z as a function of z and contributions from the different z-intervals.

When r and z go to infinity, $a(r,z)$ approaches the value

$$a_c = \frac{G_c \rho_c}{\alpha_c} \tag{23}$$

a_c, the "Machian force", is a fundamental constant playing a central role in cosmological processes. Its finite value solves the gravity paradox in a straightforward manner. However, since out of the multitude of the cosmic equilibrium processes, only the EGC and redshift effect is present in equations (21) and (22), the solution is not perfect. We shall touch on the question in Section 7.i.

In a vectorical presentation of equations (19) and (21), forces from opposite directions would cancel each other. However, particularly in view of the form of the step function in Figure 4, the actual non-isotropic nearby distribution becomes important, as it was for Mach (1893). Clearly, finiteness of a_c in equation (23) is important; if a_c were infinite, the cosmological matter distribution would be highly unstable, and the existence of quasi-stationary individual mass systems could not be understood.

iii. Effective Radius of the Universe

One half of the CBG, as well as one half of the CBR, comes from within $z = 1$, where $a_z/a_c = I_z/I_c = \frac{1}{2}$. This scale can be defined as an "effective radius of the Universe", D. For $H = 60$ km s^{-1} Mpc^{-1}, we have $D = \ln 2(c/H) = 3466$ Mpc. Of course, this does not affect the main statement about an infinite Universe.

The given D can be seen as a more rigorous scale parameter than $\alpha^{-1} = 5000$ Mpc, which occupies no special position in z or in $z/(1+z)$: $z = 1.718$, $z/(1+z) = 0.632$.

iv. Field Equation for Electrogravitational Coupling

Energy and momentum are conserved in each photon-graviton collision in the EGC scheme. The strength of redshift is known to increase as a function of density (Sections 2 and 5). This means that, within the systems, part of the outward momentum of the γ's is transferred to the infalling graviton flow. As a consequence, the G-parameter decreases with decreasing distance and increasing strength of redshift $\alpha(r)$. A field equation valid for EGC in quasi-stationary systems is as follows:

$$G(r)\alpha(r) = A = \text{constant} \qquad (24)$$

Inserting the local values, $G_o = 6.67 \times 10^{-8}$ cm^3 g^{-1} s^{-2} and $\alpha_o \approx 10\alpha_c \approx 6.33 \times 10^{-28}$ cm^{-1}, one obtains $A \approx 4.22 \ 10^{-35}$ cm^2 g^{-1} s^{-2}. Instead of emphasizing a local value, G_o, and the cosmological value $\alpha_c = H/c$, we should consider the universal behaviour of the variables $G(r)$ and $\alpha(r)$, the product of which, the electrogravitational coupling constant A, is a more fundamental constant than the special values G_o and α_c.

v. Cosmological Strength of Gravitation and Size of the Machian Force

From equation (23) and $\alpha_o \approx 10\alpha_c$, it follows that $G_c \approx 10G_o = 6.67 \times 10^{-7}$ cm^3 g^{-1} s^{-2}. The strength of gravitation per unit mass (e.g. affecting a singular atom) is thus an order of magnitude higher in the intergalactic space than locally: there exists no opposite force balancing the Machian gravitation pressure a_c.

With this value of G_c, the Machian force a_c of equation (23) becomes $a_c \approx 1.05 \times 10^{-8}$ cm s^{-2}. Actually, a_c is uncertain by a factor of a few integers. Constant a_c turns out to be a fundamental cosmological constant which determines macroscopic structure in the Universe.

vi. Interpretation of the Large-Scale Structure of Matter

Interpretation of the structural organization of matter observed on various scales is, of course, a major problem for the whole of physics. In the standard cosmology, the problem of the macroscopic structure has been dealt with, starting from either the large scale (Doroshkevich et al. 1974) or from the small scale (Peebles 1971) density disturbances in the early phases of the big bang. Resorting to hypothetical initial conditions is not very satisfactory scientifically. We should try to understand the observed Universe through the processes which exist here and now. In EC, we must consider the role of the Machian force a_c in the large-scale structural organization of matter. Table 2 gives the local gravitational acceleration

$$a_l = \frac{G(R)M}{R^2} \qquad (25)$$

at the edge of a system with radius R and mass M for various systems within a range of scales, over a factor of about 10^3; a_c is also given.

Table 2. Local acceleration $a_l = G(R)M(R)/R^2$ at the edges of systems of different scales. The Machian force a_c is also given.

Class of objects	log M (g)	log R (cm)	a (cm s^{-2})
Supergalaxies	48.7	25.5	3.4×10^{-9}
Large clusters of ellipticals	47.9	24.5	5.4×10^{-8}
Small clusters of ellipticals	47.2	24.3	2.7×10^{-8}
Small clouds of spirals	47.0	24.3	1.7×10^{-8}
Small loose groups of spirals	46.5	24.1	1.3×10^{-8}
Small dense clusters of ellipticals	46.5	23.7	8.5×10^{-8}
Compact groups of spirals	45.5	22.6	1.3×10^{-8}
Giant ellipticals	45.5	22.35	4.2×10^{-6}
M31 (Sb)	44.6	22.89	4.5×10^{-8}
Milky Way (Sbc)	44.45	22.79	5.0×10^{-8}
The Universe (a_c)	—	—	1.1×10^{-8}

It is assumed that at the edge of systems of galaxies the strength of gravitation attains the cosmological value: $G(R) = G_c = 10G_0$. For galaxies usually lying in the larger systems, and hence with further gradient of $G(r)$ for $r > R$, $G(R) = 0.2G_c = 2G_0$ is adopted. This fits through equation (24) the observation of $\alpha(r)$ in the Milky Way (Jaakkola et al. 1984). Estimates of R and M are from de Vaucouleurs (1971, Allen 1973); for M31 and the Galaxy 25 kpc and 20 kpc are adopted for R.

With the exception of supergalaxies and giant ellipticals, the a_l-values are practically equal for the systems in the list, and at the same time also equal to a_c. This remarkable fact implies that the CBG, the Machian force, controls the gross mass-radius structure of galaxies and their systems.

The high value of a_l for the E-galaxies is well understood as due to their long individual dynamical evolution. Supergalaxies are discussed in the next subsection. The origin of galaxies and systems of galaxies is examined in Section 7.iii.

vii. Supergalaxies, Passage from Hierarchic Structure to Cosmological Distribution, and Isotropy

In Table 2, a_l is for supergalaxies an order of magnitude smaller than a_c. These are therefore unstable systems being broken up by the Machian force and screening attraction by neighboring systems. The second-order clusters are not rotating ellipsoids but often chain-like configurations. The incidence of clusters belonging to the second-order systems, a few tens of percent, supports non-stationarity of the superclusters. The Local supergalaxy is probably a rotating, quasi-stationary system; it contains only one cluster with attached groups and field galaxies.

Furthermore, $a_2 < a_c$ implies that third order systems do not coalesce, since evidently $a_3 < a_2$, i.e. $a_3 \ll a_c$ would be true. A second point to note is the rapid decrease of the number of units with the level of hierarchy: there are 10^8–10^{12} stars in a galaxy, less than 10^4 galaxies in a cluster, and only a few clusters in a second-order cluster; i.e. the number of sub-units has already fallen close to the minimum.

Hence, in addition to supplying an explanation for hierarchic cosmic structure over a broad range of scales, the EC-theory, due to a fixed finite size of a_c, also offers an interpretation of the absolute scale of the hierarchic structure and a passage to the homogenous cosmological distribution. EC thus defines the dimensions of its own subject matter.

Naturally, the above also furnishes an interpretation of the observed large-scale isotropy.

viii. Global Stability

Global stability can be understood in the light of the preceding arguments, i.e. finiteness of CBG, cessation of hierarchical structuring on a certain scale, and consequent lack of any basis for larger scale motions. A rapid weakening of the contributions from the different distances (Figure 4) should also be noted. On smaller scales, explosive processes in the nuclei of galaxies balance the gravitational increase of density, as discussed in Sections 6.xi and 7.ii. Global stability is a matter of cause and effect: there is no cause for global motion, and, consequently, there is none.

ix. Equation of State for Cosmic Mass Systems

Consider a function $a(r) = G(r)\rho(r)/\alpha(r)$ at distance r from the center of a galaxy or a system of galaxies. By equation (24), it acquires the form $a(r) = A\rho(r)/\alpha(r)^2$. For $\alpha(r)$ a relation $\alpha(r) = \alpha_c + dr^{-p}$ has been found empirically, with $p = 0.8$ for average systems on different scales (Jaakkola 1978a) and $p = 1.2$ within the Milky Way (Jaakkola 1978b). Here we adopt the value $p = 1$. For $\rho(r)$, we can make a reasonable assumption that the density distribution follows the Emden isothermal gas sphere, $\rho(r) \propto Kr^{-2}$. Zwicky (1957) has applied this model successfully to clusters of stars and galaxies. We then obtain the remarkable result that $a(r)$, the local gravitational acceleration, is a constant, a_l. Evidently, $a_l \approx a_c$. This is found by writing a quasi-Newtonian force law

$$F(r) = ma(r) = mM(r)G(r)\frac{e^{-\alpha(r)r}}{r^2} \qquad (26)$$

where $e^{-\alpha(r)} \approx 1$ on galactic scales. Since $G(r) = A\alpha^{-1}(r)$ and $\alpha(r) = dr^{-1}$, as above, we obtain a $1/r$ force law for galaxies. Function $a(r)$ of equation (26) is, again, a constant if $M(r) \propto r$. With the local values $r = 10$ kpc, $G(r) = G_o$ and $M(r) \approx 10^{11} M_\odot$, $a(r) = 1.4 \cdot 10^{-8}$ cm s$^{-2} \approx a_c$. We obtain an equation, valid in quasi-stationary systems

$$a(r) = \frac{G(r)\rho(r)}{\alpha(r)} = \text{constant } a_l \approx a_c \qquad (27)$$

Therefore, a_c is a universal constant which controls both the gross M/R-structure of galaxies and systems of galaxies and their intrinsic structure. Evidently, it also controls the smooth evolution of density distribution. Equation (27) is a plausible result, since a gradient in $a(r)$ would mean an imbalance between inward gravitation pressure and outward centrifugal force and radiation pressure. Equation (27), the equation of state for the EC-theory, binds together the two long-range forces and the large-scale distribution of matter.

x. The Problems of Dark Matter and Mean Density

During the past few decades, the problem of the existence and amount of non-luminous matter, to which Zwicky was the first to pay attention in the 1930's (Zwicky 1957), has become one of the dominant themes in astrophysics and cosmology. In the author's opinion, for systems of galaxies, the problem is solvable, first, as an artifact caused by non-Doppler intrinsic redshifts in the member galaxies and in the intergalactic field; and second, due to the non-Newtonian dynamics outlined in the preceding section. The more difficult problem, that posed by the flat rotation curves of spiral galaxies was actually solved above by deducing for the galaxies a $1/r$ force law replacing Newton's law. According to Sanders (1990), a $1/r$ distance law is needed to explain constant $V(r)$ without additional dark matter.

The actual value of the cosmological mean density ρ_c is still unclear. In EC, ρ_c has a different role than in ST; it is present, together with other factors, in the formulae for α_c, a_c, G_c and in $b-\gamma-g$ equilibrium processes. After the influences of non-Doppler redshifts and non-Newtonian dynamics are clarified, an effort should be made to estimate the amount of possible dark matter in the Universe and the true value of ρ_c. Matter in the form of the background gravitation field must be included in the new value of ρ_c.

xi. Connection with Violent Motions in Nuclei of Galaxies

Parallel to the solution of the missing mass problem, EGC also offers a solution to the "missing energy problem" posed by ejection of large amounts of matter from the nuclei of galaxies, in spite of gravitation due to central condensations of 10^6–10^{10} solar masses. According to conventional dynamics, in some cases 10^{60} ergs are required for such events.

Energy required is proportional to $G(r)$, where $G(r) = Ar/d$. At $r = 1$ kpc, $G(r) \approx G_0/10$, at $r = 1$ pc, $G(r) = 10^{-4} G_0$. With these factors, outward mass motions are less energetic on these scales. Hence the missing energy problem can be approached in a totally new light. This solution also has a role in the origin of galaxies (7.iii).

xii. Discussion of Gravitational Cosmology

How do the various cosmological theories fare when confronted with the three outstanding features of the observable Universe which are intimately connected with the effect of gravitation: homogeneity and isotropy, the existence of local structure (galaxies and systems of galaxies), and the static state of the Metagalaxy?

The standard big-bang cosmology cannot explain any of them. Homogeneity and isotropy cannot be derived from fundamental theory (Peebles 1980), and causality poses serious problems: regions now 3 degrees apart were causally disconnected at the time of recombination. The origin of galaxies in the expanding frame is extremely difficult to understand. The seeds of galaxies and their systems should have existed from the very beginning of the Universe. The recently reported small-scale fluctuations of the 3° K background with scales of ~10 degrees do not relate to such seeds. Moreover, if they existed, their development into galaxies cannot be deduced from known physical principles (Peebles 1980); rather, they should have led to black holes. Third, while the big-bang theory cannot be required to interpret the staticity of the Metagalaxy (as it takes expansion for granted), one might insist on an explanation of the assumed expansion directly from fundamental theory (i.e., gravitation). However, fundamental theory points to a contraction of the Universe, and expansion is obtained only by an *ad hoc* assumption of an physically implausible initial "big-bang".

In contrast to the situation in ST, the EC theory succeeds in explaining homogeneity, isotropy and large-scale stability in a physically plausible manner, and it explains effectively the formation and the major features of local structure at the level of galaxies, groups, clusters and superclusters, sets the scale of the transition from hierarchical to homogenous structure in the correct place—and all this with parallel solutions of the major problems of the radiation cosmology. New light is thrown on the astrophysical mass and energy problems. The empirical facts can be seen to converge with different parts of theory. The status in the triangle of big problems in gravitation cosmology is shown schematically in Figure 5.

Problem	Standard Cosmology	Equilibrium Cosmology
STABILITY HOMOGENEITY ISOTROPY LOCAL STRUCTURE	CONTRACTION ↑ EXPANSION ↑ ↓ INHOMOGENEOUS ✗ HOMOGENEOUS ISOTROPIC ↓ GALAXIES CLUSTERS ✗ ANISOTROPIC ↓ BLACK HOLES	STABLE ↕ ↕ HOMOGENEOUS ISOTROPIC ↕ GALAXIES CLUSTERS

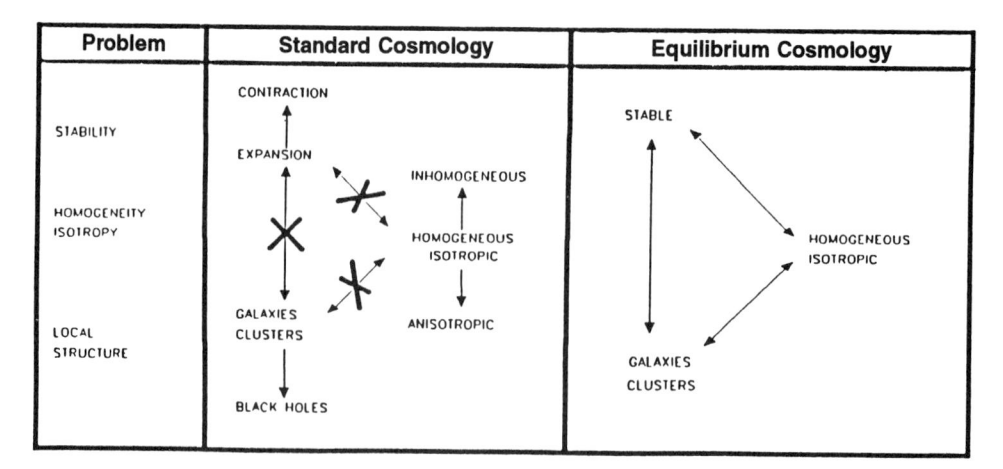

Figure 5. Comparison of two theories. Crossed arrows in the left indicate that all three major features are mutually inconsistent; the short arrows show predictions of the fundamental theory of standard cosmology. In the equilibrium theory, all three major features are consistent with one another.

7. Cosmology of Equilibrium Evolutionary Processes

The fact that the Universe as a whole does not evolve, while all individual systems do evolve, confronts cosmology with a fundamental problem which may be called the Third Cosmological Paradox: everything changes—the whole does not change.

It is appropriate to point out that the eternal, steady-state existence of the material world, as empirically discussed in Section 2 and translated into scientific terms in CP (3.i.), has been anticipated by philosophy for at least two and half thousand years (see Jaakkola 1989). I have no hesitation in saying—acknowledging the historical roots of any true advance in modern science as well as parallel efforts made in other branches of knowledge—that I find this conformity very gratifying.

Explaining equilibrium in the Universe will be the core of the new cosmology. The task is to examine, theoretically and empirically, the processes which provide for the properties of matter unchanged on the cosmological scale, while other processes change these properties in individual systems. In what follows, equilibrium evolutionary processes (EEPs) of some parameters are briefly discussed in a preliminary and qualitative manner. No solutions to the numerous theoretical problems confronted are given, nor are statistical verifications of the hypotheses attempted. The aim is merely to initiate discussion concerning a problem which will confront science with a new and extremely rich field of work.

i. Origin of Baryonic Matter: the Photon-Baryon Equilibrium

The hypothesis of a single basis for everything that dates back to the ancient Greek philosophers. The nature of the primary substance varied: water (Thales), fire (Heraclitus), air (Anaximenes), soil (Empedokles). Pythagoras and Plato, in their idealist philosophies, proposed numbers and ideas, respectively. More general notions of the fundamental substance were also proposed: the *apeiron*, an infinite, undetermined nature (Anaximander), homoiomers (Anaxagoras), and atoms (Leucippus and Democritus). In none of these hypotheses is there any notion of a birth of the Universe, which was thought to be eternal and globally unchanging.

The big-bang theory—together with certain religious doctrines—is clearly at variance with this great line of thought. It supposes an origin of everything that exists, including the baryonic matter, in the big bang, and even out of nothing, as a gigantic quantum fluctuation.

In equilibrium cosmology there is no primary origin of any kind of substance; the substances only change form into one another. As for the baryonic matter, there is an equilibrium between it and other forms of matter. If baryonic matter is converted into electromagnetic radiation via fairly well-known processes in stars and the nuclei of galaxies and quasars, for equilibrium to be valid there must be a reciprocal process of transformation of radiation energy (γ) into baryonic matter (b). In equation (18) which reduces the problem of the $\gamma - b$ number ratio to a problem of redshift and emissivity (M/L) of matter, only the CP and the interpretation of CBR are implicit; it says nothing about the $\gamma \rightarrow b$ transformation.

The redshift effect is, without doubt, one step of this process. In the EGC scheme, it transfers energy from radiation (γ) to the graviton bath (g). Part of this redshift energy re-emits as CBR. Part of the RE must in some way return to baryonic matter. Otherwise, Olbers' paradox would remain unresolved; the facts behind the paradox can be taken as a proof of the reality of the $\gamma \to b$ process. In spite of finiteness of I_c in equation (15), redshift alone is not enough for a solution, since it is an absorption effect (of energy) and as such open to the same criticism as Olbers' original solution: if nothing else happens, the redshifting medium would shine as bright as the stars. It must be said that the solutions to Olbers' paradox and the $b-\gamma$ equilibrium are parallel.

How does energy return back to baryonic matter? What is the role of gravitation, i.e. the flow of g's to mass systems? Do baryons vanish in radiation, are new baryons born in the reciprocal process? In that case, the law of conservation of baryon number would be valid only on the large scale, not locally (if the new matter is not equally of negative and positive charge). If fresh baryons are born, does this happen in galaxies, or in the intergalactic space where the redshift (z_c) occurs? Probably in both; the inflow motions (see subsection ii below) suggest that the latter does occur. The condensing of redshift energy into gravitons must be a lengthy process: high-energy gravitons ("supergravitons") and gathering of g's into existing b's, then to divide into two b's (like cell division) are possible pathways. Are the known facts enough to rule out these proposals?

The production rate of baryonic matter in the EC-model is

$$C_m = \frac{n_c L}{c^2} \qquad (28)$$

We obtain the value

$$C_m \approx 10^{21} \text{ g s}^{-1} \text{ Mpc}^{-3} \approx 10^5 \, M_\odot \text{y}^{-1} \text{ Mpc}^{-3}$$
$$\approx 1 \, M_{gal}\left(10^{16} \, y\right)^{-1} \text{Mpc}^{-3} \approx 3 \times 10^{-53} \text{ g s}^{-1} \text{ cm}^{-3}$$

or one proton in a lab of $5 \times 5 \times 4$ m in 3×10^{14} years. This is only $\approx 5 \times 10^{-4}$ times the "creation" rate in the expanding steady-state model.

It is likely not possible to create in the laboratory the conditions prevailing in space where the particles are born. However, in the future it may become possible to follow this process experimentally with an extremely intense source of radiation and the conditions for a strong redshift field. This is a matter for experimental and theoretical physicists to ponder.

On the other hand, the vanishingly small effect C_m may be within the range of observations; it is worth pointing out that C_m coincides with the energy density of cosmic rays with $E > 10^{18}$ eV, usually assumed to make up the extragalactic background that cannot be trapped by the magnetic field of the Galaxy.

An important cosmological time-scale, "a cosmological cycle" is

$$\tau_c = \frac{n}{C_m} \qquad (29)$$

With $n = 0.03$ gal Mpc^{-3}, $\tau_c = 3 \times 10^{14}$ years. This is the time-scale on which one galactic mass in a proper volume is transformed into radiation, and, correspondingly, a similar quantity of particles is born. This interval of time covers thousands of generations of stars, and probably numerous morphological cycles of individual galaxies.

When we contemplate various processes occurring in cosmological space, processes which we think are necessarily there in order to maintain the general equilibrium, but which we have never seen, we are standing on the shore of a newly discovered, uncharted ocean. I see cosmology as a deeply human, courageous enterprise. We must prepare ourselves to set out on the great adventure that awaits us.

ii. Density Distribution

The problem of a large-scale stability was discussed in the preceding section. In galaxies, gravitation tends to increase central density. Dense nuclei found in almost all galaxies prove this natural trend. Examples of large scale inward motions are given by such intensively studied events as the merging of galaxies, cooling flows and high-velocity clouds connected with the Milky Way. Other fashionable, though possibly not real, items are the Virgocentric infall and the one towards the hypothetical Great Attractor.

We have plenty of data on outward motions. Optical and radio jets from the nuclei of galaxies are most evident. Quasars and the birth of new galaxies are associated with expulsion of matter from the nuclei of existing galaxies. Outflows of gas, and even stars, are reported in the central regions of some well-known galaxies (Rubin *et al.* 1973), however, this data is distorted by non-velocity redshift gradients. Actual examples of outward motions are jets in M87 and 3C273, the Ambartsumian knot, M82, NGC 1275, and the 3-kpc arm and the counter arm in the Milky Way.

One prediction of the EC-model, which is very precise and difficult to check, is that the inward and outward flows of energy should cancel, when summed over a representative sample of galaxies.

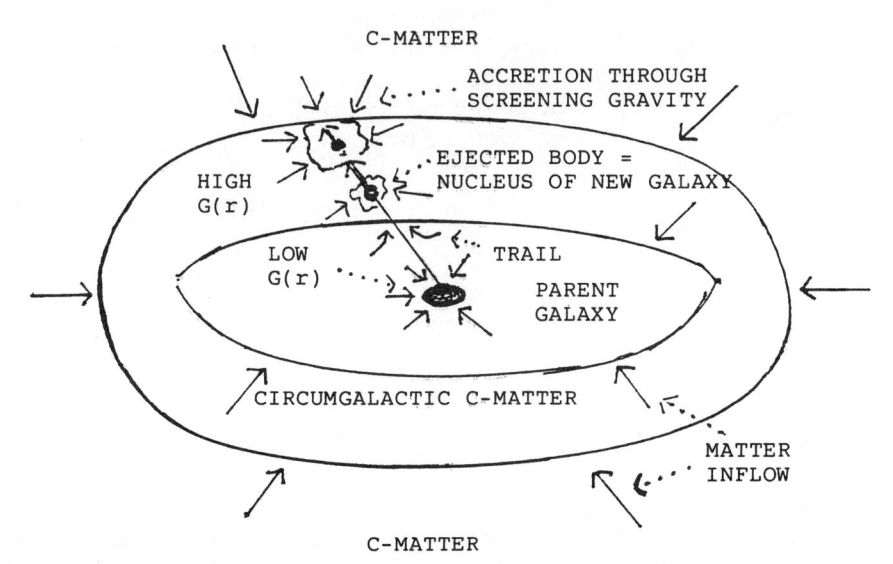

Figure 6. A schematic representation of the origin of galaxies in a process involving ejection of a compact body and accretion of galactic and newly born C-matter by screening gravity with varying $G(r)$.

iii. Origin and Fate of Galaxies

Figure 6 is a schematic representation of how galaxies may be born. Dense nuclei, observed in almost all types of galaxies, may have been the germs around which the screening gravity builds the galaxies. Originally, the germs have been ejected from the nuclei of existing galaxies.

Ejections of embryos of new galaxies are frequently observed as knotty radial jets. In the EGC scheme ejection is aided by the low $G(r)$-value in the center of the parent galaxy. The screening effect by the center may be weak in the beginning, with the field of the expelled fragment possibly being constructed only gradually; this also helps the protogalaxy to escape. When the embryo reaches the edge, both $G(r)$ rising up to G_c and screening by the galaxy come into play, and this promotes the growth of the new galaxy. This process also brakes the outward velocity, explaining the origin of double galaxies, M51-type and other type companion galaxy systems, and groups of galaxies. In the large clusters, the supergiant cD-galaxies and other giant galaxies play the role of parent galaxies. In addition, galaxy formation by fragmentation, apparent in some cases, such as Vorontsov-Velyaminov's "nest of galaxies," is allowed by the low $G(r)$ in the central regions of giant galaxies.

During the escape phase, the accreted material is drawn from the parent galaxy. This explains old population stars in irregular and young galaxies—one reason for the prevailing mistaken view that all galaxies are of similar age. The bridges to presumably newborn companions are tidal tracks formed mainly from material from the parent galaxy. Further out, material accreting into the young galaxy is supplied by the newborn matter discussed in 7.i. This must be collected into and around the parent galaxies, which further explains the group structure in the distribution of galaxies.

While the above scheme is favored by the bulk of extragalactic data, the more conventional mode of birth by condensation of low-density clouds becomes more comprehensible with the PIG gravity. The difficulty with classical gravitation is that the cloud should condense into one large object, not many smaller ones, as required for stars or galaxies in clusters. Screening gravity is able to collect material around small causal seeds. "Mock gravity" working by screening of radiation pressure has been suggested for the formation of planets (Whipple 1946) and galaxies (Hogan and White 1986). In PIG, screening gravity provides a more powerful piling mechanism, whether the center is a chance fluctuation or a splinter from an old-generation object.

In the equilibrium Universe, the rate of death of galaxies must be the same as the birth rate. Galaxies die, *i.e.* lose their identity as a single system, usually by merging into a larger galaxy, or by collective merging of galaxies of similar status to form a new giant galaxy: cD galaxies in the clusters are born in that way. Tidal disruption of a companion is also possible; the nucleus continues its life with a changed identity, possibly as a globular cluster. In a sense, the origin of galaxies by fragmentation—probably not a very common event—means the death, *i.e.* loss of identity, of the fragmented galaxy. The more common production of galaxies via the expulsion of embryos does not cause the parent galaxy to wither away; recreated by new material flowing in, its life may be very long and end only in the course of the dynamical evolution of the system of which it is a member.

iv. Morphological Distribution and Stellar Content

In the prevailing conception, the morphological types of galaxies result from initial conditions in the clouds from which galaxies are thought to have formed at a certain epoch after the big bang. Clearly, this is a contrived solution. In EC, the formation and waning of each type is continuous process, and the frequencies of the types measure the evolutionary periods corresponding to those particular types on each level of mass and status-in-system.

Contrary to the usual view, Hubble's morphological sequence is most probably an evolutionary one. The usual argument, namely, that within the Hubble period there is not enough time for a morphological evolution does not apply in the eternal equilibrium universe. It can be pointed out that those parameters which are expected to change in the course of astrophysical evolution and differential rotation (stellar types, element abundances, gas fraction, disk/bulge ratio, tightness, length, massiveness and clumpiness of spiral arms, rotation curves) change in the expected manner along the Hubble sequence from the "late" to the "early" types. Parameters

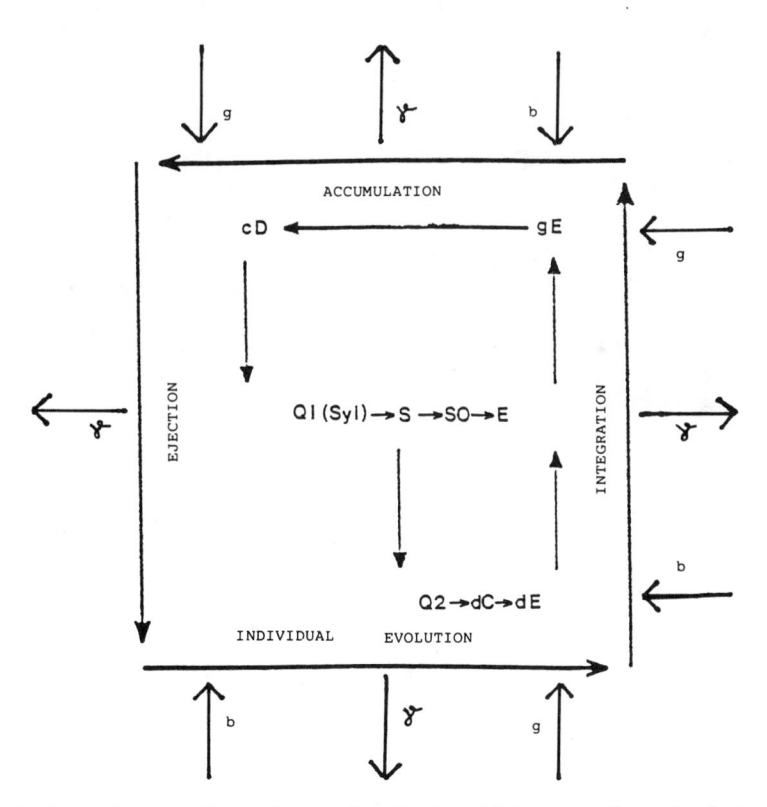

Figure 7. A schema for evolution and mutual and external interconnections of galaxies. Quasars Q1 are thought to be identical to Seyfert galaxies; Q2 are local quasars with intrinsic z. C-compact, d-dwarf, gE-giant elliptical, cD-supergiant in the center of cluster. Inward and outward arrows: b and g-baryon and graviton inflow, γ-electromagnetic emission.

which are expected not to change, such as the masses and globular cluster systems, do not change.

Evolution along the Hubble sequence means that differential rotation destroys the spiral structure. What recreates the spirals in that case? N-body computer simulations indicate that the tides by the encounters or by companion galaxies excite spiral structure, at least in the outer regions. Spiral arms extending to the galactic centre are probably due to ejections of coherent bodies from the core, and differential rotation of the tidal trails. In some cases, as in that of the 3-kpc arm and its counterarm in the Milky Way, outflows of diffuse matter may come into play. This mode of formation of spiral structure is closely connected to the galaxy cosmogony discussed in the preceding subsection; obviously the angle between the ejection and the plane of the galaxy determines the outcome. Moreover, tidal excitation of the spiral structure by the satellite galaxies eventually becomes connected with nuclear activity, in view of the origin of the satellites outlined above.

v. Stellar Population and Gas Content

The average stellar content of galaxies ages along the Hubble sequence. Periodically, new stars are formed from interstellar matter accreting onto embryos of stars (either remnants or fragments of older generation stars, or of planet sized objects) along the tidal trails of coherent bodies ejected from the galactic nuclei, or along the jets of diffuse matter. Gas content diminishes accordingly, though it is replenished by nova and supernova explosions and stellar winds, as well as outflows from the galactic nucleus.

An important factor in the quasi-equilibrium of a galaxy, greatly expanding its active age, is an inflow of inter- and circum-galactic neutral hydrogen onto the plane of the galaxy. In the Milky Way, this is observed as high velocity clouds, which in most cases exhibit negative radial velocities and are hence falling into the Galaxy with a rate which is enough to produce the observed amount of gas in the Hubble time. In this sense, the formation of a galaxy is not a single event, but a continuous process.

vi. Galactic Rotation

True to form, existing theories of rotation seek the cause for this phenomenon in conditions "at the beginning". In the vortex theory (Descartes, 1644), rotation is seen as a relic of motions in the primordial clouds from which the galaxies are said to have been born in some early phase of the big bang (von Weizsäcker 1951, Gamow 1954, Ozernoy and Chernin 1969, Harrison 1971). The tidal theory suggests that there were density fluctuations which caused these motions by their gravitation (Hoyle 1949, Peebles 1969, Harrison 1971). A combination of the two primordial causes has been suggested (Harrison 1971). Again, these explanations smack of the contrived; the problem is only pushed to such a distance that it is neither necessary to deal with it, nor possible to check it empirically. Of course, these theories become immediately invalid in the present approach which does not presuppose any global initial conditions.

Therefore, cosmology and physics still face the problem of explaining the phenomenon of rotation—of the planets, stars, galaxies. Only the moons, which always turn the same face towards their planets, do not present this problem: the tidal forces induced by the mother planets cause their observed rotation.

To discover the physics behind rotation, we should first try to find clues in the empirical data. Among galaxies, the spiral galaxies rotate very fast, with velocities of hundreds of km s^{-1}, and with typical rotation curves for the different morphological subclasses. The difficulty of finding enough mass to prevent the spiral galaxy from flying apart by rotation proves that rotation in these galaxies is not a relic phenomenon, but that the spirals are rotationally active systems. On the other hand, elliptical galaxies, which are dynamically more evolved systems, rotate with velocities of a few tens of km s^{-1} only, below the accuracy of the observations. The interesting point in the galactic data is that the rotationally active spiral galaxies are also active in other respects, *i.e.*, as regards structure, star formation, magnetic field and, most important, activity of their nuclei. The slowly rotating E galaxies do not show such activity, except some supergiant galaxies which have a special position in the centre of clusters, and, without doubt, also in the evolutionary history. Therefore, in the realm of galaxies, rotational and non-stationary activity appear to be connected. The same feature could be shown to hold true also for the stars and planets. The universal connection between the non-stationary and rotational activity adds one fundamental aspect to the non-stationary phenomena discovered and much emphasized originally by Ambartsumian (1971).

A possible way to increase angular momentum, for example, in the case of galaxies, is the following. When a galactic nucleus throws out part of its mass, this gives rise to a wave of outward motions in the outer regions. It is a well-known fact in stellar dynamics that stars which have peculiar motion in the direction of rotation most easily reach escape velocity. Thus, these stars move outward in the equatorial plane, increasing its angular momentum. The slower stars, due to encounters, drift out of the plane to form the slower-rotating halo. The total effect of this scheme is that, triggered by nuclear activity, kinetic energy in the form of random motion is transformed into rotational motion. A similar process may also play a role in the rotation of stars and planets.

Galaxies are not isolated systems, but are connected with the surrounding cosmological environment. Yet this does not explain the equilibrium feature $a_c \approx a_l$. Newly created matter collected around a galaxy is wound up to rotation by the drag of the rotating galaxy, inflows into the galaxy and increases its angular momentum, hence compensating the loss due to tidal friction by the halo and by the neighboring galaxies (some net change naturally occurs). The resulting rotation curve is such that it is in quasi-equilibrium with the external and galactic gravitational fields, as explained in the preceding section. Hence, again, the local effect, connected to the activity of galactic nuclei, and the cosmological factor are both present when equilibrium of a physical parameter, now angular momentum, is set up.

vii. Element Abundances

Cosmological conservation of element abundances must now be outlined. In EC, the observed abundance distribution should be a function of the stability of nuclei and of astrophysical conditions for further fusion, decay or violent dissolution. For hydrogen, baryon-photon-graviton equilibrium also enters into the picture.

In 1917 Harkins drew attention to the connection between the abundance of the elements and the stability of atomic nuclei—a matter naturally of extreme interest in the present context. This was followed by what was called "equilibrium theory", originated by Tolman (1932). He showed that equilibrium between H and He cannot be obtained in normal densities and temperatures within a reasonable time. This is indeed curious: in an equilibrium theory a time-scale should not be a problem, and in a model with a limiting time-scale, equilibrium should not be of interest. I have heard one astrophysicist say that given a tenfold time, the observed amount of helium could well have been produced in the stars. In the EC-model, there is certainly time enough for any astrophysical process to reach a state of equilibrium.

When hydrogen in stars burns into helium and heavier elements, what turns the nuclei back into protons? Today, we can only guess. Once again, nuclei of galaxies may be the focus of the problem, possibly working as "smelting furnaces" of matter. There is indeed some evidence that helium is deficient in the nucleus of the Galaxy (Mezger et al. 1970), as well as in quasars and Seyfert galaxies (Peimbert and Spinrad 1970a,b; Osterbrock and Parker 1966; Bahcall and Kozlovsky 1969). According to Shklovskii (1978) the metal nuclei dissociate in the extreme temperatures of supernovae into helium, and further into protons and neutrons. A similar process may be connected with the activity of the nuclei of galaxies. And, as a final point, in the equilibrium Universe with an infinite time-scale there is time enough for any nuclei to decay into protons, even gradually.

The light element abundances are cited repeatedly as a proof of the hot big bang model (Peebles et al. 1991). Actually the helium abundance of about 25 % is only known for the Sun, young stars and interstellar medium, i.e., in a very restricted range of astrophysical circumstances. Possible deficiency in galactic nuclei and quasars was mentioned above. In young star clusters He-abundance is known to increase with stellar age. Therefore an argument of a primordial, spatially constant He-content, so essential for ST, is observationally less well grounded. Hoyle (1980) points out that due to the two-to-one mass ratio of D to H, deuterium is measured where its concentration is unusually high, and that primordial ^{7}Li is easily shaken loose by non-thermal collisions. Therefore, the light element argument, for all its ingenuity, is empirically weak.

From the point of view of EC, the observational features mentioned above fit the theory. So does the fact that the metal content increases both towards the nuclear regions of galaxies, and with increasing luminosity (age) of the galaxy. The gross feature, that the abundant elements (H, He, Fe) are the ones which can be produced both by fusion (as in 7.i.) and by decay or dissolution, is quite telling in the present context. Yet the details of the process are not yet clear.

viii. Discussion of the EEPs

The preceding examples will have served, I hope, as an introduction to the new, extremely difficult, but also extremely rewarding field of science, i.e., the investigation of how the equilibrium Universe works. Indeed, this is hinted at by the etymology of the word "cosmology": studying the order of the things.

Reference to the second law of thermodynamics can be anticipated in the context of the third paradox. It should be pointed out that the cosmological system considered here is not an isolated system, but one connected to the entire Universe in a certain manner according to equations (22), (23) and (27). Further, this system, unlike a gas, is not composed of identical particles but of several qualitatively different constituents, *i.e.*, stars, objects in the nuclei of galaxies, interstellar and intergalactic gas and dust, as well as the gravitational, magnetic, and radiation fields. All this is arranged in a hierarchic manner ascending from small to large. Taken as a whole, the cosmological system is too complicated for the thermodynamical theory to apply.

It is appropriate to mention a methodological point. Constancy of the various parameters of matter over the cosmological scale and time-period, as stated by the (strong) CP, provides us with a number of new "conservation principles", which may turn out to constitute, in cosmology and astrophysics, as powerful a tool as the well-known conservation principles in physics. All local evolutionary processes can be viewed against these principles, and this may guide galactic, stellar and even planetary astrophysics into unforeseen developments. The number of the cosmological conservation principles is as large as the number of the parameters of matter, *i.e.* practically limitless.

8. Testing the Equilibrium Cosmology

The Universe may eventually prove to be non-expanding and to be in equilibrium in the other respects, but at the same time work through mechanisms other than those suggested here. Therefore, the tests of expansion and (non-)existence of cosmic evolution are now of the highest priority, possibly maintaining the fundamentals of EC, even if the above theory proves to be partially incorrect. To prove or disprove the present theory, all of its theoretical assessments, including the hypotheses of CP, EGC and PIG, and the conclusions drawn in Sections 5, 6 and 7, must be tested with care.

There is little disagreement about the spatial aspect of the CP (2.v.). Tests of its temporal aspect are those discussed in Section 2.i to iv, and in the papers referred to there. These include tests of the nature of redshifts, the global tests, and tests of cosmic evolution. For the latter, EC predicts, for any physical parameter p, determined as a statistical average in a representative sample

$$p(z) = \text{constant} \tag{30}$$

The number of such tests is very large, amounting to dozens. Checks of the EGC hypothesis should include critical analysis of the arguments in its favour, as listed in Table 1 and discussed (with references) in Jaakkola (1991), as well as developing the theory and its tests further.

Due to the nature of the PIG theory, there are, at least in principle, very many possibilities for testing it; therefore, it clearly satisfies the main criterion of scientific theories. In practice, of course, working out the experiments will be difficult, as always in non-trivial physical problems. The following list of parameters to be tested as possible factors affecting gravitation follows from the different aspects of the theory: distance (the $1/r^2$ law), mass, density, material, temperature, time, velocity, rotation, acceleration, shape, orientation with respect to the Earth, orientation with respect to the Sun and the Moon, and with respect to the distant cosmic objects (Milky Way and the CMB dipole), electric field, magnetic field, occultations, intervening matter, *etc.* Existing data on the tides represent a wealth of material for analysis in relation to the PIG-hypothesis.

Experiments in gravitation have a long history. Most parameters have been examined, and only a few minor anomalies remain unexplained in the Newtonian-Einsteinian framework (Cook 1987; Will 1987; however, see Jaakkola 1991). The aspects of the PIG theory may require special experimental setups. Many of the above factors have not yet been studied, and surprises may wait us in further experimental gravity research.

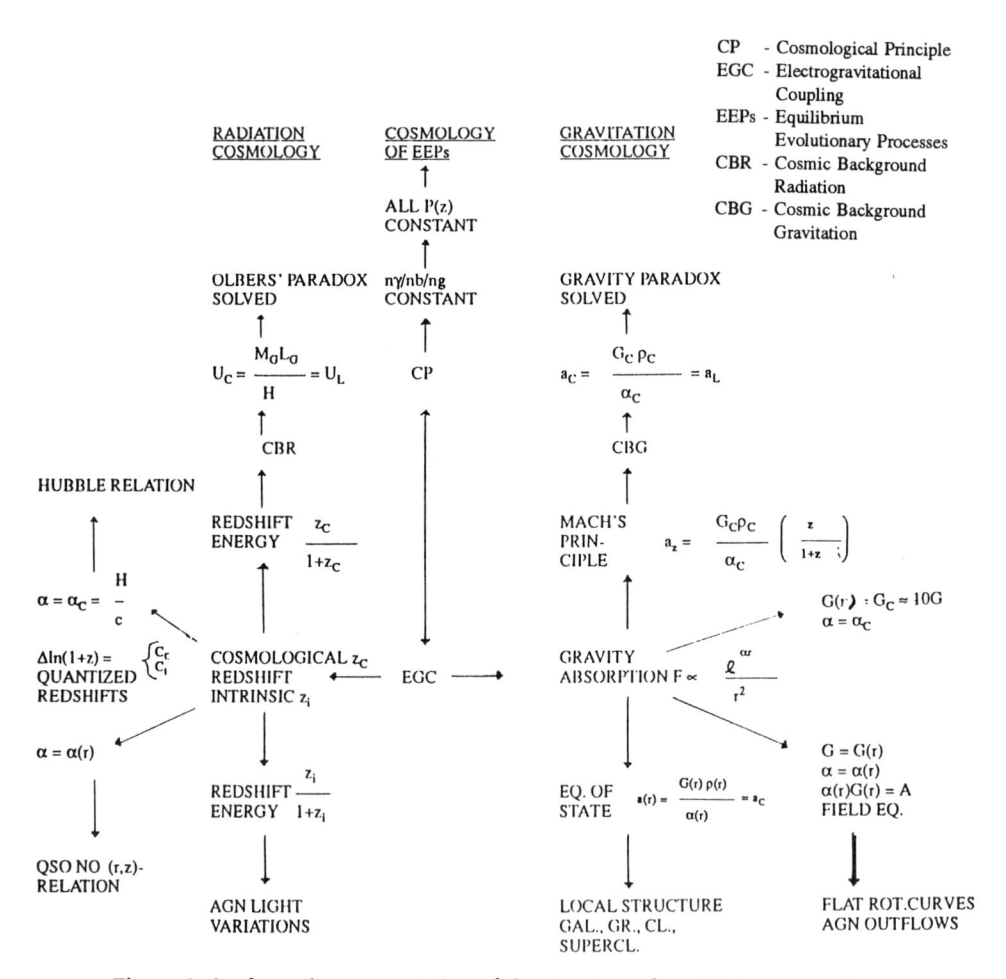

CP — Cosmological Principle
EGC — Electrogravitational Coupling
EEPs — Equilibrium Evolutionary Processes
CBR — Cosmic Background Radiation
CBG — Cosmic Background Gravitation

Figure 8. A schematic representation of the structure of equilibrium cosmology.

In radiation cosmology in particular, features F4, *i.e.* $\alpha \propto \sqrt{\rho}$, and F9, short time-scale variability, demand further verification. In Table 1, the gap between 10^{12} and 10^{18} cm should be bridged. With regard to the CBR, a strict intercomparison of U_c, T_c, α_c, n_cL and ρ_c, also involving requirements from the $b-\gamma-g$ equilibrium, has yet to be made.

In gravitation cosmology, equations (24) and (27) are the chief targets of empirical tests. The equality $a_t \approx a_c$ should be checked for a large sample of galaxies and systems of galaxies, also comparing the degree of variance from the other measures of dynamical evolution.

In the science of the equilibrium processes, thorough analysis of the problems may suffice to bring about advances in this new field of research. Consideration of possible laboratory experiments relating in particular to the $b - \gamma - g$ equilibrium is a necessity. Perhaps such conditions can never be attained; however, the possibility that something interesting can be found in measured effects should also be considered.

A cosmological test of the EEPs and the state of the Universe concerns the age dispersion δ_A of galaxies: a small δ_A at a given z supports ST, and a large δ_A, EC. Young galaxies at small z and old galaxies at large z would imply the latter case (as is actually found).

Another general idea for cosmological tests of the EEPs may be explored. For an evolutionary process, such as morphological evolution of galaxies, EC predicts

$$f_i = T_i \frac{\sum_j f_j}{\sum_j T_j} \tag{31}$$

where f_i is the observed frequency of a sub-class (e.g. Hubble sub-type), T_i its theoretical evolutionary period, $\sum_j f_j$ the total number in the whole j class (Hubble sequence within a restricted range of mass), and $\sum_j T_j$ the period of the whole cycle. Differences in the environments, as for galaxies in the clusters and in the field, at the same time complicate and refine the test. The observed f_S/f_E-difference between the field and clusters may turn out to be one further argument in favour of the equilibrium Universe, adding to the number of those given in Section 2.

9. Discussion

In order to advance discussion in the science of cosmology, it seems useful to present openly the author's view of the present status of the most important rivals of the EC theory. Following this, I shall outline my thoughts on the present state of EC.

Two well-known papers published in *Nature* (Arp et al. 1990; Peebles et al. 1991) give a good basis for estimation of the status of the two most prominent theories. As for the standard theory (ST), without going into details, it can be commented that the strongest empirical arguments cited in support of ST (expansion, cosmic evolution), are clearly not true (Section 2), and some arguments, even if empirically true (light element abundances, number of neutrino families), are theoretically so frail and vulnerable that they cannot be considered compelling. As pointed out in 7.vi., abundances are actually known only in a very narrow range of astrophysical circumstances. Observations of very distant galaxies with apparently locally normal abundances constitute a blow to ST, as do local galaxies seen in the process of formation. Internal inconsistency of ST in the cosmological test results is as bad as it could be (2.ii) To keep from collapsing, the ST needs far more epicycles than the Ptolemaic system before its collapse. But the ST may fight for survival by trying to find, *e.g.* new (false) cosmic evolutionary effects, analysis of which from the present point of view is, unfortunately, still far from complete.

For all its sophistication and extreme theoretical development, ST is actually a rather clumsy construction. Almost everything is solved by resorting to initial conditions and choosing them in an *ad hoc* manner. There is one exception: the connection between the CBR temperature and the helium abundance forms the finest chapter of the theory. My reservations were discussed above. The worst fault of the theory is its doctrine of creation of the Universe, with matter, space and time emerging from a singularity (out of nothing) at time $t = 0$. This violates the conservation principles which are so fundamental in physics, and essentially reduces the theory to sheer metaphysics.

Moreover, the ST rests on a false fundamental theory, *i.e.* general relativity. Without going into a full theoretical and empirical criticism (see Jaakkola 1991), it is sufficient to note that an interaction redshift effect invalidates both special and general relativity. The propagation of light is found to be connected with properties (*e.g.* density) of matter, and this leaves no further grounds for connecting to it properties of space and time, as is done in relativity theory.

As a model of a steady-state universe, the theory of Bondi, Gold and Hoyle (SST) captured an essential property of the Universe, expressed in the (perfect) CP. However, because it is based on belief in a universal expansion, it is open to the same crucial empirical criticism as the ST. In the SST, the origin of helium and the CBR are, in principle, similar to the ST case (Hoyle 1980); the difference is that it emerges in "little bangs" or "white holes" instead of one big bang, as in the ST; for thermalization at 2.7° K, Hoyle supposes a specific interaction with graphite particles.

The SST uses general relativity, which we have criticized above. The principal feature of the SST—violation of the physical conservation principles by creation of matter in continuous creation or in white holes, again out of nothing—is just as unacceptable as the single creation event of the big bang. This is not improved by putting into the equations a C-field describing mathematically a creation process. Nor is the problem of conservation of energy solved by saying that in the model of continuous creation energy is conserved within a volume determined by fixed rulers. Conservation principles have a certain physical meaning: they are not simply a matter of convention. The SST is thus not a satisfactory alternative to the ST, although it has the historical merit that it may have prevented the latter from growing to become an even more unshakable monolith than it is today.

The present article was intended as a summary discussion of one proposal for a theory of the equilibrium Universe. I hope it has provided at least an idea of the scope of the problems. The empirical grounds already seem very strong. As a scientific construction, the EC theory is flawless, which of course does not mean that it is true. Solutions for the redshift and the CBR, gross structure of galaxies, background paradoxes, *etc.*, achieved in the EC fairly successfully (compared to the other theories), may eventually prove to be different from those suggested in Sections 5 and 6. A core should be present in the approach to the evolution paradox in Section 7; however, research into EEPs is just beginning. Therefore, we can be content with the status of EC only when comparing it to the ST. But I am convinced that a concerted attack on the problem, with resources amounting to only a fraction of those mobilized behind the ST today, would bring about remarkable advances in cosmology.

Today many scientists are working on the physics of the non-expanding Universe and, as seen in this volume, in some areas they have advanced further than the content of the present paper. A most interesting historical antecedent of the themes discussed in this paper is to be found in the ideas of William Macmillan at a time when the redshift phenomenon had not yet been discovered, and even the identity of the galaxies was unclear. MacMillan writes (1918, 1920):

> ...cosmology, ... might be defined as the study of the transformations of energy throughout the cosmos, the study of the origins being of no more interest than the study of dissolutions.... It is natural to suppose that atoms ... are formed by the flow of energy through space.... Just as the atom and the molecule are permanent forms of physical existence, so also is the star a permanent form of physical existence, notwithstanding that the individual may pass from birth to its dissolution. There is no necessary limit to its age, and though the star itself may rise and fall, the universe as a whole is not essentially altered. The singular points may change their positions and their brilliancy, but it is not necessary to suppose that the universe as a whole has ever been or ever will be essentially different from what it is today....

To which I would only add: *It is infinite, eternal and unchangeable!*

This study was prepared as part of a research project supported by the Academy of Finland.

References

Allen, C.W., 1973, *Astrophysical Quantities*, 3rd ed., Univ. of London, The Athlone Press.
Ambartsumian, V.A., 1971, in: D.J.K. O'Connell (ed.), *Nuclei of Galaxies*, North-Holland Publ. Co., Amsterdam, p. 9.
Arp, H., 1987, *Quasars, Redshifts and Controversies*, Interstellar Media.
Arp, H., 1991, *Apeiron* 9-10:18.
Arp, H., 1992, *Mon. Not. R. Astr. Soc.* 258:800.
Arp, H., Bi, H.G., Chu, Y. and Zhu, X., 1990, *Astron. Astrophys.* 239:33.
Arp, H.C., Burbidge, G., Hoyle, F., Narlikar, J.V., and Wickramasinghe, 1990, *Nature* 346:807.
Bahcall, J.N. and Kozlovsky, B., 1969, *Astrophys.J.* 155:1077.
Bondi, H., 1961, *Cosmology*, Cambridge Univ. Press.
Broberg, H., 1982, *ESA Journal* 6:207.
Cheng, E.S., Saulson, P.R., Wilkinson, D.T. and Corey, B.E., 1979, *Astrophys.J.* 232:139.
Cook, A.H., 1987, in: S.W. Hawking and W. Israel (eds.), *Three Hundred Years of Gravitation*, Cambridge Univ. Press, p. 51.
Crane, P. and Hoffman, A.W., 1976, in: *Décalages vers le rouge et expansion de l'Univers*, IAU Coll. 37:531.
Descartes, R., 1644, *Principia Philosophiae*.
de Vaucouleurs, G., 1971, *Publ. Astron. Soc. Pacific* 83:113.
de Vaucouleurs, G., de Vaucouleurs, A., Corwin Jr., H.G. Buta, R.J., Paturel, G. and Fougue, P., 1991, *Third Reference Catalogue of Bright Galaxies*.
Faraday, M., 1855, *Expt. Res. Elec.* 3:1.
Gamow, G., 1954, *Proc. Nat. Acad. Sci.* 40:480.
Gehren, T., 1985, in: J.L. Nieto (ed.), *New Aspects in Galaxy Photometry*, Springer Verlag, p. 227.
Green, R.F. and Schmidt, M., 1978, *Astrophys.J. Lett.* 220:L1.
Guthrie, B.N.G. and Napier, W.M., 1990, *Mon. Not. R. Astr. Soc.* 243:431.
Harkins, W.D., 1917, *J. Am. Chem. Soc.* 39:856.
Harrison, E.R., 1971, *Mon. Not. Roy. Acad. Soc.* 154:167.
Hoffman, A.W. and Crane, P., 1977, *Astrophys.J.* 215:379.

Hoyle, F., 1949, in: *Problems of Cosmical Aerodynamics*, p. 195.
Hoyle, F., 1980, *Steady-State Theory Re-visited*, Cardiff Univ. College Press.
Hubble, E., 1934, *Astrophys.J.* 79:8.
Hubble, E. and Tolman, R., 1935, *Astrophys.J.* 82:302.
Jaakkola, T., 1971, *Nature* 234:534.
Jaakkola, T. 1973, *Astron.Astrophys.* 27:449.
Jaakkola, T., 1976, in: *Proc. 3rd European Astron. Meeting*, p. 488.
Jaakkola, T., 1978a, *Acta Cosmologica* 7:17.
Jaakkola, T., 1978b, *Scient. Inf. Astr. Council USSR Acad. Sci.* 45:190.
Jaakkola, T., 1982, *Astrophys. Space Sci.* 88:283.
Jaakkola, T. 1983, in: A. van der Merwe (ed.), *Old and New Questions in Physics, Cosmology, Philosophy and Theoretical Biology*, Plenum Publ. Co., p. 223.
Jaakkola, T., 1984, *Proc. Nordic Astron. Meeting*, Helsinki Univ. Obs. Rep. 4, p. 41.
Jaakkola, T., 1986, in: V. Hänni and I. Tuominen (eds.), *Proc. 6th Soviet-Finnish Astron. Meeting*, p. 190.
Jaakkola, T., 1987, *Apeiron* 1:5.
Jaakkola, T., 1989, *Apeiron* 4:9.
Jaakkola, T., 1991, *Apeiron* 9-10:76.
Jaakkola, T., 1993, in press.
Jaakkola, T., Donner, K.J. and Teerikorpi, P. 1975b, *Astrophys. Space Sci.* 37:301.
Jaakkola, T., Holsti, N., Laurikainen, E. and Teerikorpi, P., 1984, *Astrophys. Space Sci.* 107:85.
Jaakkola, T., Moles, M. and Vigier, J.P., 1978, *Astrophys. Space Sci.* 58:99.
Jaakkola, T., Moles, M. and Vigier, J.P., 1979, *Astron. Nachr.* 300:229.
Jaakkola, T., Teerikorpi, P. and Donner, K.J., 1975a, *Astron.Astrophys.* 58:99.
Karlsson, K.G., 1971, *Astron. Astrophys.* 13:333.
Laurikainen, E. and Jaakkola, T., 1985a, *Astrophys. Space Sci.* 109:111.
Laurikainen, E. and Jaakkola, T., 1985b, in: J.L. Nieto (ed.), *New Aspects in Galaxy Photometry*, Springer Verlag, p. 309.
Le Sage, G.L., 1784, *Berlin Mem.*, p. 404.
Mach, E., 1883, engl. transl. 1960, *The Science of Mechanics*, Open Court.
MacMillan, W.D., 1918, *Astrophys. J.* 48:13.
MacMillan, W.D., 1920, *Astrophys. J.* 51:309.
Megger, P.G., Wilson, T., Gardner, F.F. and Milne, D.K., 1970, *Astrophys.Lett.* 6:35.
Moles, M. and Jaakkola, T., 1976, *Astron. Astrophys.* 53:389.
Monti, R., 1991, *Proc. 2nd Conf. on Space and Time in Natural Science*, St. Petersburg (in press).
Napier, W.M., 1991, *Apeiron* 9-10:8.
Napier, W.M., 1993, this volume.
Nilsson, K., Valtonen, M.J., Kotilainen, J. and Jaakkola, T., 1993, *Astrophys. J.* (in press).
North, J.D., 1965, *The Measure of the Universe, A History of Modern Cosmology*, Clarendon Press.
Osterbrock, E. and Parker, R.A., 1966, *Astrophys. J.* 143:268.
Ozernoy, L.M. and Chernin, A.D., 1969, *Sov. Astron.* 12:901.
Pecker, J.C., 1976, in: *Décalages vers le rouge et expansion de l'Univers*, IAU Coll. 37:451.
Peebles, P.J.E., 1969, *Astrophys.J.* 155:193.
Peebles, P.J.E., 1971, *Physical Cosmology*, Princeton Univ. Press.
Peebles, P.J.E., 1980, *The Large-Scale Structure of the Universe*, Princeton Univ. Press.
Peebles, P.J.E., Schramm, D.N., Turner, E.L. and Kron, R.G., 1991, *Nature* 352:769.
Peimbert, M. and Spinrad, H., 1970, *Astrophys. J.* 159:809.
Rubin, V.C., Ford Jr., W.K. and Kumar, C.K., 1973, *Astrophys. J.* 181:61.
Rudnicki, K., 1989, *Apeiron* 4:1.
Sandage, A. and Perelmuter, J.M., 1991, *Astrophys. J.* 370:455.
Sanders, R.H., 1990, *Astron. Astrophys. Rev.* 2:1.
Shklovskii, J.S., 1978, *Stars, Their Birth, Life and Death*, W.H. Freeman and Co., p. 294.
Smirnov, Yu.N., 1964, *Astron. Zh.* 41:1084.
Smoot, G.F. et al. (27 co-authors), 1992, *Astrophys. J. Lett.* 396:L1.
Songaila, A., Covie, L.L. and Weaver, H., 1988, *Astrophys. J.* 329:580.
Staveley-Smith, L. and Davies, R.D., 1989, *Mon. Not. R. Astr. Soc.* 241:787.
Tifft, W.G., 1976, *Astrophys. J.* 206:38.

Tifft, W., 1988, in: F. Bertola, J.W. Sulentic and B.F. Madore (eds.), *New Ideas in Astronomy*, Cambridge Univ. Press, p. 173.
Tolman, R.C., 1922, *J. Am. Chem. Soc.* 44:1902.
von Weizsäcker, C.F., 1951, *Astrophys. J.* 114:165.
Will, C.M., 1987, in: S.W. Hawking and W. Israel (eds.), *Three Hundred Years of Gravitation*, Cambridge Univ. Press, p. 80.
Woodward, J.F., 1983, in A. van der Merwe (ed.), *Old and New Questions in Physics, Cosmology, Philosophy and Theoretical Biology*, Plenum Publ. Co., p. 885.
Zwicky, F., 1957, *Morphological Astronomy*, Springer-Verlag, pp. 163ff.

A Steady-State Cosmology

A. K. T. Assis*

Department of Cosmic Rays and Chronology
Institute of Physics
State University of Campinas
13081 Campinas, SP, Brazil

We analyze a steady-state cosmology based on a boundless universe which has always existed and which is homogeneous on the very large scale. As this is a stationary model without expansion, it does not require a continuous creation of matter, in contrast to the steady-state model of Bondi, Hoyle and Gold. We study the problems and properties of this model relating to inertia and gravitation (Mach's principle and the origin of inertia, the Seeliger-Neumann term), the cosmological redshift (alternatives to the Doppler interpretation of Hubble's law, the Finlay-Freundlich model), and the cosmic background radiation (predictions of a background temperature around 3° K previous to the experimental discovery by Penzias and Wilson in 1965). Some observational tests of this general model are outlined.
PACS: 98.80.-k Cosmology.
98.80.Dr Theoretical cosmology.
98.70.Vc Background radiations.
12.25.+e Models for gravitational interactions.
Key Words: Cosmology, Hubble's law of redshift, Mach's principle, cosmic background radiation, Seeliger-Neumann exponential gravity damping, cosmological principle

I. Introduction

In this paper we present a steady-state model of the universe that has grown out of two previous works (Assis 1992 a, b). Essentially, the model which we adopt complies with the perfect cosmological principle, which can be stated as follows: apart from local irregularities, the universe presents the same aspect from any place at any time (Bondi 1960, Ch. 2, p. 12). The history of this principle and its empirical foundations have been discussed by Rudnicki (1989) and Jaakkola (1989). The main properties of the universe at large assumed here are as follows: the universe ex-

---

* Also Collaborating Professor at the Department of Applied Mathematics, IMECC, UNICAMP, 13081 Campinas, SP, Brazil.

tends in all directions indefinitely; it has an infinite age; it is homogeneous on a large scale; the density of matter and energy are finite and constant, except for local irregularities; it is in a steady-state without expansion and without creation of matter; it complies with the principles of conservation of mass and energy.

In this work we will discuss this model and its relation with inertia and gravitation, the cosmological redshift, and the cosmic background radiation. Since we have analyzed Olbers' paradox (Olbers 1826) in detail in previous work (Assis 1992 a), we will not discuss it here.

II. Gravitation and Inertia

For low velocities and low energy densities, Newton's law of gravitation is known to be valid with high accuracy. However, when we try to apply it to an infinite and homogeneous universe, certain difficulties arise which can be easily visualized in Figure 1. Suppose we want to calculate the gravitational force on a particle of mass m_o located at point P due to the whole universe. If the origin of the coordinates is at O and the distance between O and P is r, the net force on m_o is given by

$$\frac{G(4\pi r^3 \rho/3)m_o}{r^2} = \frac{Gm_o \rho 4\pi r}{3}$$

pointing from P towards O. In this expression G is Newton's gravitational constant ($G = 6.67 \times 10^{-11} \, \text{Nm}^2 \, \text{kg}^{-2}$) and ρ is the uniform mass density of this hypothetical homogeneous universe. This result is due to the fact that a spherical shell of mass M and radius R exerts no force on any internal point, and attracts any external point as if its whole mass were concentrated on its center, according to Newton's law of universal gravitation, as Newton proved in the *Principia* in 1687. On the other hand,

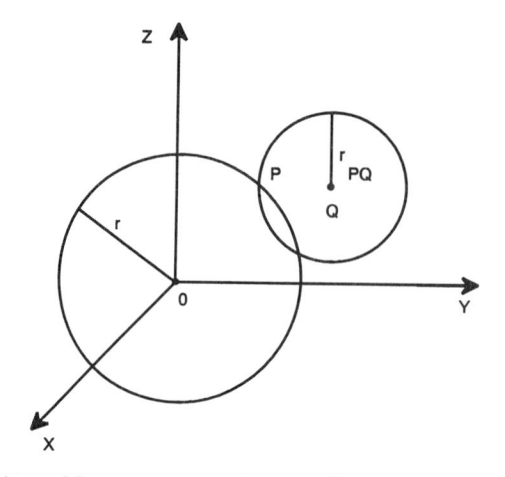

Figure 1. An infinite and homogeneous universe with constant mass density ρ. A mass m_0 is located at P. If we utilize spherical coordinates to calculate the net force on m_0 we will obtain different results if the origin of the coordinate system is at O or at Q.

if we calculate the net gravitational force on m_o from the point Q as the center of the coordinate system (see Figure 1), we find its value to be $Gm_o \rho 4\pi r_{PQ}/3$ pointing from P towards Q, where r_{PQ} is the distance between P and Q. This is different from the previous value, which shows that we can get any value for the net force on m_o depending on our arbitrary choice of the point Q.

This is certainly undesirable. In order to overcome the problem we can either assume that the universe is not infinite and homogeneous in space and time, or that Newton's law of gravitation should be modified when there is a many-body interaction. Here we will follow H. Seeliger and C. Neumann who in 1895 and 1896 proposed that the Newtonian gravitational potential should be modified by the introduction of the factor $e^{-\alpha r}$, where α is a small quantity which would only be significant for large values of r. Laplace had introduced an exponential in Newton's force law of gravitation as early as 1846 (North 1965, Ch. 2; Assis 1992 b).

One way of interpreting this exponential decay in the gravitational potential or force law is to regard this term as an absorption of gravity: instead of having a potential given by Gm_1m_2/r, it would be given by $Gm_1m_2 e^{-\alpha r}/r$, where α depends on the distribution and amount of matter in the straight line connecting m_1 and m_2. We can then look for experimental support of this proposal. Some anomalies in pendulum behavior during solar eclipses which could be due to a screening effect for gravitation have been described by Dragone (1990). To our knowledge, the best experiments specifically designed to test the idea of a gravitational absorption are due to Q. Majorana (Martins 1986, Assis 1992 b). He found a positive value with an absorption coefficient for liquid mercury of the order of $\alpha \approx 10^{-10}$ m^{-1}. Because his experiments were never repeated, we cannot have complete confidence in these results. On the other hand, the fact that they were never contested shows that there is a real possibility the effect exists, which, as we have seen, is of great cosmological significance.

A similar situation arises with inertia if we follow Mach's principle, according to which the inertia of any body is due to its interaction with the remainder of the universe (Mach 1960, Barbour 1989). For instance, recently we implemented Mach's principle quantitatively utilizing a Weber-type force law (Assis 1989 a). In order to derive Newton's first and second laws from a gravitational interaction of any body with the remainder of the universe, we postulated that the resultant force (including all kinds of interaction—gravitational, electromagnetic, nuclear, elastic, inertial, *etc.*) on any body is always zero in all frames of reference, even when the test body is in motion and accelerated. Beyond this, we also assumed that Newton's force of gravitation should be modified following the structure of Weber's force. Weber's law was introduced in electromagnetism in 1846 in order to unify electrostatics, magnetism (force between current elements) and electromagnetic induction (Weber 1872, 1892-4 and 1966; Maxwell 1954, Volume 2, Ch. 23; O'Rahilly 1965, Volume 2, Ch. 11; Wesley 1990 and 1991, Ch. 6; Phipps 1990 a, b; Assis 1989 b, 1990, 1991 and 1992 c; Assis and Caluzi 1991; Assis and Clemente 1992; Clemente and Assis 1991). The first to propose a similar modification for gravitation seems to have been G. Holzmuller in 1870, and in 1872, F. Tisserand utilized the same force law (North 1965, Ch. 3, pp. 46-47; Assis 1989 a). Recently other authors have followed the same procedure, applying a Weber-type force law to gravitation: (Eby 1977, Sokol'skii and Sadovnikov 1987). A similar idea has been followed by Ghosh, arriving at equivalent results (Ghosh 1984, 1986, 1988 and 1991). Although his force law is not exactly like Weber's, he has also succeeded in implementing Mach's principle.

Moreover his expression involves a velocity drag term, not present in Weber's force, which leads to many interesting and reasonable results.

In our previous work (Assis 1989 a) implementing Mach's ideas mathematically, we derived Newton's first and second laws from a gravitational interaction of any body with the remainder of the universe. This was done by integrating a Weber-type force law for gravitation in a finite universe of radius $R = c/H_o$, where $c = 2.998 \times 10^8$ m s^{-1} is the light velocity in vacuum and $H_o \approx 3 \times 10^{-18}$ s$^{-1} \approx 10^{-10}$ years is Hubble's constant, so that $R \approx 10^{26}$ m. If we had integrated over an infinite universe with a constant mass density compatible with the value estimated from observations ($\rho \approx 3 \times 10^{-27}$ kg m^{-3}), this model would not have worked. Accordingly, we suppose here the Mach-Weber model modified by the Seeliger-Neumann term, such that the potential energy U of the gravitational masses m_1 and m_2 is given by

$$U = -H_g \frac{m_1 m_2}{r}\left(1 - \frac{\xi \dot{r}^2}{2c^2}\right)e^{-\alpha r}. \tag{1}$$

In this equation H_g is a constant, r is the distance between m_1 and m_2, namely,

$$r = \sqrt{(x_1 - x_2)^2 + (y_1 - y_2)^2 + (z_1 - z_2)^2},$$

and $\xi = 6$ in order to obtain the correct value of the precession of the perihelion of the planets (Assis 1989 a). Moreover, $\dot{r} \equiv dr/dt = \hat{r} \cdot (\vec{v}_1 - \vec{v}_2)$, where $\hat{r} \equiv (\vec{r}_1 - \vec{r}_2)/r$ is the unit vector pointing from m_2 to m_1 and $\vec{v}_1 - \vec{v}_2 \equiv d(\vec{r}_1 - \vec{r}_2)/dt$, and α is a constant characteristic of the medium in the straight line between m_1 and m_2 ($\alpha = 0$ in a complete vacuum).

As usual, we can define the force exerted by m_2 on m_1 by $\vec{F} \equiv -\hat{r}dU/dr$ so that

$$\vec{F} = -H_g \frac{m_1 m_2 \hat{r}}{r^2}\left[1 - \frac{\xi \dot{r}^2}{2c^2} + \xi\frac{r\ddot{r}}{c^2} + \alpha r\left(1 - \frac{\xi \dot{r}^2}{2c^2}\right)\right]e^{-\alpha r} \tag{2}$$

where

$$\ddot{r} \equiv \frac{d^2 r}{dt^2} = \frac{(\vec{v}_1 - \vec{v}_2)\cdot(\vec{v}_1 - \vec{v}_2) + (\vec{r}_1 - \vec{r}_2)\cdot(\vec{a}_1 - \vec{a}_2) - \dot{r}^2}{r}$$

and $\vec{a}_1 - \vec{a}_2 \equiv d^2(\vec{r}_1 - \vec{r}_2)/dt^2$.

Weber's original expressions for the potential energy between the electric charges q_1 and q_2, and the force exerted by q_2 on q_1 are the same as equations (1) and (2) with the replacements $-H_g m_1 m_2 \to q_1 q_2/4\pi\varepsilon_o$, $\xi \to 1$ and $\alpha \to 0$.

We now follow the same procedure as in our previous work (Assis 1989 a) to calculate the force on a test mass m_o due to an isotropic distribution of mass extending to infinity. We suppose an observer at the origin of a coordinate system, such that for this observer the universe (which is supposed to have a constant mass density ρ) as a whole is spinning with an angular frequency $\vec{\omega}$. Integrating equation (2) utilizing spherical coordinates yields

$$\vec{F} = -Am_o \left[\vec{a}_o + \vec{\omega} \times (\vec{\omega} \times \vec{r}_o) + 2\vec{v}_o \times \vec{\omega} + \vec{r}_o \times \frac{d\vec{\omega}}{dt} \right] \quad (3)$$

In this equation A is a dimensionless constant given by

$$A = \frac{4\pi}{3} H_g \frac{\xi}{c^2} \rho \int_0^\infty r e^{-\alpha r} dr = \frac{4\pi}{3} H_g \frac{\xi}{c^2} \frac{\rho}{\alpha^2},$$

where α would be the mean absorption coefficient of the universe. By analogy with the cosmological redshift, which will be described in the next section (and which can be interpreted as an absorption of light by intergalactic matter), we propose that $\alpha = H_o/c$.

As in our previous work (Assis 1989 a) we can then derive Newton's first and second laws, with a very important development: these two laws are now derived in an infinite and homogeneous universe. The main idea is that the Seeliger-Neumann term both solves the gravitational paradoxes arising from Newton's law of gravitation, and leads to a derivation of inertia in full compliance with Mach's principle in an infinite and homogeneous universe. A complete derivation of these results and an analysis of its consequences is given by Assis (1992 b).

III. Cosmological Redshift

Our next subject is the cosmological redshift. When the spectra of the extragalactic nebulae or external galaxies are observed, what we measure are the apparent luminosities of nebulae and shifts in their spectra (Hubble 1958, p. 3). It has been observed that the fainter the nebula, the larger the redshift (with the exception of the nearby galaxies, most galaxies present a shift toward the red instead of the blue). By assuming the faintness of the nebulae to be related to their distance by a certain function, Hubble was able to conclude that there exists a linear relation between redshifts and distances (Hubble 1929). This relation can be written as

$$z(r) \equiv \frac{\lambda(r) - \lambda_o}{\lambda_o} = \frac{H_o}{c} r \quad (4)$$

In this relation $z(r)$ is the fractional spectral shift, λ_o is the wavelength of a certain line as observed in the laboratory (when the source and detector are at rest relative to the earth), $\lambda(r)$ is the wavelength as observed in the earth's detector of the same line which had been emitted (presumably at a wavelength λ_o) by a galaxy which is at a distance r from the earth, and H_o is Hubble's constant.

Usually this redshift is interpreted as a Doppler shift. This interpretation is what leads to the idea of the expansion of the universe, the big bang, etc. If this were the case we would have (Sciama 1971, p. 71)

$$1 + z(r) = \frac{\lambda(r)}{\lambda_o} = \sqrt{\frac{c+v}{c-v}} = 1 + \frac{v}{c} + O\left(\frac{v^2}{c^2}\right) \quad (5)$$

This would mean

$$\frac{v}{c} = \frac{(1+z)^2 - 1}{(1+z)^2 + 1} \tag{6}$$

In 1929 the largest value of v/c given by Hubble was $v/c \approx 6 \times 10^{-3}$ (Hubble 1929, Table 2), and by 1936 when he wrote *The Realm of the Nebulae* this value had increased to $v/c = 0.13$ (Hubble 1958, plate VII). In 1971 the record redshift was 2.88, which implies $v/c = 0.87$ (Sciama 1971, p. 70). This applies to quasi-stellar objects (quasars) as well as some galaxies. For instance, Arp described some galaxies which according to the Doppler interpretation of redshifts would be moving away from us at $v/c \approx 0.1$ (Arp 1987, Ch. 6). These extremely large recession velocities are a source of doubt for the interpretation of the redshift as a Doppler effect. The reason is that all other velocities of astronomical objects known to us are much smaller. For instance, the orbital velocity of the earth around the sun is approximately 30 km s^{-1} ($v/c \approx 10^{-4}$); the orbital velocity of the solar system relative to the center of our galaxy is approximately 250 km s^{-1} ($v/c \approx 10^{-3}$); and the random or peculiar motion of galaxies is of this same order of magnitude.

There are other problems with the big-bang model: the age of some structures in the universe (some galaxies and agglomerates of galaxies) is supposedly greater than the "age of the universe" derived from the big-bang model; there are intrinsic redshifts of quasars and some galaxies which are clearly not due to a Doppler effect, *etc.* (Arp 1987, Arp and van Flandern 1992). A general criticism of big bang cosmological models has been given by Kierein (1988).

If the redshift is not due to a Doppler effect, what is its origin? We prefer to assume that a photon loses energy in its journey from the surface of a star to the earth. The energy which is lost by the photon would be acquired by the matter with which it interacts in its journey. If this is the case, there would be two components in the redshift of any astronomical object: one intrinsic, due to the interaction of the photon with the atmosphere of the astronomical body (a star and its atmosphere, for instance) and with the matter surrounding it (interstellar matter), and one external component due to intergalactic matter. Both redshifts may be due to the same mechanism, but only the latter (the cosmological or Hubble component) would obey the redshift-distance relation (equation (4)). The first component should be independent of our distance to the source if the source is an external galaxy or a star belonging to this external galaxy. If the source is a star in our own galaxy, then its redshift should have an intrinsic component (due to interaction of its light with its own atmosphere and immediately surrounding matter) and a component which should depend on our distance to the star due to the interaction of its light with the interstellar matter.

This interpretation of the redshifts is usually called the tired light model. We discussed this model in our previous work (Assis 1992 a). In simple terms, the cosmological or Hubble component utilizes Einstein's expression for the energy E of a photon related to its frequency v and wavelength λ: $E = hv = hc/\lambda$, $h = 6.6 \times 10^{-34}$ Js being Planck's constant. This expression is coupled to the energy lost by the photon to the matter with which it interacts in its journey from the surface of a star to the earth: $E(r) = E_o e^{-\alpha r}$, E_o being the initial photon energy at the surface of the star, or the equivalent energy of a photon of the same frequency in the laboratory.

$E(r)$ is the energy of the photon upon its arrival at the Earth, and α is the mean absorption coefficient of light in the line of sight connecting the source and the earth. From these two relations we obtain

$$z(r) \equiv \frac{\lambda(r) - \lambda_o}{\lambda_o} = e^{\alpha r} - 1 \approx \alpha r + \frac{\alpha^2}{2} r^2 \qquad (7)$$

Comparison with (4) yields $\alpha = H_o/c$. A good discussion of the tired light model can be found in Reber (1983) and LaViolette (1986). LaViolette, in particular, has shown that the tired light model fits the data better than the big bang model in four important tests: the angular size-redshift test, the Hubble diagram test, the galaxy number count-magnitude test, and the differential $\log N - \log S$ test.

This explanation is a very simple one and avoids the problem of the high velocities mentioned above. It was advocated, for instance, by de Broglie (1966). After discussing the Doppler interpretation of the redshift, he said explicitly:

> Cependant je ne suis pas personnellement persuadé que l'interprétation des déplacements spectraux observés par un effet Doppler lié à une expansion de l'univers s'impose réellement. A mon sens, l'effet observé pourrait être dû à un "vieillissement du photon", c'est-à-dire à une perte progressive d'energie par le photon au cours de son long parcours intersidéral. Cet effet, jusqu'ici inconnu de toutes les théories de la lumière même ce tenus de l'existence des photons, pourrait résulter d'une cession continue d'énergie par le photon à l'onde qui l'entoure.*

In a previous paper he had explained in more detail how this loss of photon energy might occur (de Broglie 1962):

> Un photon venant à nous d'une nébuleuse très lointaine pourrait voir son onde u s'affaiblir par suite d'un étalement lent ou d'une absorption par les milieux absorbants extrèmement ténus qui existent, on le sait aujourd'hui, dans les espaces intersidéraux.... Il y aurait ainsi une diminution progressive du quantum $h\nu$, donc un déplacement vers le rouge, par un mécanisme tout à fait différent de l'absorption forte par le photon et de l'effet Comptom, mécanisme relié à l'affaiblissement 'faible' et continu de l'onde u.**

What are the main arguments which have been raised against this explanation? Recently Arp pointed out five observational tests tired light theories must confront in order to become useful theories (Arp 1990): (1) absence of blurring in the optical images of extragalactic objects; (2) existence or not of a correlation between

---

* Translation (Keys 1991): Personally, however, I am not convinced that the interpretation of the observed spectral shifts as due to a Doppler effect connected with expansion of the universe is really necessary. In my opinion, the observed effect could be due to a 'photon aging', *i.e.* a gradual loss of energy by photons during their long intergalactic voyage. This effect, hitherto unknown in any theory of light, even theories that admit photons, could be due to a continuous loss of energy by the photon to its surrounding wave.

** Translation (Keys 1991): A photon arriving from a very distant nebula could have its wave u weakened through a slow attenuation or absorption by the extremely tenuous absorbing matter that we now know exists in interstellar space.... This would result in a gradual decrease of the quantum $h\nu$, and hence a redshift, through a mechanism quite different from strong photon absorption or the Compton effect. The actual mechanism would be the continuous 'weak' absorption of the wave u.

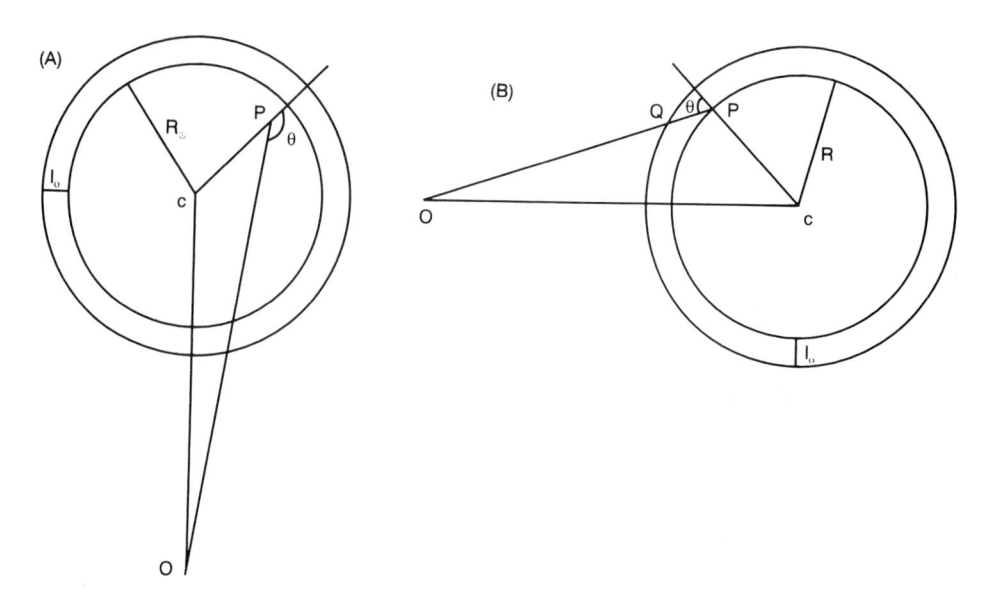

Figure 2. (A) An observer O looking at a point P in the surface of the sun (centre C, radius $R_\odot$, atmosphere with a thickness $l_0 \ll R_\odot$).
(B) The same situation in a plane cutting the sun and containing the points O, C and P. The path traversed by a photon in the sun's atmosphere is given by $l = PQ = l_0/\cos\theta = l_0 \sec\theta$.

redshifts of stars and the column density of gas (hydrogen or molecular clouds, ionized gas clouds) in front of each star; (3) the fact that two galaxies may be interacting so that they are at the same distance but may have much different redshifts; (4) spiral galaxies embedded in a redshifting medium should show a gradient of redshift from center to edge; (5) there should be severe redshift dislocations along the edge of objects due to the redshifting medium. All five tests have been specifically analyzed by Jaakkola, who showed that observational evidence complies with the tired light model (Jaakkola 1990).

The main criticisms against the tired light model are its *ad hoc* assumption of a beginning in time and lack of a suitable mechanism which could account for the observed phenomena. In this work we will not analyze any possible mechanism in depth, but will discuss an important example which gives considerable support to the tired light idea, namely the center-to-limb variation of solar lines (the sun being the star that is best known to us). The fact that the redshift in the solar lines changes from the center of the sun to the limb has been known since the turn of the century. In Figure 2 (A) (not drawn to scale) we present the main parameters which describe this phenomenon. An observer O on the earth observes a point P on the surface of the sun. The sun has a radius $R_\odot = 7 \times 10^8$ m and an atmosphere of thickness $l_0 \ll R_\odot (l_0 \approx R_\odot/2000)$. We represent by θ the angle between the line of sight and the solar radius to the point where the line of sight cuts the solar surface. In Figure 2 (B) we have the same situation in a plane which cuts the sun but contains the sun's center C, the observer O and the point P. Observations in the visible spectrum ($\lambda \approx 6100$ Å) show a redshift which changes from center to limb according to Finlay-Freundlich (1954)

$$\lambda_{obs} - \lambda_{lab} = (2.72 + 1.85 \sec\theta) \times 10^{-3} \text{ Å} \qquad (8)$$

which leads to a fractional change of

$$z = \frac{\lambda_{obs} - \lambda_{lab}}{\lambda_{lab}} = (4.5 + 3.0 \sec\theta) \times 10^{-7} \qquad (9)$$

The observed points and relation (8) are represented in Figure 3, taken from Finlay-Freundlich (1954).

Let us now try to understand the origin of this observed redshift. Is it a Doppler redshift? The answer seems to be no, because equation (9) was obtained after taking into account the known Doppler shifts resulting from the relative motion of the sun and the earth (Marmet 1989). The rotation of the sun has no influence, since equation (8) is observed everywhere on the surface of the sun (Marmet 1989). The spots on the sun's surface have a period T of rotation of approximately 21 days. This would lead to a Doppler shift of $z = v/c = 2\pi R_\odot/Tc = 8.1 \times 10^{-6}$, suggesting a redshift at the points on the surface of the sun which are moving away from us and a blue shift at the points which are coming towards us. This is not what is described by equation (9).

Is it a gravitational redshift? According to Einstein's theory of relativity, light emitted with frequency v_1 from a place of gravitational potential Φ will arrive at a place of relative gravitational potential zero with frequency v_2 such that (North 1965, p. 53):

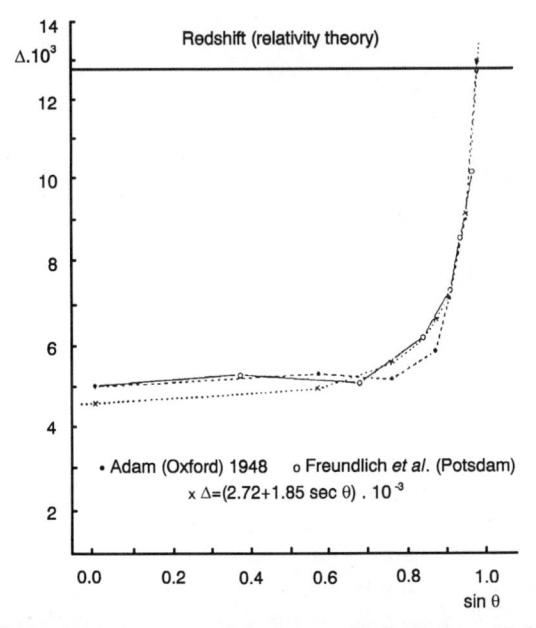

Figure 3. The redshift from the centre of the sun to the limb (from Finlay-Freundlich 1954). Full and open circles represent two sets of observations. The dotted line with ×'s is the least square solution given by (8) which best fits the data. The horizontal line is the gravitational redshift according to Einstein's theory of relativity.

$$\frac{v_1 - v_2}{v_2} = \frac{\Phi}{c^2} \qquad (10)$$

Applying this to the sun would yield $z = \Delta\lambda/\lambda = -\Delta v/v = GM_\odot/R_\odot c^2 = 2.12 \times 10^{-6}$, where $M_\odot$ is the sun's mass. As we can see from Figure 2, this is near the observed value at the limb of the sun but is much larger than the value in the center. As the gravitational potential is a constant over the surface of the sun, the redshift should not vary from center to limb. So the conclusion is that the redshift of the sun is not wholly due to a gravitational redshift.

An alternative interpretation of the cosmological or Hubble redshift has been presented by Arp (1991). According to this model, the cosmological redshift, rather than a Doppler effect, would depend on the epoch of creation of the astronomical object. This model explains many anomalous and intrinsic redshifts of some galaxies made of younger matter but cannot explain the variation of the redshift from center to limb in the sun, as the age of the matter at the sun's surface should be the same everywhere.

What explanation is left? Figure 2 (B) shows that the length that light travels across the sun's atmosphere is given by $l \equiv PQ$. From the triangle OCP it can be easily seen that (as $l_o \ll R_\odot$) $l \equiv PQ \approx -l_o/\cos(\pi - \theta) = l_o \sec\theta$. Consequently, an explanation which almost forces itself upon us is that the redshift is due to the interaction of light in its passage through the atmosphere of the sun. The redshift would then be proportional to the length of travel, which as we have seen, is proportional to $\sec\theta$. Finlay-Freundlich has successfully explained the redshift of the solar lines, as well as those anomalous redshifts of O, B and A stars, supergiant M stars, Wolf-Rayet stars and the cosmological redshift, with a simple formula (Finlay-Freundlich 1954)

$$\frac{\Delta\lambda}{\lambda} = -\frac{\Delta v}{v} = AT^4 l \qquad (11)$$

where $A = 2 \times 10^{-27}$ K^{-4} m^{-1} is a constant, T is the temperature of the radiation field where light is moving (it is not necessarily the temperature of the source) and l is the length of path traversed through the radiation field. This formula explains reasonably well the second term on the right hand side of equation (8).

This redshift of the sun's line is a clear proof that the tired light proposal meets Arp's condition (2) due to the proportionality between the redshift and the length of path across the sun's atmosphere, namely, z is proportional to $\sec\theta$, which is proportional to l.

The dependence of the solar redshift on the length of path through its atmosphere is remarkable and lends support to a tired light model. If this is the case, what is the physical mechanism responsible for this effect? When presenting equation (11), Finlay-Freundlich suggested that it might be due to a photon-photon interaction. A variation of this proposal was given by Pecker et al. (1972) Another mechanism based on an inelastic collision of the photon with atoms or molecules has been given by Marmet, and he has also successfully explained the redshift of the solar limb (Marmet 1988, 1989). Another possibility is an interaction of photons with free electrons (Kierein 1990). There are many other proposals; however, we will not discuss them here. The main difficulty which I see with regard to Marmet's proposal (or any other of this kind) is that atoms and molecules have extremely well defined levels of energy, and therefore they can only absorb and emit in these frequencies. On the other hand, an interaction between photons and free electrons would appear more plausible, since free electrons can absorb and emit photons of any frequency. Another problem is that to explain Hubble law of redshifts by the same mechanism, Marmet (1988) needed to assume an average density of hydrogen atoms throughout the universe of 2.5×10^4 atoms m^{-3}. But as we have seen, the estimated mass density in the universe (based on observations of visible galaxies, etc.) is only $\rho \approx 3 \times 10^{-27}$ kg m^{-3}, which is equivalent to only one hydrogen atom per cubic meter, four orders of magnitude smaller than what is required in Marmet's proposal.

Whatever the nature of the mechanism (there may even be several mechanisms at work simultaneously), a tired light model does seem to satisfactorily account for the data on redshifts. The center-to-limb variation of the solar lines shows clearly that there exist redshifts which are not due to a Doppler effect or a gravitational redshift, and are also not age dependent. This constitutes a proof that another mechanism is at work to create this redshift. This is the case at the sun's surface. Why should the same mechanism not work in other stars and in interstellar and intergalactic space?

IV. Cosmic Background Radiation

In 1965, working with a horn-reflector antenna at 4080 Mc s^{-1} ($v = 4.08 \times 10^9$ Hz, $\lambda = c/v = 7$ cm), Penzias and Wilson discovered an excess temperature of $3.5 \pm 1.0°$ K (Penzias and Wilson 1965). They found this temperature to be isotropic, unpolarized, and free from seasonal variations. It was soon interpreted by Dicke *et al.* as cosmic black-body radiation, a relic of a hot big-bang (Dicke *et al.* 1965). The idea of a hot big bang had been developed by Gamow, Alpher, Herman and others in the period 1948-54 (see for instance Alpher *et al.* 1948). Later, a dipole anisotropy was found in the cosmic background radiation which is usually interpreted as being due to the earth's motion through the radiation field. The value of the dipole anisotropy is well known nowadays, and allows a precise determination of our motion relative to this radiation background (see, for instance, Lubin *et al.* 1985).

Here we want to emphasize certain other parallel developments which are not so well known. First of all, in 1954 (prior to the discovery by Penzias and Wilson) Finlay-Freundlich developed his alternative interpretation of the cosmological redshift on the basis of a tired light model (see equation (11)). As a corollary he predicted a mean temperature of intergalactic space between 1.9 and 6.0° K (Finlay-Freundlich 1954), stating that: "One may have, therefore, to envisage that the cosmological redshift is not due to an expanding universe, but to a loss of energy which light suffers in the immense lengths of space it has to traverse coming from the most distant star systems. That intergalactic space is not completely empty is indicated by Stebbins and Whitford's discovery (1948) that the cosmological redshift is accompanied by a parallel unaccountable excess reddening. Thus the light must be exposed to some kind of interaction with matter and radiation in intergalactic space." Max Born discussed Finlay-Freundlich's ideas, and indicated his support for them (Born 1954).

Less well-known is an important text by Regener which was published in 1933—well before Gamow's paper (1948). Regener utilized the Stefan-Boltzmann law, equating it to the measured value of the flux energy of the night sky (due to light, heat and cosmic radiation), and obtained a mean temperature for interstellar space of 2.8° K (Regener 1933, Monti 1988). Regener's work was taken up by Nernst in his model of a boundless universe, homogeneous on the large scale and without expansion (Nernst 1937, Monti 1988). Nernst is another advocate of the tired light model.

Even before Regener, a temperature of interstellar space of 3° K had been given by Eddington in his famous book first published in 1926 (Eddington 1988, p. 371).

The conclusion we draw is that the existence of a background radiation of 2.7° K cannot be used as a proof of the big bang theory. This is true not only because other models are compatible with it, but also because other models had even predicted its existence prior to its discovery.

V. Predictions of the Model and Conclusion

Here we would like to present some consequences of the model described in this paper. These are not predictions to be tested in the laboratory, but something to be expected in future observations.

We assume that the universe is isotropic, homogeneous and boundless (extending indefinitely in all directions with a constant, finite mass density). This means that, in principle, there should be galaxies at all distances from us. Consequently, one prediction is that with the development of observational instruments we should find galaxies at an ever increasing distance from us, without limit (the only limitations are the resolving power of the instruments and the range of propagation of electromagnetic radiation).

The universe is also assumed to be homogeneous in time (on a macroscopic scale, the same in the past, now and in the future). A second prediction is thus that in any large region of space, no matter how far from us, we should find approximately the same number of galaxies dying out and being created. In the big-bang model, on the other hand, all the galaxies were formed at approximately the same time. Therefore any young galaxy found at large distance from the earth lends support to a model of the universe which is in a steady state on a large scale.

Another prediction of the model can be obtained by comparing equations (4) and (7). According to equation (4), H_o should be a constant independent of the distance. But if we write equation (7) in the same form as equation (4) we obtain a Hubble "constant" which should depend on the distance, namely

$$z(r) = e^{\alpha r} - 1 \equiv \frac{H(r)}{c} r = \alpha r + \frac{\alpha^2}{2} r^2 + \frac{\alpha^3}{3!} r^3 + O(\alpha^4 r^4) \tag{12}$$

As we have already identified α with H_o/c, we obtain:

$$H(r) = H_o + \frac{H_o^2 r}{2c} + \frac{H_o^3 r^2}{6c^2} + \ldots$$

This shows that Hubble's "constant" should in fact increase with the distance according to the tired light model. Is there any indication that this is the case? If we measure the distance of galaxies by a method which does not depend on their redshift we find that the redshift does not in fact seem to be a linear function of distance (Arp 1988; Arp and Van Flandern 1992). Furthermore, the slope of the curve $H(r) \times r$ seems to agree with equation (12). With an improvement in the observations and an increase in the number of observed galaxies, it will be possible to test this prediction in more detail in the near future.

We may thus conclude that an absorption of light offers an explanation of the cosmological redshift in an infinite and homogeneous universe which is essentially static and without expansion. A similar phenomenon in the realm of gravitation yields inertia as due to a gravitational interaction with the remainder of the universe, in compliance with Mach's principle. Moreover, this is obtained in the framework of the simplest of all models of the universe: a universe which has always existed, in which there is no expansion or creation of matter, and which is essentially uniform and homogeneous in all directions and at all distances.

Acknowledgments

The author wishes to thank FAPESP and CNPq (Brazil) for financial support during the past few years. He thanks also Dr. Thomas E. Phipps Jr. and C. Roy Keys for useful discussions and suggestions.

References

Alpher, R. A., Bethe, H. and Gamow, G., 1948, The origin of the chemical elements, *Physical Review* 73:803-804.
Arp, H., 1987, *Quasars, Redshifts and Controversies*, Interstellar Media.
Arp, H., 1988, Galaxy redshifts, in: *New Ideas in Astronomy*, Cambridge University Press, pp. 161-171.
Arp, H., 1990, Comments on tired-light mechanisms, *IEEE Transactions on Plasma Science* 18:77.
Arp, H., 1991, How non-velocity redshifts in galaxies depend on the epoch of creation, *Apeiron* 9-10:18-29.
Arp, H. and van Flandern, T., 1992, The case against the big bang, *Physics Letters A* 164:263-273.
Assis, A. K. T., 1989 a, On Mach's principle, *Foundations of Physics Letters* 2:301-318.
Assis, A. K. T., 1989 b, Weber's law and mass variation, *Physics Letters A* 136:277-280.
Assis, A. K. T., 1990, Deriving Ampère's law from Weber's law, *Hadronic Journal* 13:441-451.
Assis, A. K. T., 1991, Can a steady current generate an electric field?, *Physics Essays* 4:109-114.
Assis, A. K. T., 1992 a, On Hubble's law of redshift, Olbers' paradox and the cosmic background radiation, *Apeiron* 12:10-16.
Assis, A. K. T., 1992 b, On the absorption of gravity, *Apeiron* 13:3-11.
Assis, A. K. T., 1992 c, Deriving gravitation from electromagnetism, *Canadian Journal of Physics* 70:330-340.
Assis, A. K. T. and Caluzi, J. J., 1991, A limitation of Weber's law, *Physics Letters A* 160:25-30.
Assis, A. K. T. and Clemente, R. A., 1992, The ultimate speed implied by theories of Weber's Type, *International Journal of Theoretical Physics* 31:1063-1073.
Barbour, J., 1989, *Absolute or Relative Motion?, Volume 1: The Discovery of Dynamics*, Cambridge University Press.
Bondi, H., 1960, *Cosmology*, Cambridge University Press, 2nd edition.
Born, M., 1954, On the interpretation of Freundlich's redshift formula, *Proceedings of the Physical Society A* 67:193-194.

Clemente, R. A. and Assis, A. K. T., 1991, Two-body problem for Weber-like interactions, *International Journal of Theoretical Physics* 30:537-545.
De Broglie, L., 1962, Remarques sur l'interprétation de la dualité des ondes et des corpuscules, *Cahiers de Physique* 16:425-445.
De Broglie, L., 1966, Sur le déplacement des raies émises par un objet astronomique lointain, *Comptes Rendues de l'Academie des Sciences de Paris* 263:589-592.
Dicke, R. H., Peebles, P. J. E., Roll, P. G. and Wilkinson, D. T., 1965, Cosmic black-body radiation, *Astrophysical Journal* 142:414-419.
Dragone, L., 1990, The gravitational magnetic field, *Hadronic Journal Supplement* 5:309-334.
Eby P. B., 1977, On the perihelion precession as a Machian effect, *Lettere al Nuovo Cimento* 18:93-96.
Eddington, A. S., 1988, *The Internal Constitution of the Stars*, Cambridge University Press. First issued in 1926.
Finlay-Freundlich, E., 1954, Red shifts in the spectrum of celestial bodies, *Philosophical Magazine* 45:303-319.
Ghosh, A., 1984, Velocity-dependent inertial induction: an extension of Mach's principle, *Pramana Journal of Physics* 23:L671-L674.
Ghosh, A., 1986, Velocity-dependent inertial induction and secular retardation of the earth's rotation, *Pramana Journal of Physics* 26:1-8.
Ghosh, A., 1988, A possible servomechanism for matter distribution yielding flat rotation curves in spiral galaxies, *Astrophysics and Space Science* 141:1-7.
Ghosh, A., 1991, Velocity-dependent inertial induction: a possible tired-light mechanism, *Apeiron* 9-10:35-44.
Hubble, E., 1929, A relation between distance and radial velocity among extra-galactic nebulae, *Proceedings of the National Academy of Sciences* 15:168-173.
Hubble, E., 1958, *The Realm of the Nebulae*, Dover.
Jaakkola, T., 1989, The cosmological principle: theoretical and empirical foundations, *Apeiron* 4:9-31.
Jaakkola, T., 1990, On reviving tired light, *Apeiron* 6:5-6.
Keys, C. R., 1991, Preface: Festschrift Vigier, *Apeiron* 9-10:i-v.
Kierein, J. W., 1988, A criticism of big bang cosmological models based on interpretation of the redshift, *Laser and Particle Beams* 6:453-456.
Kierein, J., 1990, Implications of the Compton effect interpretation of the redshift, *IEEE Transactions on Plasma Science* 18:61-63.
LaViolette, P. A., 1986, Is the universe really expanding?, *Astrophysical Journal* 301, 544-553.
Lubin, P., Villela, T., Epstein, G. and Smoot, G., 1985, A map of the cosmic background radiation at 3 millimeters, *Astrophysical Journal* 298:L1-L5.
Mach, E., 1960, *The Science of Mechanics—A Critical and Historical Account of its Development*, Open Court, La Salle.
Marmet, 1988, A new non-Doppler redshift, *Physics Essays* 1:24-32.
Marmet, P., 1989, Red shift of spectral lines in the sun's chromosphere, *IEEE Transactions on Plasma Science* 17:238-244.
Martins, R. de A., 1986, Pesquisas sobre absorção da gravidade, *Anais do Primeiro Seminário Nacional de História da Ciência e da Tecnologia*, Mast (Rio de Janeiro), pp. 198-213.
Maxwell, J. C., 1954, *A Treatise on Electricity and Magnetism*, Dover.
Monti, R., 1988, *Progress in Cosmology According to Walther Nernst and Edwin Hubble*, Società Editrice Andromeda (Bologna).
Nernst, W., 1937, Weitere Prüfung der Annahme eines stationären Zustandes im Weltall, *Zeitschrift der Physik* 106:633-661.
North, J. D, 1965, *The Measure of the Universe-A History of Modern Cosmology*, Clarendon Press.
Olbers, H. W. M., 1826, On the transparency of space, *Edinburgh New Philosophical Journal* 1:141-150.
O'Rahilly, A., 1965, *Electromagnetic Theory-A Critical Examination of Fundamentals*, Dover.
Pecker, J. C., Roberts, A. P. and Vigier, J. P., 1972, Non-velocity redshifts and photon-photon interactions, *Nature* 237:227-229.

Penzias, A. A. and Wilson, R. W., 1965, A measurement of excess antenna temperature at 4080 Mc/s, *Astrophysical Journal* 143:419-421.
Phipps Jr., T. E., 1990 a, Toward modernization of Weber's force law, *Physics Essays* 3, 414-420.
Phipps Jr., T. E., 1990 b, Weber-type laws of action-at-a-distance in modern physics," *Apeiron* 8:8-14.
Reber, G., 1983, Intergalactic plasma, *IEEE Transactions on Plasma Science* PS-14:678-682.
Regener, E., 1933, Der Energiestrom der Ultrastrahlung, *Zeitschrift für Physik* 80:666-669.
Rudnicki, K., 1989, The importance of cosmological principles for research in cosmology, *Apeiron* 4:1-7.
Sciama, D. W., 1971, *Modern Cosmology*, Cambridge University Press.
Sokol'skii, A. G. and Sadovnikov, A. A., 1987, Lagrangian solutions for Weber's law of attraction, *Soviet Astronomy (Astronomical Journal)* 31:90-93.
Weber, W., 1872, Electrodynamic measurements-Sixth memoir, relating specially to the principle of conservation of energy, *Philosophical Magazine* 43:1-20 and 119-149.
Weber, W., 1892-1894, *Wilhelm Weber's Werke*, Springer-Verlag, 6 Volumes.
Weber, W., 1966, On the measurement of electrodynamic forces, in: R. Taylor (editor), *Scientific Memoirs*, Volume 5, Johnson Reprint Corporation, pp. 489-529.
Wesley, J. P., 1990, Weber electrodynamics, part I: General theory, steady current effects, *Foundations of Physics Letters* 3:443-469; Weber electrodynamics, part II: Unipolar induction, Z-antenna, ibid. pp. 471-490; Weber electrodynamics, part III: Mechanics, gravitation, ibid. pp. 581-606.
Wesley, J. P., 1991, *Selected Topics in Advanced Fundamental Physics*, Benjamin Wesley (Blumberg, Germany).

Cosmological Principles

Konrad Rudnicki

Jagiellonian University Astronomical Observatory
ul. Orla 171, 30-244 Krakow, Poland

The most important cosmological principles are presented and briefly discussed. This lecture is a summary of a long monograph on cosmological principles being prepared for publication. A more developed presentation of problems mentioned here and more complete references to original papers can be found in that work.

1. Astronomy and Cosmology

Astronomy is an observational and theoretical science dealing with celestial bodies and celestial phenomena accessible to human eyes as well as to astronomical and physical scientific instruments. Cosmology is the science of the entire Universe. Cosmology would be just another name for astronomy, if the total area of the Universe were accessible to observations. And in fact there are some cosmological models, called "small Universe models" (cf. Ellis 1987, Dyer 1987), which represent the Universe as a region entirely accessible to earth-based observations. These models, however, are not developed in the main stream of cosmological ideas. They were devised more to illustrate some methodological possibilities than to depict reality.

In most contemporary conceptions and contemporary models of the Universe, the so called cosmological horizon is accepted, i.e. a surface located at a certain distance, from beyond which no signal can reach the observer. Points located beyond the cosmological horizon cannot have any causal connection with the observer. There are various definitions of cosmological horizons. For our purposes all of them are of the same value. Because of the cosmological horizon, the entire Universe is divided for every observer into two parts: one part that is accessible and another that is not accessible to observations.*

One can discuss whether cosmology, the field of human knowledge about the observable as well as unobservable regions of the Universe, can be accepted as a

---

* For a Universe born all at the same time (Ed. note).

science, or should instead be considered as a part of metaphysics. But in other exact sciences statements are also made which are not in direct relation to observations. In mathematics no geometrical figure can be seen or measured by physical devices. All that can be constructed in the physical world may only roughly remind us of mathematical reality, which is accessible as such for the mind only. Physical laws, as such, are likewise not observable. We can observe only their consequences. Still nobody takes this fact as an argument for considering mathematics or physical laws as belonging to metaphysics. Similarly, nobody has the right to demand that everything in cosmology should be observable. The human mind can supplement what cannot be reached by sense perception.

2. The Building Stones of the Universe

In addition to one or more cosmological principles, any theory of the Universe, and any model of the same, has to contain some ideas about the main constituents of the Universe. In various epochs of the development of cosmology, different celestial bodies or their agglomerations were accepted as such building stones. To mention only the most important ones in the time interval from Copernicus till today, they were: planets, stars, galaxies, clusters of galaxies, bubbles of "Voronoy Foam" as well as abstract agglomerations of matter called "fundamental bodies". Besides these, the structure of spacetime as such is often (not always!) taken as a main "building stuff" of the Universe. These "building materials" have, however, only secondary influence on the property of theories or models of the Universe. I will not discuss the problems connected with choosing among these. The participants at the School or the reader of these proceedings can easily supplement for himself all that I will say with comments on constituents of the Universe used in various models and theories.

3. Extrapolation: Mach's Principle as a Basic Cosmological Principle

The simplest way to describe the regions not accessible directly to observations is to extrapolate the observable into unobservable regions.

There exists in mathematics a wide class of analytical functions possessing the characteristic property that out of the features of a function in even a very small neighbourhood of a point, the features of the function in the entire domain of its arguments can be derived. This is the basic theorem of such functions. There exists a certain kind of a counterpart to this theorem in the realm of physical principles, called Mach's Principle. This can be formulated as follows: out of the features (positions and velocities of all physical bodies, qualitative and quantitative properties of all physical fields and of the spacetime itself) even in the immediate cosmological vicinity of an observer, the features of the entire Universe can be derived. Mach's Principle has many different formulations *e.g.* "local physical properties are uniquely determined by the features of the entire Universe and vice versa." Mach's Principle does not belong to the main physical principles. Nobody, till now, has given any significant theoretical arguments in its favor. It is rather a postulate which, with the authority of Albert Einstein, who was a fan of it, has gained an important place in cosmological considerations. The first strict formulation of the rather general ideas of Ernst Mach as well as the name Mach's Principle itself was given by Einstein. Most modern cosmological principles (and some historical ones) are just specifications of this general conviction of Mach, that out of intrinsic knowledge of local features of the World, all features of the entire Universe can be deciphered.

Mathematics deals with ideal situations. One can think about perfect knowledge of properties of an analytical function in the vicinity of a point. It is, however, much more difficult to explain what it means "to know perfectly" all properties of the physical world in the vicinity of an observer. It is difficult to define, and even more difficult to achieve, even in the macrocosmic sense, perfect cognizance of physical reality. The problem becomes even more complicated when the problem of the microcosmic uncertainty principle has to be taken into consideration. Nevertheless, Mach's idea, which admits the possibility of reconstructing the totality of existence out of a fragment is accepted today by most cosmologists. In a subconscious way, it was also accepted in past epochs.

4. The Genuine Copernican Cosmological Principle

There exist today about 30 different cosmological principles in the cosmological literature. I want to limit myself to the four main streams of cosmological principles in this compact presentation.

The first stream was originated by Nicolaus Copernicus when he dared to postulate that the Universe observed from every planet looks roughly the same. This statement, called today the Genuine Copernican Cosmological Principle, was a basis for the model of the Universe created by Copernicus himself, as well as for two other historical models, the most important of them being the model of Johannes Kepler.

5. The Generalized Copernican Cosmological Principle

Einstein and his contemporary cosmologists introduced the natural generalization of the Genuine Copernican Principle when they assumed by solving the General Relativity equations that the Universe observed from every point and in every direction looks roughly the same. This principle has various names. Besides the Generalized Copernican Cosmological Principle, it is also called the Ordinary, the Narrow, or the Weak Cosmological Principle. In fact this was the first cosmological principle accepted as such by cosmologists. As long as other cosmological principles were not known, it was called simply "the Cosmological Principle," and sometimes it is called this even today.

This principle leads to the Hubble Law of spatial expansion. All models based on this principle allow in the Universe only radial systematic motions with the velocity proportional to the distance:

$$V_r = Hr$$

where V_r is the velocity of radial motion, r—the distance between two points (fundamental bodies) of the Universe, H—the so called Hubble Constant, which can be positive, negative or equal to zero. The Hubble Law was first considered to be due to General Relativity. Much later it was proved that it is not connected with

General Relativity or with any physical theory or assumption, but simply with the Generalized Copernican Cosmological Principle, which restricts the possibility of systematic velocities exclusively to radial and proportional-to-distance. After the discovery of this fact, the name "Weak Cosmological Principle" is not used any more because in fact it is a powerful principle which produces the Hubble Law.

The generalization of the Genuine Copernican Principle has to be understood in a specific sense. In fact the genuine principle is more general than the generalized one. Every model fulfilling the Generalized Copernican Principle likewise fulfills the Genuine Copernican Principle, but not vice versa. If the Universe looks (roughly) the same from every point, it looks (roughly) the same from every planet. But it is not enough to look (roughly) the same from every planet in order to look (roughly) the same in every direction and from every point in the Universe.

Most contemporary models of the Universe (*e.g.* the Friedman type models, or the majority of the inflationary models) are based on the Generalized Copernican Principle.

If we accept the Generalized Copernican Principle with a positive value for the Hubble Constant, we obtain a theory (a model) of the Universe with an initial singularity. A subclass of such models are Big Bang models.

6. The Perfect Cosmological Principle

A further modification of Copernican ideas leads to the Perfect or the Strong Cosmological Principle: The Universe observed from every point in every direction at any time looks roughly the same. Again, this is a kind of generalization because it concerns not only space but also time, yet it also narrows the possibility of being fulfilled by cosmological models. Every model which fulfills the Perfect Principle also fulfills the Generalized Copernican Principle but not vice versa. What looks the same "from every point and at any time" looks the same "from every point." However, the Universe could also look the same from every point, but look different at various times.

The models fulfilling the Perfect Principle must, of course, obey the Hubble Law. If we accept the Hubble Constant to be not equal zero, we could obtain a Universe where the density of matter increases or diminishes constantly due to expansion or to contraction of the Universe. This would be contradictory to the assumption that the Universe looks the same at any time. To avoid this contradiction one has to introduce the production of matter out of nothing. In the case of positive value of the Hubble Constant (expansion), one gets the famous "Steady State Universe". In the case of contraction, the vanishing of matter is required.

7. The Lucretian Cosmological Principle

One can avoid these contradictions and preserve the principle of preservation of mass, by accepting zero for the Hubble Constant. With this assumption we obtain a static Universe. Models of such a Universe were popular in 19th century and in the early 20th. A contemporary picture of such Universe has been proposed by Jaakkola (1989). The assumption that the Universe looks roughly (on the large scale) the same in every direction and at every point in time from every point and admits no systematic motions is sometimes called the Lucretian Cosmological Principle.

8. The Generalized Perfect Cosmological Principle

Universe models based on the Genuine Copernican Principle are not connected with any simple geometrical symmetry in the space. The subclass of these models fulfilling the Generalized Copernican Cosmological Principle possesses simple space symmetry. The General Relativity theory employs the concept of spacetime and thereby requires certain properties in spacetime, not in space only. In this sense the Perfect Principle is much more relativistic because it requires uniformity from both. Still it limits the requirement of isotropy to space only, while not requiring isotropy of time.

At first glance, the requirement of isotropy in time seems to be impossible in a real Cosmos. Even in the stationary Universe of Jaakkola, isotropy in time can be considered in a metaphorical sense only. To be sure, the general view of the Universe is the same when we move in the positive and the negative direction of time, but local phenomena are not reversible. The gravitation pulling forces and the explosive dispersing forces are working in opposite directions in time, but, even in Jaakkola's Universe, by no means produce the same kind of phenomena, when changing only the time direction. The same is true with electromagnetic radiation. The arrow of time still exists.

But one cannot exclude the possibility that, in the course of further search for similarities and identities in various physical interactions, some theories may emerge with perfect symmetries of phenomena according to time. Then, a subclass of models based on the Perfect Principle and fulfilling the narrower requirement of uniformity as well as isotropy of space and time can be envisioned. This requirement might perhaps be called the Generalized Perfect Cosmological Principle.

9. The Fully Perfect Cosmological Principle

But even such a Generalized Perfect Cosmological Principle is not completely relativistic in the sense that it requires homogeneity, but only separately from space and separately from time. One can propose a Fully Perfect Cosmological Principle, establishing one requirement for spacetime as such. At first glance, such total isotropy seems impossible, at least as long as we keep the ordinary notion of the relativistic spacetime, because the metric signature distinguishes time with the opposite sign to the signs of spatial dimensions. But there do exist mathematical concepts of other spacetimes, and it is possible to introduce imaginary time ix_0 instead of x_0 in simple relativistic spacetime. Then the metric formula:

$$ds^2 = +dx_0^2 - dx_1^2 - dx_2^2 - dx_3^2$$

becomes fully symmetrical with respect to all four dimensions.

The Fully Perfect Principle seems useless, at least with today's physical and astronomical concepts, but may still be explored with some more exotic cosmological ideas.

The principles discussed thus far—Genuine Copernican, Generalized Copernican, Perfect, Lucretian, Generalized Perfect and Fully Perfect Principle—form a series where the class of cosmological models fulfilling every consecutive principle is a subclass of the previous one. According to most of today's cosmologists, the Genuine Copernican Principle is too broad to be useful for understanding the Universe, and the Fully Perfect Principle is too narrow for this purpose. These extreme principles are not popular. Models considered similar to reality are today based on principles lying somewhere in the diapason from the Generalized Copernican (the standard model, the inflationary Universes) to Lucretian principles (the Universe of Jaakkola).

In any case, the series of principles that originated with Copernican ideas is the most popular in today's cosmology.

10. The Ancient Greek Cosmological Principle

The basis for most cosmological models constructed in the time from Eudoxios* to Tycho Brahe was the Ancient Greek Cosmological Principle, which can be formulated as follows: Our Earth is the natural center of the Universe. Today this principle is not popular.

11. The Generalized Ancient Greek Cosmological Principle

The following statement can be considered as a generalization of the former principle: The Universe possesses its distinguished center. This principle is sometimes used today. For example, Ellis, Maartens and Nel (1978) proposed a model of spherically symmetrical Universe with our Galaxy in its center. They showed that this model is not contradictory to observations. Furthermore, the Schwarzschild point solution of General Relativity theory, when considered as a model of the Universe, fulfills this principle.

It can be mentioned that all known heliocentric models fulfilled at the same time the Genuine Copernican and the Generalized Ancient Greek Principle. It is important to remember that cosmological models can fulfill several cosmological principles, not necessarily only one of them.

12. The Ancient Hindu Cosmological Principle

The ancient Hindu cosmological views can be expressed in contemporary terms this way: The Universe is infinite in space and time and is infinitely heterogeneous. The principles discussed in sections 4–11 were certain realizations of a general Mach's Principle. All of them give some possibility to extrapolate the local observational facts to the unobservable parts of the Universe. This is not the case with the Ancient Hindu Cosmological Principle. Here the Universe is considered as infinitely heterogenous. With this principle, an extrapolation can, in best of cases, be validated only in a small restricted domain of time and space.

In contemporary cosmology, the Ancient Hindu Cosmological Principle is in fact accepted in all theories (models) of domain-universes, and in all arguments along the line that what is observed here is not necessarily valid everywhere. Take,

---

* With the notable exception of the early period, during the time of Aristarchos of Samos (Ed. note).

for example, the case where the expansion of the Universe can be true in a domain of space in which all observable parts of the Universe are contained within the entire surface of our cosmological horizon. But in an other domain only a general contraction can take place, while others can posses a different number of dimensions or different signatures of their spacetime.

13. The Cosmological Anthropic Principle

The anthropic principle, formulated in 1973, has aspired to the status of a cosmological principle in the last decade. This principle states that the physical properties of the Universe have to be taken as a logical conclusion from the premise that real observers are present in some parts of the Universe's spacetime. Here I shall not discuss the entire, very complex problem of the anthropic principle (*cf.* Barrow and Tipler 1986). I only wish to say (Rudnicki 1989) that this principle, when considered as a cosmological principle, is a very weak one, and can produce specific cosmological models only with the help of other cosmological principles.

14. Final Remarks

The sole possibility of crossing the cosmological horizon is given by the power of human thinking. Cosmological principles form only one possible pattern, are just one example of such thinking. I would resist overestimating this particular pattern. My personal conviction is that the Universe is too complicated to grasp by means of one sentence called a cosmological principle. Since, however, cosmological principles are in common use in toady's cosmology, I think that it is prudent to undertake a logical discussion and arrive at a classification of principles. This short lecture may be understood as a preliminary introduction to the problem.

References

Barrow, J.D. and Tipler, F.J., 1986, *The Anthropic Cosmological Principle* (first edition), Clarendon Press.
Dyer, C.C., 1987, *Theory and Observational Limits in Cosmology*, Ed. W.R. Stoeger, Specola Vaticana.
Ellis, G.F.R., 1987, *Theory and Observational Limits in Cosmology*. Ed. W.R. Stoeger, Specola Vaticana.
Ellis, G.F.R., Maartens, R. and Nel, S.D., 1978, *Mont. Not. Roy. Astr. Soc.* 184:439
Jaakkola, T., 1989, *Apeiron* 4:9.
Rudnicki, K., 1990, *Astronomy Quarterly* 7:117.

The Meta Model: A New Deductive Cosmology from First Principles

T. Van Flandern

Meta Research
6327 Western Ave. NW
Washington, DC. 20015
EMail: METARES@WELL.SF.CA.US

A model of the universe can be derived, entirely deductively, from first principles. The advantage of such a process is that deductive reasoning is unique (induction is not); so the resulting model is severely constrained in its degrees of freedom. The disadvantage of this approach is that finding valid starting premises is extraordinarily difficult. A starting point which seems to work well is a universe consisting of nothing at all—no space, no time, no scale, no light, no gravity, no matter or energy, no implicit structure, no directions or orientations. From there we introduce units of substance one at a time, deriving their properties as we go. We soon must confront Zeno's paradoxes, which lead to new understandings of the meaning of space, time, and matter. The model also demands that the four standard dimensions be infinite, as must a fifth dimension of scale, which has profound implications. Substances can interact only through collisions, which create what we call "forces". One of these interactions at our scale gives rise to the phenomenon we call gravity, which behaves just as in General Relativity with three exceptions: no singularities arise, the field has finite range, and the flux entities travel faster than light. A new understanding of Special Relativity arises from these concepts. Finally, the model makes specific predictions about the nature of cosmological redshift, the cosmic microwave radiation, quasars, dark matter, and both the large-scale and quantum universes.

Why not the Big Bang?

The standard model for the origin of the universe in current cosmology is the Big Bang, in which the universe originated in an explosion of space and time between 10 and 15 billion years ago. The two cornerstones of the Big Bang are the observed galaxy redshift-brightness relationship, which seems to imply that the universe is expanding, and the microwave blackbody radiation, which appears to be a fireball remnant of the Big Bang explosion. But both of these interpretations are assumed, and are unconfirmed by observations.

Recent data has cast considerable doubt on both assumptions. If galaxy brightness and distance were related in the way that the Big Bang requires, then the angular diameters of radio galaxies should diminish rapidly with increasing redshift. That is not seen. Instead, there appears to be a minimum angular diameter cut-off for high-redshift objects. That is inconsistent with the redshift-implies-expansion assumption (Hewish et al. 1974).

Over twenty years ago, the American astronomer Tifft found that redshifts of galaxies were "quantized"; that is, they tended to occur more frequently in multiples of about 36 km s^{-1}, for example. Since this is impossible in a Big Bang universe where redshift indicates velocity, it has been assumed that Tifft was mistaken. But just recently, his result was confirmed with an independent sample of galaxies by independent investigators (Guthrie and Napier 1992).

The interpretation of the microwave blackbody radiation as coming from beyond the galaxies has also run into trouble. Intergalactic absorption has been shown to be sufficiently severe to prevent the direct visibility of microwave radiation at a redshift of over 10,000, as the Big Bang requires (Lerner 1990). Moreover, it has been shown that the temperature of that radiation, 3 degrees Kelvin, is about the same as the equilibrium temperature of most nearby interstellar and intergalactic material, suggesting that the radiation sources are nearby also.

Still another anomaly is that the bulk of the galaxies in our part of the universe appear to be streaming in one direction when their peculiar velocities are measured with respect to the microwave radiation (Lindley 1992). The alternative is that the radiation is asymmetric with respect to us. So we are forced to choose between a flowing stream of galaxies in our part of the universe with respect to a fixed microwave background, or an asymmetric source of microwave radiation and relatively fixed galaxies. Occam's Razor certainly favors the latter view, that contradicts the Big Bang. But even the former picture is difficult for the Big Bang to accommodate if the universe is homogeneous, as the Big Bang hypothesizes.

Starting Assumptions

To build a model one must have a starting point—one or more observations or assumptions from which one may deduce or generalize. One starting assumption for the Big Bang is called the Cosmological Principle. It is assumed that the universe is homogeneous (matter density is everywhere the same over large enough scales) and isotropic (the universe looks the same in all directions to all observers). But such assumptions are motivated by aesthetics. In effect, we are telling the universe how it must be, instead of asking how reality is.

A second problem is that we use inductive reasoning to take observations and guess their interpretation, so that we can guess the models which will explain them. Inductive reasoning backwards from observations to causes is non-unique and often well off the mark. Experience has shown that it is not a good method for arriving at successful models of reality.

Deductive reasoning, by contrast, does not suffer from this handicap. If our starting point is correct and our reasoning valid, we may trust our deductions quite absolutely. Moreover, every successful comparison between deductions and observations is a new validation of our starting point. Models developed deductively are severely constrained in their degrees of freedom, because introducing any *ad hoc* helper hypotheses along the way violates the methodology. The chances of such a model resembling reality are vanishingly remote unless the starting point is valid. The disadvantage of this approach is that finding a valid starting point is extraordinarily difficult.

The Meta Model is the first cosmology which is completely a product of deductive reasoning. After the model was developed, the name was chosen from the dictionary meaning of the prefix "meta": "...later or more highly organized or specialized form of; more comprehensive; transcending; used with the name of a discipline to designate a new but related discipline designed to deal critically with the original one."

A starting point which seems to work well is a universe consisting of nothing at all—no space, no time, no scale, no light, no gravity, no matter or energy, no implicit structure, no directions or orientations.

Introduce one unit of substance into this universe. ("Substance" means whatever exists and influences its environment. It might be more general than matter and energy, the two known forms of substance.) Note that it has no determinable dimensions, and there are still no properties to the universe. This unit cannot move, because any possible state of motion would be indistinguishable from non-motion.

Introduce a second unit of substance. This gives meaning to direction and scale, which can be measured relative to the two substances. But all possible orientations of the two substances are equivalent. And if the substances are dimensionless, then all possible separations are equivalent also, except zero separation.

We can distinguish zero separation of two substances from non-zero separation. The first coincidence of two substances is an observable change in the state of the universe, and marks the beginning of a time interval. The second coincidence completes the time interval. The length of such an interval is indeterminate.

With a third unit of substance, time intervals, distance intervals, and scale intervals can be measured relatively. For example, a time interval involving A and B is relatively longer than other time intervals involving B and C if there are more B-C coincidences than A-B coincidences in it. This gives meaning to the concept of "time", which is necessarily a relative construct.

The Nature of Space, Time, and Matter

In one of Zeno's paradoxes, it is argued that one cannot traverse an interval because an infinite number of half-the-remaining-distance steps is required. But if there were a smallest possible unit of space, then straight line motion through such units in arbitrary directions would be impossible, since at some level lines would have to pass between units of space.

In another paradox, one arrow in flight cannot catch another arrow because an infinite number of half-the-remaining-time steps is required. But if there were a smallest possible unit of time, then change between time units would be discontinuous, requiring constant re-creation of the universe.

In a Zeno-like paradox for mass or scale, it may be argued that nothing can ever touch anything else because each thing is composed of smaller entities which are mostly empty space and not touching. But if there were a smallest possible unit of substance, it could not be further composed; so it could not deform on collision or rebound or otherwise react to other substance; nor could its interior contribute in any way to the external universe. Such things are properties of substances which are further composed.

Such paradoxes must be resolvable. There exists a 1-to-1 correspondence between points in a space or time interval and the rational numbers between zero and one. The same may be said of the logarithm of a scale interval. This implies that infinite divisibility of finite amounts of space, time, and substance is possible. It is therefore required, since finite divisibility still leads to the logical paradoxes already mentioned.

By logical extension, our universe must have five, and only five, dimensions: three in space, plus time and scale. It must be infinite in all five dimensions, because no possible boundaries in space or time or scale could be stable. This Meta Model differs from most other theoretical universes by being infinite in scale; that is, the smallest imaginable particle in it is an entire universe on a still smaller scale; and the largest imaginable construct is but an elementary particle on a still larger scale.

As in the case of the Cosmological Principle, we would expect that all parts of the universe are ultimately made up of assemblages of similar smaller substances, so the universe looks similar in all places and at all times and from the vantage point of all scales. But since it is infinite in all five dimensions, no one vantage point may be expected to be "average." That is, local differences may always be expected to outweigh distant similarities. We might conclude, therefore, that all the universe we see is but structure in the "ocean" of a huge assemblage on a vaster scale. But not far away on that great scale must lie a "shore" and an "atmosphere" and "space" beyond, implying that the character of the universe surrounding us must change greatly in its details if we look far enough away.

The Nature of Force

The only means by which the substances in this model universe can interact is by collision. Collisions produce a reaction, always of the "pushing" type, which we might call a "force". Different forces are simply interactions between substances on different scales. There must therefore be an infinite number of fundamental forces, of which only a few will manifest themselves as viewed from any one scale.

Consider two large substances, A and B, immersed in a sea of much smaller, rapidly-moving substances (see Figure 1). The constant bombardment on the surfaces of A and B from the smaller substances produces a downward-pushing force, making it appear that A and B pull things toward themselves. Moreover, both A and B shadow each other from some impacts by smaller substances, resulting in a net force toward each other, as if they "attracted" one another. On an astronomical scale, we call the smaller substances "C-gravitons" ("C" for "classical," since these are not the same as the hypothetical spin-2 gravitons of quantum physics). And we call the resulting apparent force between A and B "gravity."

Let us further conjecture that the CGs are so small that the large substance (say, a mass) is nearly transparent to them. Then only a small fraction of the CGs contribute to the force, the rest passing through without effect. Under such circumstances the force would be proportional to the total matter content in the mass, since every bit of its three dimensional interior is equally likely to reflect CGs.

Indeed such a construction has all the properties of Newton's Universal Law of Gravitation: Any two masses immersed in a high-velocity sea of CGs will appear to attract one another with a force directly proportional to the product of their masses and inversely proportional to the square of the distance between them.

On other scales, other forces of nature arise with properties that depend on the relative size, mean density, and velocity of the medium in which particles are

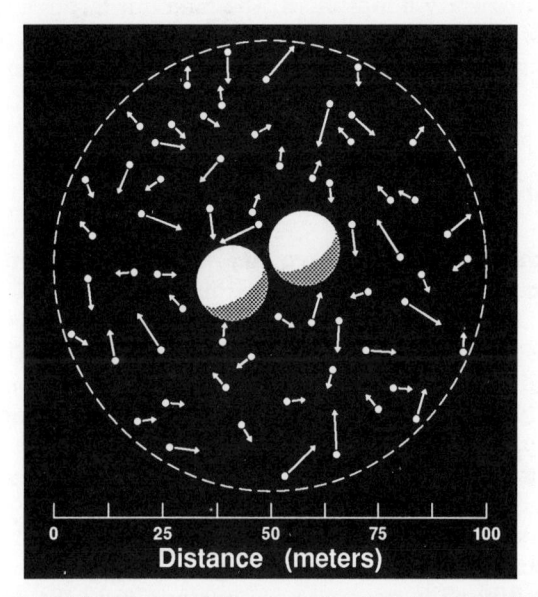

Figure 1. If two large particles are much closer together than the mean distance between collisions of the smaller particles, they will feel a net push toward one another.

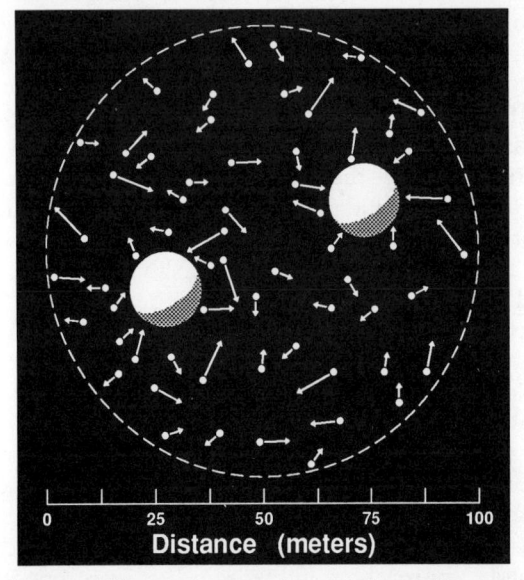

Figure 2. If the two large particles are far enough apart, the smaller particles backscatter into the space between them, and the large particles feel no net force.

immersed. External media give rise to attractive forces, and entities that are emitted from inside particles give rise to repulsive forces.

Waves may propagate through any of these media at any scale. Each medium will have some characteristic velocity; for example, that of lightwaves. The medium that carries lightwaves must be as different from the medium of CGs as oceans are from atmospheres. The medium of CGs consists of rapidly-moving separate entities, much like a planetary atmosphere. The medium through which light propagates must consist of contiguous, relatively stationary entities, much like water in the ocean, since lightwaves are transverse. We call this light-carrying medium the LCM.

Special Relativity

Anything sensitive to this LCM would also propagate through it, as light does. Consider the effect this would have on electrons orbiting atoms. They must propagate through the LCM at less than the speed of light. As their speed of propagation approached lightspeed, the orbits of electrons would necessarily all become compressed in their direction of motion, just as sound waves must bunch up together as the speed of sound is approached in an atmosphere. If the dimensions of electron orbits, or rulers made of assemblages of such electrons, served as the unit of length, it would appear to external observers that those moving rulers had contracted in the direction of motion.

In like manner, we know it takes longer to make a round trip in a moving stream than in a stationary one. So electrons will take longer to complete their orbits when moving rapidly with respect to the medium they are propagating in than when stationary in it. If the circuit time for the electron or clocks based on electrons serves as the unit of time, it will appear to external observers that those clocks have slowed. The amounts of these length contractions and time dilations are given by the well-known Lorentz formula.

If the moving observer could communicate via faster-than-light signals, he would have no difficulty detecting his own motion and judging that his rulers and clocks were the affected ones. But if his communications are limited to signaling through the LCM at the speed of light, the moving observer will be unable to detect his own motion. This is because the speed of light will appear to him to be the same in all directions. Moreover, the length contractions and time dilations will make it appear to him that the rulers of other relatively moving observers are contracted, and that relatively moving clocks are dilated. This illusion is invincible as long as an observer is limited to communications at the speed of light.

These properties, the relativity of motion for all observers and the constancy of the speed of light for all observers, are the fundamental postulates of Einstein's theory of Special Relativity. These postulates are obeyed in the Meta Model. But we can see that, contrary to the conventional interpretation, those postulates do not necessarily imply that faster-than-light communication is impossible. They only imply that the universe will appear that way to observers who do not have faster-than-light signals at their disposal.

This point is important. In the standard interpretation of Special Relativity, time slows for a body approaching lightspeed. If it were possible for an entity to exceed lightspeed, time would appear to move backwards for that entity. Hence, it has been concluded that crossing the light barrier is impossible. But in the Meta Model interpretation of the same postulates, electrons approaching the speed of light would take longer and longer to complete their orbits. At faster-than-light speeds, those electrons would reverse the direction of their orbital motion. Clocks which measured time using those electrons might indeed "run backwards." But the meta-time of the universe neither slows nor reverses. Observers with faster-than-light communications would be able to measure this meta-time. And in principle, they could shield themselves from the LCM, thereby theoretically permitting matter and observers to travel faster than light in forward time.

General Relativity

CGs do not propagate through the LCM, and so do not experience Special Relativistic effects as ordinary matter does. They are far too small to be affected by the LCM in any bulk way, although individual CGs can occasionally be deflected by the LCM. Near large masses, the net downward pressure toward the mass produced by the CGs in its vicinity also collects more LCM near the mass. The increased density of the LCM near mass then affects light and clocks in precisely the manner described by the theory of General Relativity (GR). But GR ascribes these same effects to a "curvature of space-time." In the Meta Model, this curvature is actually ordinary refraction in the LCM. The extra density of LCM near mass refracts photons, which slows their travel (as in the radar-delay test of GR), causes extra bending near the Sun's limb, and slows clocks in a gravitational potential. In like manner, the increased density of LCM near the perihelion of Mercury's orbit, as compared with its aphelion, produces an effectively stronger radial force at perihelion that progressively advances the perihelion. Hence, all classical tests of GR are obeyed exactly, but by virtue of refractive curvature in a medium of varying density instead of curvature of space-time.

It is well-known but not often mentioned that the direction of the Sun's gravitational field and of its arriving photons are not parallel for an Earth observer because of "aberration," an effect arising from the finite speed of light. Gravitational forces, by contrast with radiation-pressure forces, always appear to operate without detectable delay. This phenomenon is explained in GR by postulating that space-time curvature at a distance instantly conforms to the new velocity and acceleration state of a body. In the Meta Model, this apparent instantaneous action of gravity is explained by CGs moving at mean speeds of at least 10^{10} times the speed of light (as experimentally required). As we have seen, such superluminal speeds are allowed in the Meta Model, and no causality violations occur.

Properties of the Universe

As lightwaves propagate over great distances through the LCM, they gradually lose energy because of resistance from the medium of CGs, just as ocean waves gradually lose energy from friction with the atmosphere. This slow, gradual energy loss would be observed as a redshift that is proportional to the distance traveled by the lightwave through the CGs. Unsuspecting observers, finding that all arriving lightwaves have less energy than when they were emitted, might interpret this "redshift" as due to an expansion of the universe around them.

Although ordinary matter is largely transparent to CGs, having only occasional interactions with them, collapsing matter would eventually become so dense that CGs can no longer penetrate at all. Such a superdense body would exert the strongest "gravitational field" it is capable of, since it would reflect all CGs and allow none to pass through. Hence the gravity field would get no stronger if more matter were added to the superdense body's interior in the same volume. Infinite gravitational fields and collapse to a singularity are not possible. Hence there are no black holes in the Meta Model. When matter exceeds a certain density, additional gravitational force becomes shielded. Such an effect may be measurable when three bodies line up and the middle one is dense enough to produce significant gravitational shielding. For example, a very slight effect might be seen in an Earth satellite such as Lageos during eclipse seasons, when the Sun's gravity would be mildly shielded by the Earth's dense core.

Instead of forming black holes, large collapsed masses would simply produce ultra-strong gravitational fields, and the light escaping from them would be highly redshifted by this intense gravitation. Such "quasi-stellar" objects would have properties which resemble quasars—principally a high redshift from an object with stellar dimensions. Regular high-speed ejection of such objects from galaxies in supernova explosions is to be expected. But since their gravitational redshift would be much stronger than their doppler shift, blue-shifted quasars would be unlikely to occur.

Gravitational shielding effects would change stellar models for high-mass stars significantly. Because the interior matter content and densities would be greater than the external gravity field indicated, neutrino fluxes (for example) would be smaller than predicted. Stable stars could exist over much longer times with masses well in excess of 100 solar masses. In fact, there would be a red and a blue extension of the stellar H-R diagram into the high-luminosity regions. The blue branch would produce many supergiant stars with ultraviolet excesses. Nearby galaxies containing these would look normal because ultraviolet light is blocked by the Earth's atmosphere. But distant galaxies with many extended-blue-supergiant-branch stars would have the ultraviolet excess redshifted into the blue part of the spectrum, and we would observe an excess of blue galaxies at those redshifts.

No matter how transparent matter may be to CGs, there will be a characteristic distance that CGs will travel before collisions with one another. At much greater distances than this "mean free path" for CGs, the CG medium will begin to act like a "perfect gas." Over such large scales CGs can still produce pressure, but not an apparent inverse square force, just as is true for air molecules. The force of gravity would apparently change its character from inverse square to inverse linear (the actual form of the force law would be more complex—see Figure 2). If this characteristic distance were about 2000 parsecs, this would then explain the observed "non-inverse-square" behavior of rotation velocities of matter in galaxies, and the behavior of clusters of galaxies, without need for inventing hypothetical "dark matter." The actual behavior of these systems on scales over 2 kpc is more nearly inverse linear than inverse square with distance, just as this feature predicts.

Pencil-beam surveys show at least a dozen equi-spaced "walls" of galaxies stretching across the visible universe. Such large-scale structure is difficult for the

Big Bang to explain. In the Meta Model, we would be seeing waves in an enormous medium in which galaxies are the constituent entities.

The Meta Model is not vulnerable to "heat death" from the increase in entropy, as the Big Bang is. Forces operating at some scales, such as electromagnetism, always increase entropy. Forces operating at other scales, such as gravity, always decrease entropy. The net over an infinite number of fundamental forces operating over an infinite range of scales is that entropy is perfectly conserved.

Properties of the Quantum World

The Meta Model also has many implications for quantum physics, especially in light of its requirement that the universe be infinitely divisible in scale. It implies new interpretations for particle-wave duality, Heisenberg uncertainty, observer-created reality, and the Bell Interconnectedness Theorem. Since this is an extensive topic in its own right, we will develop it at length elsewhere (Van Flandern 1993).

Summary

From first principles, this model predicts a universe in which wave phenomena will suffer a redshift proportional to distance, creating the illusion of expansion in a universe that is actually infinite. It predicts the existence of superdense stellar-class bodies with high gravitational redshifts located relatively nearby, whose properties resemble the objects we call quasars. It insists that gravity and all forces must eventually lose their inverse square character and have a finite range. In the case of gravity, if that failure sets in at a characteristic distance of about 2000 parsecs, the behavior of galaxy rotations and clusters of galaxies is explained without the need to invoke missing "dark matter". Indeed, the model is completely devoid of such *ad hoc* helper hypotheses, and has very few adjustable parameters (*e.g.*, the characteristic distance between CG collisions). Finally the model illustrates one way in which entities may propagate faster than light in forward time without the slightest contradiction of the vast body of experimental evidence supporting special relativity.

A model universe built up from first principles, with each step deductively required by the previous steps and without helper hypotheses to make it "fit" reality, turns out to resemble reality in remarkable ways; yet it has somewhat different interpretations of many common phenomena. Since the model seems not to be obviously contradicted by observations, and since it makes many predictions and is therefore a "falsifiable" model, it seems to have met the criteria of a viable model that may be considered and tested further by the scientific community.

The model's chief benefit is its intuitive quality, providing many new insights into the nature of things. Although this brief exposition may not contain enough detail to see how the model's features follow from its first principles in every case without *ad hoc* manipulation (see Van Flandern 1993 for more particulars), the fact that it does arrive at descriptions of all the main properties of the observable universe from those principles without the author needing to think up explanations and add them to the model is its greatest strength.

References

Guthrie, B. and Napier, W.M., 1992, *Sky and Telescope* 84:129.
Hewish, A., Readhead, A.C.S. and Duffet-Smith, P.J., 1974, *Nature* 252:657.
Lerner, E.J., 1990, *Astrophys. J.* 361:63.
Lindley, D., 1992, *Nature* 356:657.
Van Flandern, T., 1993, *Dark Matter, Missing Planets and New Comets*, North Atlantic Books.

Dark Matter, Spiral Arms and Giant Comets

S.V.M. Clube

Department of Physics
University of Oxford, UK

On the scales of cosmological, galactic and planetary systems, we come across prominent categories of matter whose relevance, to one another, is not usually perceived as being particularly close: dark matter, spiral arms and giant comets. Our understanding of these categories is discussed here in the context of two assumptions regarding the nature of the physical substratum. *i.e.*, whether it is expanding or stationary. In particular, it is emphasized that these assumptions are, themselves, dependent on the adopted process by which electromagnetic radiation is transmitted, leading to a certain arbitrariness at present in our understanding of the origin of the cosmological redshift. Accordingly, the halo-disc discrepancies and quantization effects amongst cosmological redshifts should now be taken into consideration. If these new physical effects prove to be real, a logical basis for returning to Lorentzian theory would now appear to exist.

1. Introduction

By the beginning of the 20th century, straightforward laboratory experiments had failed to reveal the Earth's absolute motion through space. This finding was a surprise, and placed unexpected constraints on the physical nature of the substratum *and* the mode of electromagnetic communication between material bodies in space and time. The resolution of the problem became an issue for both physics *and* astrophysics.

1.1 Cosmological and Galactic Considerations

Conventionally, the massless vacuum (Einstein 1905b) was associated with new proposals regarding the transmission of electromagnetic radiation through space (Einstein 1905a) and gravitational action (Einstein 1916), as a result of which the redshifts of cosmological photons were interpreted as a universal effect of the expanding substratum (Eddington 1931). This interpretation was not achieved, however, without a fundamental assumption regarding the nature of the commonest building blocks of the universe, namely, spiral galaxies and their environments. Thus, the contemporary debate excluded any possible role for active galactic nuclei (not then observed) in the production of spiral arms, since it also eliminated any possibility that the associated extreme physics might be so extreme as to cause perturbations of the "Hubble flow" within distances ~ 100 Mpc (also not then observed), i.e., on timescales not significantly greater than the circulation time, based on the observed few turns, of typical spirals $<10^9$ years.

Gravity theory was thus constrained so that galaxies were understood in each case as having a time-invariant mass and an axisymmetric circulation of gas and dust released by normal evolved stars, this circulation also being a *continuous* source of new stellar condensations and planetesimal accretions in an appropriately "grand design" spiral: the crucial point here being that the spiral pattern results from the passage through the disc of a density wave whose lifetime is substantially greater than the circulation time of typical spiral arms $>10^9$ years. The theory was to remain unchanged in spite of the discovery, subsequently, of (i) extreme activity in a proportion of galactic nuclei, frequently associated with the production of primitive arms or jets, and (ii) significant departures from the underlying cosmological flow at distances within ~ 100 Mpc.

Any theory involving this degree of arbitrariness clearly cannot expect to enjoy universal support amongst physicists and astrophysicists. Alternative proposals regarding the nature of the substratum have never, therefore, been excluded. Thus, the material aether (Lorentz 1904) is associated with alternative proposals regarding the transmission of electromagnetic radiation (waves) and gravitational action (Dicke 1961; see also Atkinson 1962), as a result of which the cosmological redshift may be understood as an effect of the stationary substratum (this paper). The gravitational theory, in this instance, is evidently constrained so that the extreme physical conditions in central galactic concentrations are responsible for the (repeated) plasma outflows (*e.g.* Jeans 1928) which, due to subsequent cooling and backpressure by the disc, are also the natural source of stellar through to planetesimal condensations in spiral arms. It follows that we can expect these extreme conditions to be associated with active phases in the evolution of galactic nuclei *and* departures from uniformity in the underlying cosmological flow, which is usually regarded as revealing the presence of substantial dark matter.

Essentially, therefore, a situation has developed in late twentieth century astrophysics where an unknown characteristic of the substratum (*i.e.*, whether it is expanding or stationary) is associated with alternative perceptions regarding (i) the source of dark matter in galactic environments, and (ii) the physical nature of planetesimals. So far as dark matter is concerned, the choice lies between unidentified primordial material *continuously* present in galactic halos, and *intermittent* central galactic concentrations displaying symptoms of extreme physics. So far as planetesimals are concerned, the choice lies between (post-stellar) accretions and (pre-stellar) condensations. In their case, the solar system provides an accessible environment for detailed study, and it may well be that the true nature of the substratum has a good prospect of being revealed, if not settled, through this line of enquiry. We consider this point first.

1.2 Stellar and Planetary Considerations

In accordance with conventional theory, supposing that there is a big bang to explain universal expansion (Sandage 1961), and a process of star formation associated with halo collapse and spiral density waves in the disc to explain galactic evolution (Eggen, Lynden-Bell and Sandage 1962), the present short-period comet population is thought of as being predominantly supplied by the *residual* solar nebula, the latter necessarily containing planetesimal accretions of dust and gas drawn from the galactic disc (Kuiper 1951, Whipple 1963, Cameron 1962). Planetesimal accretions in such nebular accretions are believed, in fact, to be the building blocks for larger objects, *i.e.*, meteorite parent bodies and proto-planets, whose differentiated state is also materially assisted by radioactive heating. The later fragmentation and reconstitution of some of these objects creates something of a theoretical distinction between the comets and asteroids . This distinction may be artificial, however, for we cannot *a priori* exclude the possibility that the range of bodies covered by planetesimals through to proto-planets embraces rapid condensations on different scales in a hot, rapidly cooling proto-stellar medium which is suitably compressed, in which case we may start with differentiated planetesimals (comets with asteroidal cores), and there need not be a nebula prior to the formation of planets and stars (*cf.* McCrea 1978).

The solar nebula was considered to be an essential component of conventional theory when the latter was set up during the 1960's. The nebula, however, is an aspect of theory which is more readily tested than most by reason of its accessibility, and it is by no means fully endorsed by cometary scientists. Thus, the short-period comet population is also widely thought of as being dominated by possible planetesimal condensations associated with the isotropic cometary cloud (Oort 1950; see Section 2.1), since it is recognised that prograde, low inclination orbits are especially favoured amongst those captured when Jupiter deflections are involved (Everhart 1972). The observed source and nature of short-period comets, in other words, have not been settled, and continue to be critical to the cosmological debate.

The question, then, is whether the only product of the star formation process we are able to study in detail, namely the Sun and its planetary system, is obviously formed from a solar nebula containing accreted planetesimals, the accepted natural adjunct of the cosmological and galactic theory currently preferred; or whether it is more plausibly associated with an isotropic cloud containing condensed planetesimals, the perceived natural adjunct of the cosmological and galactic theory regarded here as being the likely alternative. From the most recent studies of the evolution and dynamics of giant comets such as Chiron and the Taurid progenitor (Bailey *et al.* 1993, Steel *et al.* 1993), it would seem that the isotropic cloud known as the Oort cloud is, in fact, the more probable source of short-period comets.

However, it is not merely the apparent absence of the residual solar nebula that forces this change of view. There is, in addition, a new category of evidence arising from the catastrophism studies of the recent decade which also supports the Oort cloud as the source. This is the so-called eschatological record, the fact that repeated enhancements of the global fireball flux (Hasegawa 1992) which are associated with revivals of the astrological doctrine of "last times" during the last 2,000 years, are also associated with the disintegrations of the latest giant comet in sub-Jovian space, which have produced the so-called Taurid stream (Clube 1993). The point here is that the disintegration of comets and asteroids by splitting, marked also by the formation of dust-trails, is probably the principal process by which captured Oort cloud bodies deposit new material in sub-Jovian space, while the cometary *and* asteroidal constituents of the Taurid stream undergoing a hierarchy of disintegrations, are apparently the dominant source (over millennia) at the present time. It seems, therefore, from the piecemeal structure of the original giant comet, which allows this kind of hierarchical breakdown, that we are dealing with a Chiron-like, differentiated planetesimal which was itself built up from lesser condensations already formed.

If so, the evidence to hand is favouring substantial bodies which are produced through a hierarchy of rapid condensations and aggregations, such as are associated with a formation regime undergoing rapid compression and terminating with Jeans collapse. Indeed, the conventional slow accretion of planetesimals no longer appears particularly likely, and it follows from considerations such as these that we now have a scientific basis for preferring plasma outflows from galactic nuclei and, hence, cosmological redshifts in a stationary substratum. Such redshifts may then be attributed to the so-called quadratic Doppler effect (this paper), as a result of which there are (i) halo-disc discrepancies in extragalactic sources arising from stellar mass function differences due to mean age (Arp, these proceedings), and (ii) "quantisation" effects in accordance with the statistical distribution of cosmic material in spiral galaxies throughout the universe (Napier and Guthrie, these proceedings).

The purpose of the present paper is to provide an overview of some of these developments. The factors influencing the growing eschatological debate are considered first (Section 2). Matters relating to galactic theory are then briefly described prior to a recapitulation of Lorentzian theory (Section 3), with particular emphasis on the quadratic Doppler effect and its consequences (Section 4).

2. Short-period Comets and the Eschatological Debate

2.1 The Current View

The favoured world-view of the 1990's is essentially that of the 1960's (Sandage 1961; Eggen, Lynden-Bell and Sandage 1962). The theory has been considerably embellished, of course, as a result of detailed studies in those areas of science which have subsequently attracted most technological support. Much of this effort, however, has had the undesirable effect of polarising the enquiry towards areas of knowledge which seem particularly critical so far as the favoured world-view is concerned, with the paradoxical result of hindering any movement towards an early resolution of the underlying substratum debate. Moreover, the 1960's worldview, as we have seen, presupposes (i) the existence of a residual solar nebula which, rather than the Oort cloud, is the prime source of all short-period comets; and (ii) the existence of an accretion process during the formation of the solar nebula which is responsible for the presumed homogeneity, non-asteroidal constitution and extremely rapid comminution of short-period comets in the vicinity of the Sun.

The latter is not a precisely quantified aspect of the theory, though it is interpreted in such a way as to provide a precise description of how most of the sub-cometary and sub-asteroidal material is continuously breaking down in the inner solar system environment. As part of this description, we note the flux to Earth of cometary fireballs (diameter < 10 m; Ceplecha 1992) is currently observed to be ~ 10–10^2 times their long-term flux inferred from the impact cratering record on the Moon (Shoemaker 1983), consistent with ~ 10^4–10^5 year flux enhancements at ~ 10^6 year intervals. Enhancements of this kind are usually though of as taking place as a consequence of the hierarchical splitting of intermittent large comets (Kresak 1981), such as may be invoked to explain the recent production of the present zodiacal cloud (Whipple 1967).

Nevertheless, it is *not* anticipated that the fireball excess extends in the same proportion to intermediate objects (diameter ~ 10–1000 m), such as may be associated with the incompletely disintegrated products of a large comet in the inner solar system *at the present time*. The accepted picture arising, therefore, from the intermittent fireball flux is that of random missiles in the Tunguska class (diameter ~ 100 m) which are currently being deflected from the asteroid belt at the appropriately steady long-term rate (*e.g.* Brandt and Chapman 1982), thereby positively discounting coherent streams of cometary-*cum*-asteroidal debris of ~ 10–1000 m dimensions, such as may be produced during the disintegration of a differentiated giant comet.

However, it is hardly realistic now to discount giant comets and disintegrating streams. In fact, the assumptions (i) and (ii) above, for short-period comets, are now more of the nature of a bold hypothesis for which recent studies provide little support. Thus, dynamical considerations based upon the observed splitting of Oort cloud members deflected towards inner solar system space have long indicated the progressively increased resilience and reduced volatility of the dominant material species from the originating isotropic cloud, more readily associated with condensations than with accretions (Oort 1950). The implied differentiation and break-up of the larger planetesimals would then lead us to expect a population of short-period comets which is more numerous than it would otherwise be, as observed, and which probably includes large cometary bodies of asteroidal appearance, also as observed. The discovery of Chiron during the 1970's and other similar bodies since, along with the absence of any strong evidence for the residual solar nebula, have now seriously weakened the idea of planetesimal accretions.

Indeed, the 1960's world-view is increasingly difficult to maintain in the face of the most recent discoveries in sub-Jovian space. The critical development has been the substantial evidence for the large cometary-*cum*-asteroidal stream, the Taurids, resulting from the breakup of the latest Chiron-like body to be deflected therein (Steel *et al.* 1991, Asher *et al.* 1993). It appears that the stream includes a dense, coherently moving swarm of recent debris from the progenitor, indicating after all that the current excess in the cometary fireball flux may well extend to Tunguska class objects as well, these latter being mostly in mean motion resonance with Jupiter close to the orbit of Comet Encke (Asher and Clube 1993). These objects cannot be readily seen at present, but their interactions with the Earth in the past and in the future are now expected, introducing a degree of immediacy and realism to cosmological and cosmogonic speculation which would otherwise be absent. Altogether, it seems, therefore, that there may be considerable elements of wishful thinking in the principal theoretical structures underpinning the popular view of comets, leading one to suspect the cosmological and evolutionary considerations relating to the present world-view are by no means secure.

2.2 The Displaced World-View

The recovery of the Oort cloud as the primary source of short-period comets coupled with the observed frequency of Chiron-like planetesimals implies the repopulation of sub-Jovian space by the disintegrating debris of these planetesimals every 10^5–10^6 years or so. Indeed we expect separate orbital classes of debris corresponding to particular ranges of values for certain orbital parameters (semi-major axis, orbital inclination) of the source, namely $(a,i) \geq$ and $\leq$ (2.5 AU, 10°), corresponding to so-called "Halley-types" (*e.g.* the so-called Kreutz group) and "Encke-types" (*e.g.* the Taurids) respectively. Since the debris in highly elongated orbits is the more widely dispersed, it follows that we expect the Encke-types to dominate the supply of inner solar system dust (the zodiacal cloud) and the incidence of physical encounters with the Earth (through the fireball flux). It also follows that the most recent glacial and interglacial periods (*i.e.*, the late Pleistocene and the Holocene), characteristic of glacial-interglacials generally, may be plausibly associated with the expected periods of erosion and fragmentation of the latest (progressively structured) giant comet to reach sub-Jovian space (*i.e.* ~ 20,000 BP) and undergo hierarchical splitting. The importance of these developments in cometary science is the fact that Chiron-like planetesimals are now probably dominant and commonplace throughout the history and evolution of the Earth (*i.e.*, the dinosaurs were victims of an earlier arrival and disintegration in sub-Jovian space, some 65 million years ago); furthermore, they are probably numerous in the spiral arms of the Galaxy (being the natural building blocks of planets and stars).

The significance of the evidence for an overabundance at present of Tunguska class objects in sub-Jovian space (Steel *et al.* 1993) resides in its potentially serious implications for the eschatological and fireball flux records. The latter record was in fact available to western civilization early in the nineteenth century (Biot 1848) but was essentially set aside during the rise of uniformitarianism. The 1860 British Association Meeting at Oxford marks the occasion of the Huxley-Wilberforce debate, which was subsequently represented as being the point in time when the uniformitarian view of geological and biological evolution was established (Gould 1991). Nowadays, of course, we are accustomed to the very long timescale that is required for the major geological and biological processes on Earth (which was Huxley's principal concern), and we tend to overlook the much shorter timescale of the cosmic interventions which were considered to be the possible cause of catastrophic processes during the historical period (which were Wilberforce's principal concern). Wilberforce, in fact, had very little to say regarding the Earth's great age or the immense scale of cosmic phenomena which had already begun in his time to extend far beyond the planetary system. Rather, the issue at stake, so far as he was concerned, was the historical evidence for the recognised cosmic interventions in terrestrial (and state) affairs, which he and his non-secular contemporaries thought of as having divine implications as well.

In this respect, of course, he was not so very different from earlier luminaries, such as Ussher and Whiston, or indeed historians today who continue to ponder the evidence for our repeatedly eschatological past. Thus the problem which historians have always sought to resolve is why our ancestors were afflicted from time to time with the deeply held eschatological belief that the "last times" were imminent. Usually nowadays, especially as a result of the Huxley-Wilberforce debate and its nineteenth century aftermath, this belief is regarded as having a purely (crowd-) psychological basis. However, the correlation in time (Figure 1) between the known effects of eschatological impulses (Ball 1975, Hill 1980) and the enhanced fireball flux runs counter to the assumption, and it is now likely that the belief was promoted by truly observed celestial phenomena which, in the light of historical knowledge, were liable to be interpreted as phenomena which had sometimes proved

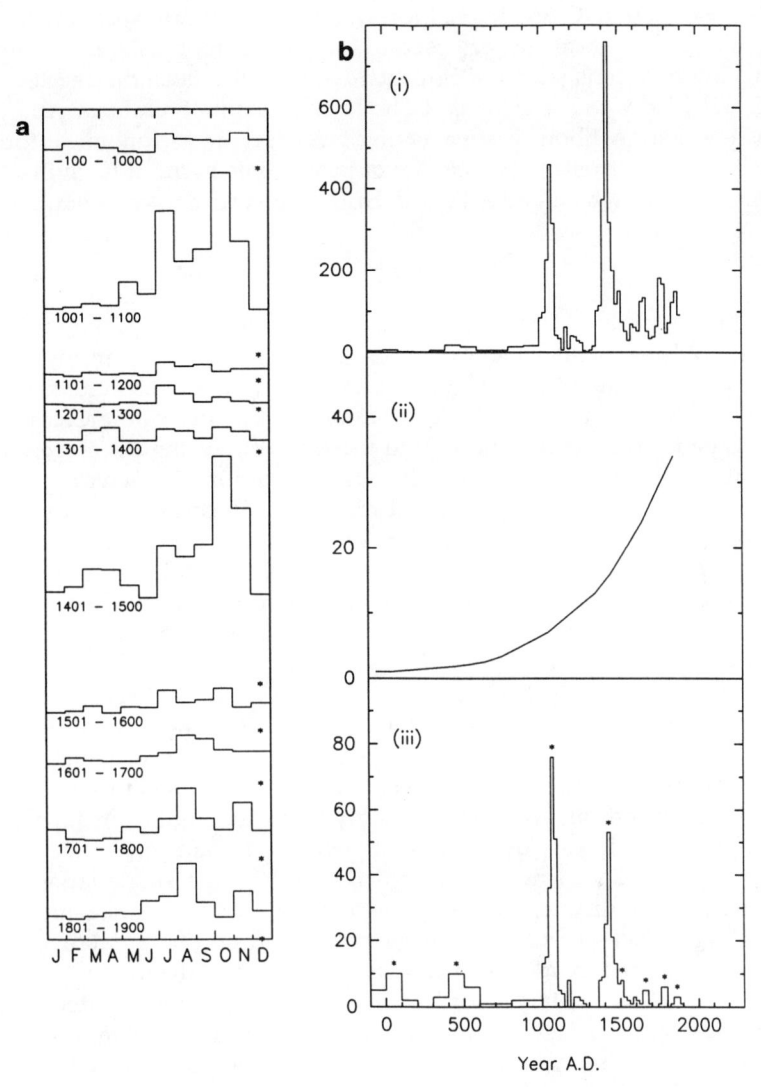

Figure 1. Combined monthly (a) and annual (b) variations of the large meteor (fireball) flux during the last two millennia, based on Chinese records (cf. Clube 1993). Note the persistent July-August and October-November peaks in the millennial and centennial fluxes consistent with the broad Taurid stream association. Note also the major enhancements of the annual flux, due apparently to partial or total fragmentations of individual Taurid members, *e.g.* the progenitor now much reduced, each of which coincides also with periods of supposed "divine providence" and intensified eschatological concern. These are associated, in fact, with significant revolutionary epochs, *e.g.*, the early Christian period, the European dark age, the split with Islam, the great schism, the reformation, the English revolution, the French revolution (contemporary with the American war of independence) and the years of European revolutions. By taking note of the recorded background flux, so far as this is possible, to estimate (b{ ii }), the recorded total flux (b{ i }) may be appropriately rectified and normalized in order to obtain a realistic estimate of the relative fragmentation fluxes (b{ iii }). It is not expected that these are particularly well calibrated, one with respect to the other, especially before ca. 1000 AD, since it is likely then that only the very largest fireballs were recorded.

harmful to civilization (Clube 1993). Indeed, the uniformitarian belief that evolution on the geological timescale (as opposed to the biblical timescale) is also unaffected by cosmic interventions (asteroid impacts) has likewise been discredited. In other words, it is by now rather clear that Huxley's precipitate uniformitarian view in 1860 was endorsed without it ever having had a secure astronomical foundation. The possibility thus exists that accounts of past events harmful to civilization may well relate to crises which had a Taurid Tunguska swarm as their source.

2.3 The Possible Outcome of the Current Debate

The historical record, then, is characterized by eschatological impulses. These correspond to those relatively short periods (say about 50 years) during the last two millennia when the attention given to apocalyptic studies was at an exceptionally high level. Such periods occurred, for example, at the turn of the Christian era, at the end of the fourth century through to the seventh, during the eleventh century and then again, several times, during the past millennium. In each case, there was a significant association with political and religious upheavals. Thus, by the time of the 30 years' war in Europe (1618–1648) and its civil war counterpart in Britain (1642–1649), the issue at stake was the precise (astrological) nature of the cosmic interventions in terrestrial (and state) affairs, the original question at the catholic-puritan divide being whether or not they indicate a forthcoming apocalyptic judgment on human behaviour.

From this particular period, however, we also learn that the apocalyptic revivals were associated with a perceived recurrence time (orbital revolution) of ~ 3.5 years (*e.g.* Mede 1672, Brightman 1644, Alsted 1627) during the decadal-to-centennial enhancements of the fireball flux at centennial-to-millennial intervals. These enhancements are ostensibly due to the underlying helion-antihelion stream of meteoroids and asteroids in sub-Jovian space known as the Taurids (Figure 1) which, as we have seen, is believed now to have originated from a single large progenitor through hierarchical disintegrations during the last ~ 2×10^4 years. Since the stream includes Comet Encke (period ~ 3.30 years), in addition to a putative cometary-*cum*-asteroidal source in 7:2 mean motion resonance with Jupiter (period ~ 3.39 years), the basically cometary nature of the progenitor seems assured. Also, since the stochastically recurring eschatological impulses are in accordance with a continuing hierarchical disintegration of the progenitor, the fundamental question has now been raised whether a source cluster of low-velocity debris (*i.e.*, a putative cometary-*cum*-asteroidal source containing objects as large as 0.1–1 km in size) is penetrated by the Earth in the wake of significant cometary splits (Asher and Clube 1993). Such events clearly provide a valid physical basis for particular cosmic interventions with the character of multiple-Tunguska bombardments which seem also to be implied by the well known (Christian) apocalyptic.

The outcome of the eschatological debate is unlikely now to be long delayed. Following the decision by US Congress in 1990 to assess and consider ways of ameliorating the celestial hazard to civilization (Morrison *et al.*, 1992; Canavan *et al.* 1992), an issue now exercising the attention of at least three government agencies, the potential role of the previously unsuspected source cluster and its mode of origin are now under increased scientific scrutiny (Clube 1993, Bailey *et al.* 1993, Steel *et al.* 1993). The emergence of a viewpoint which admits the past exclusion of eschatological impulses from the historical record whilst recognizing that Oort cloud planetesimals of the same type as Chiron dominate the cometary mass function, marks a significant turning point in our perception of the astronomical environment. Thus we envisage large comets which feed into sub-Jovian space at ~ 10^5–10^6 year intervals where they break up under the influence of the sun over timespans of several 10^4 years, providing massive inputs to the Earth of heat-processed cometary material (meteoroids, dust, chemicals) which essentially determine the evolution of civilization, climate and life. While this implied new focus on terrestrial evolution may not be accepted immediately by theologians, historians, biologists, climatologists and geologists, members of the disciplines which came to be strongly guided by the Darwin-Huxley tradition of uniformitarianism, neither the US Congress (concerned with the civilization hazard) nor astrophysicists (concerned with the origin of planetesimals) can be expected now to ignore the full implications of "eschatological impulses" in the historical record.

3. Spiral Arms and the Cosmological Debate

3.1 Galaxies in Perspective

To the extent that the comets we observe are frozen planetesimals of the type expected to form in proto-stellar discs, namely, *homogeneous dust-volatile accretions*, the lack of evidence confirming the existence of any residual solar nebula from which most short-period comets are likely to be derived, proves, as we have seen, not to be adequately explained. On the other hand, we do observe a top-heavy mass distribution of comets from the Oort cloud and dynamical-evolutionary products such as Chiron and the Taurids which are consistent with hierarchical splitting and fading in the vicinity of the Sun. This is the sort of behaviour we expect if there is a history of rapid planetesimal formation which is likewise hierarchical, involving both condensation and aggregation. Indeed, if a formative medium which is the ultimate source of stars is the source of lesser gravitational condensations as well, then it is to be expected that the linear dimension and amount of (heat) processing to be associated with the components that emerge as a result of subsequent splitting will be correlated in accordance with their reduced volatility and increased resilience. Any such hierarchy would suggest we are dealing with a primordial medium in comet forming regions of low temperature and high molecular weight which undergoes rapid compression terminating in Jeans collapse. If so, we are bound to take seriously the possibility of a formation process, applied to planetesimals through to stars in the general location of galactic Keplerian discs, which has the perceived plasma injections producing successive spiral arms also undergoing adiabatic cooling and expansion prior to their containment and recompression by the disc. A general mechanism of this kind, accounting for the most conspicuous physical properties of the comets and asteroids we observe, involving the rapid hierarchical condensation of intermittent outflows from the galactic centre, evidently presupposes the repeated gathering up and release of part (or all) of the Keplerian disc material into and from compact galactic cores. The question arises therefore whether there is a natural process responsible for producing galactic cores on timescales significantly less than the spiral circulation period.

Under the normal conditions of stellar mass loss in relaxed galaxies, we anticipate the emergence of supermassive stars from time to time in galactic nuclei (*e.g.*

Bailey and Clube 1978). This is interesting not only for the periods of their gestation, which are ~ 10^6–10^8 years, depending on their mass and their environment, but for the fact of their subsequent evolution, which causes exceptionally large amounts of cosmic material to assume a macroscopic physical state, extreme gravitational potential, which is not normally encountered in the laboratory or in the solar system. It follows that the environments of galactic nuclei are then of interest on account of their providing the circumstances par excellence wherein it is possible to examine *empirically* what happens when ordinary matter advances to states of extreme gravitational potential. The simplest practical expression of this consequence, in accordance with weak field dynamical facts, is that given by the Lorentz-Dicke theory of matter (Dicke 1961) in which the inertial-*cum*-gravitational rest mass of ordinary material (m_{oo}) and the corresponding velocity of light (c_{oo}) are elementary functions of the local Newtonian potential ϕ, namely,

$$m_o = m_{oo} \exp\left(-\frac{3\phi}{c_{oo}^2}\right) \tag{1}$$

$$c_o = c_{oo} \exp\left(\frac{2\phi}{c_{oo}^2}\right) \tag{2}$$

It is immediately apparent from these relationships that supermassive stars can be understood as evolving to hypermassive states in near-relativistic regimes in which $|\phi| \sim c_{oo}^2$ though, of course, the extrapolation from the weak-field regime where the relationships are established is one which has to be guided in practice as much by the empirical facts as by any underlying theory. Similar expressions for the material rest-mass and the velocity of light are also associated with forms of relativistic theory which allow these physical parameters to vary with gravitational potential (Atkinson 1962), but the general principles prescribing the space-time metric are then no longer self-evident, and it is clear that, here too, we need to be guided by empirical facts as much as by underlying theory.

Be this as it may, it is apparent that we must also expect highly evolved supermassive stars in galactic nuclei to undergo further growth, probably through the medium of an enhanced nuclear accretion disc involving enhanced nuclear activity. The evolution here depends, however, on the precise balance between the mass outflow (due to normal (super) stellar processes) and the mass inflow (due to hypermassive effects on the environment) and may not be easy to predict. Nevertheless, in principle, there is an obvious limiting factor to this growth, namely, the distribution of material in the associated parent galaxy itself. In other words we can expect to observe cosmic masses at least as great as

$$m_L \sim m_{oo}\{\text{spiral galaxy}\} \exp\left(\frac{-3\phi_{extreme}}{c_{oo}^2}\right) \tag{3}$$

where $\phi_{extreme}$ is a physical parameter depending on the structural properties of the material aether associated with the Lorentz-Dicke theory. Arbitrarily assuming such cosmic masses are amongst the most luminous quasars known, whilst admitting perhaps that the gravitational and cosmological components of typical large redshifts are not readily separated at present, we then anticipate (approximate) "great attractor" masses of $m_L < 10^{12} e^{10} M_\odot$, or

$$m_L \leq 10^{16} M_\odot \qquad (4)$$

Whether or not this is a realistic upper limit, given also that fundamental particles are thought of in this framework as extreme states of very small aether aggregates which undergo bifurcation when suitably perturbed (*i.e.*, the process of matter—antimatter creation), it does seem that we also have reason to expect that the extreme states of much larger aether aggregates rendered unstable may likewise undergo bifurcation, giving rise in effect to the symmetrical outflows which, it seems, may be responsible for spiral arms. A fully developed theory for this process is not, of course, available at present, but the more obvious properties of comets, spiral arms and dark matter, taken together, do now appear to provide evidence that a familiar macroscopic process (spiral arm formation) and a familiar high energy process (pair production) may be understood in similar fundamental terms. They also provide a *prima facie* physical basis for examining further the basic implications of Lorentz's theory.

3.2 Lorentzian Theory in Perspective

To some extent, it is now a self-serving justification of relativistic physics that the paradigm shift accepting the principle of relativity is generally considered to have occurred more or less at the time of the original publication (Einstein 1905b). In practice of course, the continuity between relativistic physics and nineteenth century physics was capable of being maintained (*e.g.* Lorentz 1904, Cunningham 1914), the implications for the material aether being abandoned only as recently as the 1930's following the recognition of cosmological redshifts (Hubble 1929). By the 1930's, however, the quantum character of the physical processes involved in the absorption and emission of electromagnetic radiation were accepted as playing a fundamental role in physics and there was a greater readiness than before to dispose of the material aether and admit the possibility that electromagnetic radiation is *transmitted* by zero-mass photons, as proposed by Einstein.

Under these circumstances, the existence of cosmological redshifts necessarily implied an expanding substratum. This fact was recognised more or less immediately by astrophysicists (Eddington 1931), but the enthusiasm for zero-mass photons was not apparently universal. Thus, several decades were to pass before this development was accepted by physicists generally. Nevertheless, the wave transmission of electromagnetic radiation remained a viable concept, possibly associated with a material aether, and it was clearly necessary to keep open this theoretical option so far as the cosmological redshift was concerned. Even Hubble himself was strongly inclined to insist on this viewpoint, though there is a tendency nowadays to associate him exclusively with the expanding substratum concept.

"Inertial frames of reference" are essentially a twentieth century concept describing coherently moving distributions of Euclidean space (x, y, z) in association with co-moving (inertial) masses. Newtonian gravitational fields, on the other hand, are essentially a concept of the nineteenth century (and earlier) describing coherently moving distributions of gravitational potential $\phi(x, y, z)$ within appropriate volumes of Euclidean space containing co-moving material. The principle of equivalence permits a degree of congruence between these concepts, so that the nineteenth century identity between gravitational and inertial mass may continue to be allied with both moving fields and moving inertial frames.

This alliance implied that otherwise identical fields in different states of absolute motion were originally regarded as referring to "corresponding states" (Lorentz 1899). At the microscopic level, these corresponding states would appear to have been commonly envisaged as instantaneous, topologically identical $\phi(x, y, z)$-dependent distributions of (visible) material and (material) aether. At the same time, however, the aether as a whole was evidently regarded as a topologically invariant substratum comprising an infinite array of intrinsically identical fundamental particles (i.e., invisible material in bound motion), whilst visible material in its varied manifestations corresponded to physical states of the (material) aether which depended on their potential. It follows that the "corresponding states" were perceived as instantaneous configurations of (visible) material and (material) aether reflecting the passage of potential fields, a picture which seemed, then, to result in the idea that fields were in some sense more basic than the instantaneous material states they described. It also follows from this line of thinking that if an underlying material substratum is physically interlocked, then (visible) matter in motion is itself a succession of aether states involving annihilation-creation events which are coordinated, not only in accordance with an underlying universal motion (or universal clock), but also in accordance with the advancing field. Subsequently it was a reassuring development that the material aether apparently participated in isolated annihilation and creation events involving symmetrical particle pairs suggestive of bound motion in the aether.

At the beginning of the twentieth century, therefore, a general understanding of the physical properties of (visible) matter was sought (e.g. Larmor 1900, Lorentz 1904, etc.) with a particular perception of the (material) aether in mind, according to which identical-dependent distributions of (visible) material and (material) aether in different states of absolute motion experienced correspondingly orchestrated effects, namely, Fitzgerald contraction and clock retardation. The connections amongst the bound motions of the common substratum were inevitably considered to be preserved through (i) a universally regular phase, and (ii) a common value for its bound motion in the ground state. After a certain amount of exploration during the last two decades of the nineteenth century, these orchestrated effects were eventually reduced to a description in terms of event sequences and physical relationships (the Lorentz transformations) between corresponding states S and S'.

In their most familiar form, due to Einstein, these relationships between corresponding states are:

$$x' = \beta(x - vt) \tag{5}$$

$$t' = \beta\left(t - \frac{vx}{c^2}\right) \tag{6}$$

where $\beta = (1 - v^2/c^2)^{-1/2}$ and the other symbols have their usual meanings. The second expression may also be written in the form

$$t' \approx \beta^{-1}\left(t - \frac{vx}{c^2} + \frac{tv^2}{c^2}\right) \tag{7}$$

and hence

$$t' = \beta^{-1}\left(t - [x - vt]\left[\frac{1}{c} - \frac{1}{c+v}\right]\right) \qquad (8)$$

provided $v/c \ll 1$ and we limit the expansion of series to no more than quadratic terms in v/c, making it clear that the second bracketed term in equation (8), assuming (x,t) coordinates relate to an event in absolute space and time, arises at this level of approximation as a consequence of the absolute time difference in the reception of signals from remote events by stationary and moving clocks. It follows that (x,t) and (x',t') correspond to remotely observed event coordinates; specifically, t was known as "local time" in Lorentz's analysis; also that equation (6) is an approximation if equation (8) is regarded as exact. Thus, whilst expecting the universality and constancy of c to be a realistic requirement for the material aether, corresponding to the universal application of physics, the choice amongst these particular relationships between corresponding states was seen as an issue which had to be decided by experiment alone. In the supposed absence of experiments demonstrating the existence of the material aether, Einstein elected to interpret equations (5) and (6) as exact, whence he attributed their derivation to a fundamental 'principle of relativity'. In the supposed absence of experiments demonstrating the non-existence of the material aether on the other hand, Lorentz elected to interpret these equations in terms of material undergoing Fitzgerald contraction (the β term in (5)) and clock retardation (the β^{-1} term in equation 8), treating the Einstein (1905b) principle, therefore, as non-fundamental.

More or less in parallel with the proposed relativity principle, apparently assuming that he had independent support for the inferred non-existence of the material aether, Einstein (1905a) provided an explanation of the photoelectric effect in terms of the transmission of electromagnetic radiation by photons rather than by waves. Be that as it may, the predicted Doppler effect based on the Lorentz transformations for radiation passing between emitters and absorbers associated with relatively moving gravitational fields is

$$v' = v\beta^{-1}\left(1 + \cos\theta \frac{v}{c}\right)^{-1} \qquad (9)$$

where θ is the angle between the line of transmission and the line of relative motion and the other symbols have their usual meaning. It follows that the dispersion free spectral shifts along and perpendicular to the line of transmission, assuming $v/c \ll 1$, are, respectively

$$cz_\| = v + \frac{1}{2}c\frac{v^2}{c^2} \qquad (10)$$

$$cz_\perp = \frac{1}{2}c\frac{v^2}{c^2} \qquad (11)$$

the longitudinal and transverse quadratic terms being subsequently confirmed by experiments with positive rays (Ives and Stilwell 1938, 1941). In the absence of the material aether, photons are generally envisaged as travelling unaltered between source and distant receiver, the value of $cz_\parallel$ then being close to their relative velocity $v (\ll c)$. Correcting for the absolute motion of the sun and assuming fields with random motions, the expected relative velocity is zero, whence it follows that the observed cosmological redshift $cz_\parallel$ is largely due to the expanding substratum (Eddington 1931). In the presence of a material aether however, with successive absorbers and emitters associated with intervening gravitational fields in relative motion v, having components u along the radiation path, the observed spectral shift is given by

$$cz_\parallel = \sum u + \frac{1}{2} c \sum \frac{v^2}{c^2} \qquad (12)$$

whence, correcting for the absolute motion of the sun and summing over cosmological paths dominated by a sufficiently large number of intervening fields in random absolute motion (i.e., $\sum u \to 0$), the spectral displacement (cosmological redshift) is given by

$$cz_\parallel = \frac{1}{2} c \sum \frac{v^2}{c^2} \qquad (13)$$

For the purposes of discussion, we shall assume that the successive relative velocities along any line of sight "between galaxies" and "between stars" are comparable, in which case $cz_\parallel$ is dominated by the stellar component alone. To this degree of approximation, for an object at distance r, it follows that

$$cz_\parallel = Hr \qquad (14)$$

the quadratic Doppler expression for the Hubble constant being $H \sim \frac{1}{2} n_g \left(\frac{3}{2} n_s d_g \frac{v_s^2}{c} \right)$, where n_g is the number of typical (spiral) galaxies per megaparsec, d_g is a representative depth for a typical galaxy halo intercepted by the line of sight, n_s is a representative number density per unit distance for halo stars, and v_s is one component of the typical "between stars" relative velocity. Evidently for a cosmological distribution of intervening gravitational fields which are isotropic, random and of similar dimensions and mass, we anticipate a common value $Q \sim \left(\frac{3}{2} n_s d_g \frac{v_s^2}{c} \right)$, and the Hubble law is "quantized". This quantization is meaningful in the sense that individual galaxies apparently have redshifts of the form

$$cz_\parallel = kQ + w \qquad (15)$$

where k is integer and w is the line of sight velocity component of a galaxy with respect to the sun. In general this may be re-expressed in the form

$$cz_\parallel = (k + k')Q + \varepsilon \qquad (16)$$

in which k' is integer and $|\varepsilon| < Q$, with the awkward consequence that dynamical effects (reflecting the influence of both ordinary and dark matter) may be masked by quantization.

On the other hand, for near equivalent lines of sight originating from the same galactic or supergalactic system, we anticipate halo-disc $cz_\parallel$ discrepancies $\sim \Delta \left(\frac{3}{2} n_s d_g \frac{v_s^2}{c} \right)$, depending on typical near-halo and near-disc values of (n_s, v_s). The preponderance of red rather than blue shifts is consistent with a stellar relative velocity effect which is stronger amongst more aged stars, in accordance with the galactic-dynamical properties of stellar populations generally. It would appear that the quantization effect and halo-disc discrepancies among cosmological redshifts are both potential discriminants, therefore, so far as stationary and expanding versions of the underlying substratum are concerned.

4. The Cosmological Redshift

The approach of this paper has been to demonstrate the disadvantages with which we are saddled by the present "top-down" theoretical attitude to astrophysics. Thus, throughout the twentieth century, there has been a general desire amongst physicists to explore to the fullest the "relativity principle" introduced by Einstein and its remarkable implications for massless photons and the massless vacuum. However, sight has never been lost of the fact that we could be dealing with photons of finite mass interacting with massive bosons presumed to be present in the material vacuum (*e.g.* Vigier 1990). This has kept open the possibility of a (material) aether traversed by waves, as previously envisaged by nineteenth century physicists. On the other hand, astrophysicists have been rather single-minded in their enthusiasm for the relativity principle. Indeed, through their premature acceptance of its implications for the cosmological redshift (Eddington 1931), they have risked abrogating their responsibility for a *balanced* understanding of the full range of known astrophysical effects at lesser scales.

Their enthusiasm appears now to have been least bridled during two separate postwar periods which can be seen as marking the implicit acceptance (during the 1920's) of the massless vacuum, fixating ideas about the general state of the substratum and spiral galaxies, and (during the 1950's) of the star-forming process in spiral arms, fixating ideas about an accreting source for comets and planets. It is true of course that the discovery during the 1960's of the cosmic microwave background—leaving aside the coincidental value of its energy density and the discovery subsequently of additional backgrounds (X-ray, IR)—helped to entrench further the universal expansion based upon the supposed big bang. However, this assumption plainly set aside any prediction that cooled plasma is injected into intergalactic space, such as may be expected from the many jets which escape from galactic discs. Moreover, it has taken the full paraphernalia of the space age to uncover the broad character of the cometary and asteroidal populations within the solar system, incidentally affecting the Earth, to begin revealing that our cosmologically and cosmogonically fixated ideas about comets not only defied the facts in 1960, but do so still in 1990. A *balanced* understanding of the full range of known astrophysical effects must, of course, include cometary phenomena, and we must seek to avoid any tendency of the "top-down" approach to subordinate our knowledge of comets to the demands of an imagined cosmos simply on the grounds of their relative mass!

Progress along these lines will, of course, require us now to examine fully the implications of a stationary aether and the quadratic Doppler effect. In Section 3, we considered in particular two predicted effects involving cosmological redshifts. In each case, so far as the observed quantum ~ 37 km s^{-1} (Napier and Guthrie *loc. cit.*) and the observed halo-disc discrepancies ≤100 km s^{-1} (Arp *loc. cit.*) are concerned, we appear to be dealing predominantly with the dynamics of typical halo stars. For some time now, it has also been known that the Galactocentric velocity relative to the local standard of rest (l.s.r.) has an apparent symmetry around + 40 km s^{-1} rather than 0 km s^{-1}. This suggests an outward motion in the Galaxy for the l.s.r. of ~ 20 km s^{-1} to which is added the effects of a typical cosmological redshift ~ 20 km s^{-1}, say, roughly half the quadratic Doppler quantum. For the Galactic centre line of sight, the disc population contribution to the redshift may be estimated on the basis of ~ 10^3–10^4 intervening fields and typical motions ~ 20 km s^{-1} (say), yielding an approximate value ~ 10–20 km s^{-1}.

The outward motion of the l.s.r. in the Galaxy is independently determined at present with respect to galaxies (*e.g.* Napier and Guthrie *loc. cit.*), globular clusters (*e.g.* Clube and Waddington, in preparation) and the relaxed disc (formerly identified as Stream II; Clube 1985), each these standards of rest being consistent with a significantly non-zero value of ~ 20 km s^{-1}.

5. Concluding Remarks

The main emphasis of this paper has, of course, been on the possibilities that arise through considering some of the latest findings in cometary science, not least through taking note of the (excluded) eschatological record. However it should also be noted that the present "top-down" theoretical approach to astrophysics, which essentially determines the assumed weight of such findings, would probably not be supported were it not for the continuing positivistic tendency in the study of physics—the predilection for conceptions of nature which treat 'nothing' as if it were 'something'.

Consider, for example: Which has most legs, a horse or no horse ? This riddle (Rees-Mogg 1993) goes back at least to the first half of the eighteenth century and may be far older. The answer is: "No horse, for a horse has but four legs, and no horse has five". This kind of logical fallacy was first noticed in ancient Greece. When Odysseus put out the eye of Cyclops, for example, he told him that his name was "Nobody"; when Cyclops complained, he went round telling his friends that Nobody had put his eye out. In the meantime Odysseus was thus able to escape!

Twentieth century physicists often appear to fall into the trap provided by this old fallacy which treats negatives as though they were merely another kind of positive. To explain the myriad properties of the vacuum, for example, it is customary to credit the aether and the no-aether, so far as possible, with corresponding physical attributes. Thus to give meaning to descriptive equations which evidently work so far as the phenomena are concerned, both the aether and the no-aether are understood as being capable of carrying electromagnetic 'waves'. At the same time, so far as the *longitudinal transmission* by the intervening medium is concerned, we apparently consider it to be a matter of convention whether we have a no-aether (or absent aether) affected by local gravitational fields. Obviously, by invalidating equation (12), we disclaim any influence on cosmological photons as we establish the expanding substratum. On the other hand, there is no logical basis for the disclaimer if the horse has four legs.

Acknowledgments

It is a pleasure to thank the editors of these proceedings: Halton C. Arp, Konrad Rudnicki and C. Roy Keys, for their forbearance as the account of this presentation was prepared and finalized.

References

Alsted, J.H., 1627, *Diatribe de mille annis apocalypticis*, Francofurti.
Asher, D.J., Steel, D.I. and Clube, S.V.M., 1993, *Mon.Not.R.Astr.Soc.* in press.
Asher, D.J. and Clube, S.V.M., 1993, *Quart.Jl.R.Astr.Soc.* in press.
Atkinson, R.d'E., 1962, *Proc.Roy.Soc.Lond.* A272:60.
Bailey, M.E. and Clube, S.V.M., 1978, *Nature* 275:278.
Bailey, M.E., Clube, S.V.M. Hahn, G., Napier, W.M. and Valsecchi, G.B., 1993, in: *Hazards due to Comets and Asteroids*, ed. Matthews, M.S. Gehrels, T., University of Arizona Press, in press.
Ball, B.W., 1975, *A Great Expectation: Eschatological Thought in English Protestantism to 1660*, E.J. Brill.
Brandt, J.C. and Chapman, R.D., 1982, *Introduction to Comets*, Cambridge University Press.
Brightman, J., 1644, *The Works*.
Cameron, A.G.W., 1962, *Icarus* 1:13.
Canavan, G.H. Solem, J.C. and Rather, J.D.G., 1992, *Proceedings of the Near-Earth Object Interception Workshop*, NASA.
Ceplecha, Z., 1992, *Astron.Astrophys.* 263:361.
Clube, S.V.M., 1985, *I.A.U. Symp.* 106:145.
Clube, S.V.M., in: *Hazards Due to Comets and Asteroids*, ed. Matthews, M.S.T., University of Arizona Press, in press.
Clube, S.V.M. and Waddington, W.G., 1993, in preparation.
Cunningham, E., 1914, *The Principle of Relativity*, Cambridge University Press.
Dicke, R.H., 1961, in: *Enrico Fermi Course XX*, ed. Möller, C., I, Academic Press.
Eddington, A.S., 1931, *Mon.Not.R.Astr.Soc.* 91:412.
Eggen, O.J., Lynden-Bell, D. and Sandage, A., 1962, *Astrophys.J.* 136:748.
Einstein, A., 1905a, *Ann. der Phys.* 17:132.
Einstein, A., 1905b, *Ann. der Phys.* 49:769.
Einstein, A., 1916, *Ann. der Phys.* 78:3.
Everhart, E., 1972, *Astrophys.J.Lett.* 10:131.
Gould, S.J., 1991, *Bully for Brontosaurus: Reflections in Natural History*, Norton.
Hasegawa, I., 1992, *Cel. Mech. and Dyn. Astron.* 54:129.
Hill, C., 1980, *Change and Continuity in Seventeenth Century England*, Weidenfeld.
Hubble, E., 1929, *Proc.Nat.Acad.Sci.(USA)* 15:168.
Ives, H.E. and Stilwell, G.R., 1938, *J.Opt.Soc.Amer.* 31:369.
Ives, H.E. and Stilwell, G.R., 1941, *J.Opt.Soc.Amer.* 28:215,
Jeans, J.H., 1928, *Astronomy and Cosmogony*, Cambridge University Press.
Kresak, L., 1978, *Bull.Astron.Inst.Czechosl.* 29:135.
Kuiper, G.P., 1951, *Proc.Natl.Acad.Sci.(USA)* 37:1.
Larmor, J., 1900, *Aether and Matter*, Cambridge University Press.
Lorentz, H.A., 1899, *Proc.Roy.Acad.Amsterdam* 1:427.
Lorentz, H.A., 1904, *Proc.Roy.Acad.Amsterdam* 6:809.
McCrea, W.H., 1978, in: *The Origin of the Solar System*, ed. Dermott, S.F., 75, Wiley.
Mede, J., 1672, *The Works*.
Morrison, D., 1992, ed., *The Spaceguard Survey: Near-Earth Object Detection Workshop*, NASA.
Oort, J.H., 1950, *Bull.Astron.Inst.Neth.* 11:91.
Rees-Mogg, W., 1993, *The Times*, London 15 Mar.:14.
Sandage, A., 1961, *Astrophys. J.* 133:355.
Shoemaker, E.M., 1983, *Rev. Earth Planet Sci.* 11:461.
Steel, D.I., Asher, D.J. and Clube, S.V.M., 1991, *Mon.Not.R.Astr.Soc.* 251:632.
Steel, D.I., Asher, D.J., Napier, W.M. and Clube, S.V.M., in: *Hazards Due to Comets and Asteroids*, ed. Matthews, M.S. and Gehrels, T., University of Arizona Press, in press.
Vigier, J.P., 1990, *I.E.E.E. Trans.Plasma Sci.* 18:64.
Whipple, F.L., 1963, in: *The Solar System IV*, ed. Middlehurst, B.M. and Kuiper, G.P., 639, University of Chicago Press.
Whipple, F.L., 1967, *NASA-SP* 150:409.

Active Galactic Nuclei:
Their Synchrotron and Cerenkov Radiations

P. F. Browne

Department of Pure and Applied Physics
University of Manchester
Institute of Science and Technology
Manchester M60 1QD

A previously suggested acceleration mechanism generates ultrarelativistic charged particles which are collimated into a beam. For a typical active galactic nucleus (AGN), the mechanism yields charges of energy 10^{14} eV, assuming a magnetic vortex tube (MVT) of radius $R \approx 10^{13}$ cm with "wall" magnetic field $B_\varphi(R) \approx 10$ mG. A column density $\approx 10^{21}$ cm^{-2} generates luminosity 4×10^{45} erg s^{-1} typical of AGNs with power-law spectra $v^{-1+\delta}$ when injected into the nonuniform field $B_\varphi(r) \propto r^{-1}$, where electron density varies as $n_e(r) \propto r^{-\delta}$. The emission is forward-directed into a solid angle depending on charge collimation.

It is shown that Cerenkov radiation develops at resonance-line frequencies if $\gamma^2 n_s$ exceeds a threshold value, where n_s is the density of atoms/ions of species s and γ is the Lorentz factor of the exciting charge. For H(Lα) the condition is $\gamma^2 n_H > 5 \times 10^{19}$ cm^{-3}, assuming thermal broadening of 150 km s^{-1}. For $\gamma = 2 \times 10^8$, this requires that $n_H > 7500$ cm^{-3}. Outflow velocity increases with distance due to continued acceleration, reaching relativistic values. Self-shielding leads to shells of gas of optical thickness unity which are individually tuned to different driving fields, explaining the richness of the Lα absorption spectrum. Continued acceleration may separate the velocities of shells, explaining different absorption redshifts associated with a single source.

Cerenkov fields at the Lβ, Lα, etc. resonance lines of H generate strong broad emission lines by stimulated Raman transitions, i → j → k, the direct transition i → k being forbidden. For strong pumping of the resonance line (i → j) stimulated emission (j → k) dominates spontaneous emission. The emission lines are broadened by AC Stark effect. AGN broad emission lines have the properties expected for stimulated Raman scattering. They originate in an outflow through the core of a magnetic vortex tube (MVT) along the rotation axis, as do other lines.

For strong pumping, the broad profile develops two peaks with separation $2\Omega_{ij} = p_{ij} E_0 / \hbar$, where p_{ij} is the transition dipole moment and E_0 is the amplitude of the pump field. Only the redshifted peak is present for Cerenkov pumping: a Cerenkov field cannot develop blueward of ω_{ij} because the refractive index $\mu < 1$. Intrinsic redshifts and asymmetrical line profiles are predicted for the broad emission lines, superimposed on Doppler blueshifts for outflow. From observed $\Delta \omega_{ij}$ one infers E_0, which yields a value for source dimension R. We obtain $R \approx 10^{12}$ cm for 3C 273.

1. Introduction

Previously (Browne 1986) I have proposed a mechanism for generating beams of ultrarelativistic charged particles in various astrophysical sources (Browne 1985, 1986, 1987, 1988a). The mechanism has been applied also to acceleration in a laboratory device, the plasma focus (Browne 1988b, 1988c) prompted by a discussion with E.J. Lerner (see Lerner 1986). Here the mechanism will be applied specifically to active galactic nuclei (AGNs), including quasi-stellar objects (QSOs), BL Lac objects, Seyfert 1 nuclei and broad-line radio galaxies.

The acceleration mechanism is summarised in Section 2. A cylindrical electrostatic double layer of radius R and thickness ΔR carries an axial electric current J which flows across lines of B_φ because of a radial electric field E_r in the double layer. The annular double layer therefore carried the axial current J which is the source of the pinching magnetic field $B_\varphi(r) = 2J/rc$. When differential rotation winds a loop of field line around the MVT, the outward and return branches of the loop become opposite helices with opposite axial source currents $\pm J$. The opposite currents are arrested by accumulation of space charge, whose axial electric field eventually acquires all of the energy in $B_\varphi(r)$. The current then reverses, and energy oscillates between $E_{\tilde{z}}$ and B_φ, but the oscillation is rapidly damped because charges acquire energy in $E_{\tilde{z}}$ which allows them to escape from the region, no longer contributing to J. A field $E_{\tilde{z}}$ can also arise when a region $\Re$ of the current channel becomes denuded of free charges due to an instability, in which case induction drives a displacement current density in $\Re$ as in a charging capacitor.

For an AGN with $R \approx 10^{13}$ cm and $B_\varphi(r) \approx 10$ mG, charges are accelerated to typically 10^{14} eV, which implies Lorentz factor $\gamma = 2 \times 10^8$ for electrons. A typical AGN of luminosity $\approx 3 \times 10^{45}$ erg s^{-1} is obtained for radiating charge column density $\approx 10^{21}$ cm^{-2}, implying perhaps electron density $n_e \approx 3 \times 10^6$ cm^{-3} and column length $Z \approx 3 \times 10^{14}$ cm. For electrons of constant γ injected into the nonuniform field $B_\varphi(r) \propto r^{-1}$, the synchrotron spectrum is $L_\nu \propto \nu^{-1+\delta}$, assuming $n_e(r) \propto r^{-\delta}$. The emission is anisotropic, with strong forward beaming determined by collimation of the electron trajectories. A bunch of relativistic electrons which are confined to the core of an MVT by a helical wall magnetic field can generate synchrotron radiation whose direction of polarization rotates by more than 360°, explaining a feature of blazars. The requirements for superluminal expansion are also met by a bunch of relativistic electrons moving outward within an MVT core, and changing direction if the MVT bends.

The accelerated charges with velocity βc generate Cerenkov radiation in a medium of refractive index μ provided that $\beta\mu > 1$, or $\gamma^2\beta^2(\mu^2 - 1) > 1$. This condition is most easily satisfied at resonance-line frequencies. In particular at the Lα line of H the condition becomes $\gamma^2 n_H > 5 \times 10^{18}$ cm^{-3}, assuming thermal broadening of 15 km s^{-1}. For Lβ, $\gamma^2 n_H > 3 \times 10^{19}$ cm^{-3}. For $\gamma = 2 \times 10^8$ these conditions are easily met. Strong Cerenkov fields at resonance line frequencies have two consequences. Firstly, radiation pressure drives strong gas outflows. Self-screening causes outflowing gas to form shells of optical thickness unity—each Doppler tuned to a different radiation band, explaining the richness of the Lα absorption line spectrum. Outflow velocity increases with distance due to continued acceleration, so that shells may, eventually, become well separated in velocity; this explains the existence of more than one redshift system in rich absorption spectra. Outflow velocity can become relativistic; for example, the observation of z_{em} = 1.95 and z_{abs} = 0.61 in PHL 938 implies outflow velocity 0.5 c. However, all absorption systems are associated with the source, as the idea of intergalactic absorbers has now been abandoned.

Secondly, Cerenkov pumping of the n = 2, 3, etc. levels of hydrogen generates strong stimulated Raman emission to levels k lying between the resonance levels i and j. This process is well known in electrooptics, where laser pumping of transitions between vibrational states of molecules is accompanied by stimulated emission from transitions between rotational levels of the upper vibrational state. The cross section for the two-photon single transition i $\rightarrow$ j $\rightarrow$ k is comparable to that of a resonance scattering i $\rightarrow$ j $\rightarrow$ i. Because the rate of the two-photon single transition is proportional to $I(\omega)I'(\omega')$, where $I(\omega)$ is the intensity of the pumping radiation at frequency $\omega \simeq \omega_{ij}$ and $I'(\omega')$ is the intensity of the emitted radiation at frequency $\omega' \simeq \omega_{jk}$, the single-transition stimulated emission dominates the two-transition spontaneous emission for strong pumping. The broad emission lines of active galactic nuclei have all of the properties of such emission (Browne 1983, 1985). Their broadening and asymmetric profiles are not caused by Doppler effect, but by AC Stark effect. They originate in the same outflow as the narrow emission lines, the only difference being that they have a Cerenkov field at the frequency of the i $\rightarrow$ j transition. The threshold condition for the Cerenkov resonance line field determines which emission lines will be broad and which will not.

Anomalous dispersion causes a Cerenkov field to develop only for the redward wing of the resonance line at ω_{ij}, the reason being that the threshold condition $\beta c > 1$ requires $\mu - 1$ to be positive, which is achieved only for $\omega < \omega_{ij}$. For strong pumping, both the ω and ω' transitions develop two peaks separated by $2\Omega_{ij} = p_{ij}E_o/\hbar$, where p_{ij} is the transition dipole moment and E_o the amplitude of the pump field. In the case of Cerenkov pumping, the blueward peak is suppressed, yielding an intrinsic redshift Ω_{ij} ($\simeq 10,000$ km s^{-1}) which is superimposed on a Doppler blueshift due to the outflow. Both the broad emission lines and the absorption lines should show a Doppler shift to the blue, but the emission lines have a superimposed intrinsic redshift which usually exceeds the blueshift. Also, the emission lines originate close to the source of relativistic charges, where outflow velocity will be less than for the absorption lines produced at long range. Now all spectral lines, emission or absorption, broad or narrow, with the same or different redshifts, originate in a single gas outflow through the core of an MVT. The emissions are forward beamed.

It is possible to infer the source dimension R from the observed value of $\Delta\omega_{ij}$, yielding a value for E_o at the source which can be compared with a value at Earth. The result obtained, $R \simeq 10^{12}$ cm, agrees with the dimension inferred from the most rapid X-ray variability, assuming of course that the distance is known from the Hubble effect.

2. Acceleration in Pinched Magnetic Vortex Tubes

(i) Axial Current Channel

In the core of a magnetic vortex tube (MVT) (Browne 1985, 1986, 1988a) charges are trapped by reflection in the $B_\varphi(R)$ field which defines the core "wall". As the wall pinches with velocity $\dot{R}$, trapped charges gain velocity at each reflection. Let an ion have radial velocity $\beta_i c$. If this velocity is reversed from $+\beta'_i$ to $-\beta'_i$ in the rest frame of the wall, then in the laboratory frame it is changed from $+\beta_i$ to $-(\beta_i + \Delta\beta_i)$, where

$$\Delta\beta_i = -\frac{2(1-\beta_i^2)\dot{R}/c}{1-2\beta_i \dot{R}/c + \dot{R}^2/c^2} \approx -2(1-\beta_i^2)\frac{\dot{R}}{c} \tag{1}$$

The approximate form assumes $\dot{R}/c \ll 1$. The number of reflections per unit time is $\beta_i c/2R$, so that

$$\dot{\beta}_i = -\beta_i(1-\beta_i^2)\frac{\dot{R}}{R} \tag{2}$$

which has the integral

$$\beta_i \gamma_i R(t) = \text{constant} \tag{3}$$

Thus, contraction of the MVT from radius $R(0)$ at time zero to $R(t)$ at time t causes the proper velocity of trapped charges to increase by a factor $R(0)/R(t)$. This factor is the same for electrons and ions, so that the wall pressure is exerted almost entirely by ions, being given by

$$p_i = (2\beta_i \gamma_i m_i c)(n_i \beta_i c) = 2\beta_i^2 \gamma_i n_i m_i c^2 \tag{4}$$

Pinching is caused by force density $\mathbf{j} \times \mathbf{B}/c$ which acts on the electrons. If $\mathbf{j}$ is nonzero in an annular layer of thickness ΔR and radius R (see below), then inward pressure is $\Delta R(\mathbf{j} \times \mathbf{B})/c$. Eliminating $\mathbf{j}$ by Maxwell's equations, and using the identity $\vec{\nabla}^2 \mathbf{B} = 2(\mathbf{B} \cdot \vec{\nabla})\mathbf{B} + 2\mathbf{B} \times \vec{\nabla} \times \mathbf{B}$, one finds (neglecting any change of $\mathbf{B}$ in the direction of $\mathbf{B}$)

$$p_i = (\mathbf{j} \times \mathbf{B}) \cdot \Delta \mathbf{R}/c = (4\pi)^{-1}\left[(\vec{\nabla} \times \mathbf{B}) \times \mathbf{B}\right] \cdot \Delta \mathbf{R}$$
$$= -(8\pi)^{-1}\vec{\nabla}B^2 \cdot \Delta \mathbf{R} = \Delta\left(\frac{B^2}{8\pi}\right) \approx \frac{B^2}{8\pi} \tag{5}$$

Since pinching force is exerted by electrons, whilst pressure opposing pinch is exerted by positive ions, there develops radial charge separation which is limited by electric field E_r extending over the thickness ΔR of an annular layer between the charge surfaces. The force per unit area restraining separation of surfaces of charge density σ is $p_r = \sigma E_r/2 = E_r^2/8\pi$, where one notes $E_r = 4\pi\sigma$. If contraction is steady, the forces balance, yielding

$$\frac{B_\varphi^2(R)}{8\pi} = \frac{E_r^2}{8\pi} = 2\beta_i^2 \gamma_i n_i m_i c^2 \tag{6}$$

It follows that $E_r \simeq B_\varphi(R)$. Hence, charges with gyroradii less than ΔR will acquire in the annular channel electric drift velocity $cE_r/B_\varphi \simeq c$. For electrons with gyroradius r_{ge}, we can show that $r_{ge} < \Delta R$, but for ions with gyroradius r_{gi}, we find that $r_{gi} > \Delta R$.

The thickness ΔR is determined by the distance required to bring to rest a proton with energy $\gamma_i m_i c^2$:

$$eE_r \Delta R = (\gamma_i - 1) m_i c^2 \tag{7}$$

Expressing the gyroradii in terms of ΔR, we have

$$\begin{aligned} r_{ge} &= \frac{\beta \gamma m_e c^2}{eB_\varphi(R)} = \frac{\beta \gamma m_e \Delta R}{(\gamma_i - 1) m_i} \\ r_{gi} &= \frac{\beta_i \gamma_i \Delta R}{\gamma_i - 1} \end{aligned} \tag{8}$$

where (6) and (7) have been used. The condition $r_{ge} < \Delta R$ reduces to $\beta_i > 2m_e/m_i = 1/918$, where we assume non-relativistic motions for the Fermi-accelerated charges and where we note that $\beta_i = \beta$ for these charges. Assuming this condition to be met, electrons acquire electric drift velocity $cE_r/B_\varphi \simeq c$, but ions do not acquire the full electric drift. Thus the annular channel carries current J. Current J is the source of the field $B_\varphi(r)$; that is, $B_\varphi(r) = 2J/rc$ for $r > R$ and $B_\varphi(r) = 0$ for $r < R$.

The annular channel approximates a current sheet when $\Delta R \ll R$. A current sheet tends to evolve into an array of parallel current filaments due to tearing mode instability (Gekelman and Pfister 1988). The magnetic field due to a cylindrical array of such filaments has been calculated by Syrovatskii (1966). Such filamentation of the current clearly occurs when the mechanism operates in the plasma focus device (Nardi 1978, Nardi et al. 1980).

(ii) Displacement Current

Winding of a loop of magnetic field line around the axis of the MVT by differential rotation produces a pair of helices of opposite parity, one for each arm of the loop (Figure 1). Because the loop has an outgoing and returning arm, the azimuthal field component B_φ reverses across the apex of the loop at $z = 0$, so that

$$B_\varphi(r, z) = -B_\varphi(r, -z), \quad j_3(r, z) = -j_3(r, -z) \tag{9}$$

The reversal of B_φ at $z = 0$ implies a reversal of j_3. Then opposite currents meet at $z = 0$. Charge of one sign accumulates at $z = 0$, and charge of the opposite sign accumulates at $z = \pm \Delta z$ (Figure 1). The equivalent circuit for this situation is a pair of LC circuits as shown in Figure 2a.

The accumulating space charge at $z = 0$ generates an axial electrostatic field $E_3(t)$ directed away from $z = 0$ to the opposite space charges at $z = \pm \Delta z$. The maximum strength of $E_3(t)$ is attained when all of the energy in the magnetic field $B_\varphi(t)$ $[= 2J(t)/rc]$ has been transferred into E_3, at which stage $J(t)$ has fallen to zero. Thereafter the energy oscillates between E_3 and B_φ, until it is dissipated.

Because the current is confined to a well defined channel, it is possible to model the oscillation with an equivalent circuit containing a self-inductance L, a capacitance C and a resistance R_Ω (Figure 2a). The energy is transferred from B_φ to $E_{\hat z}$ on the reactive time scale $t_R = (LC)^{1/2}$. The time scale for Ohmic dissipation is the magnetic energy $LJ^2/2$ divided by the Ohmic dissipation rate $R_\Omega J^2$, yielding $t_\Omega = L/2R_\Omega$. If accelerated charges escape from the vicinity of the acceleration region they will no longer contribute to J, which is to say that their energy represents dissipative loss. Since $t_R \ll t_\Omega$, this process greatly increases the rate of dissipation of magnetic field energy. At present the process is not recognized in either astrophysical or laboratory plasma work.

Beams of electrons and ions are ejected from opposite sides of each of the double acceleration region in Figure 3a. The charges which are directed inwards from either the $+E_{\hat z}$ or the $-E_{\hat z}$ region are decelerated as they pass through the other region. The charges which are directed outwards escape. During one half-cycle, positive charges escape, and during the next half-cycle negative charges escape because the polarity of the central space charge in Figure 3a alternates. The beams from either end of the double acceleration region in Figure 3a are identical and macroscopically neutral. One notes that bunches of alternately electrons and positive ions will produce periodic polarization of the beams which is maintained for some distance if both types of particle have velocity $\simeq c$, but which will eventually relax.

There is another possibility. An instability of drift velocity can denude a region $\mathfrak{R}$ of the current channel of free charges. Let $\mathfrak{R}$ be bounded by surfaces S_1 and S_2 through which charges enter $\mathfrak{R}$ and leave $\mathfrak{R}$, and suppose that the number of charges in $\mathfrak{R}$ is slightly reduced. Induction maintains constant total current (conduction plus displacement). Thus the drift velocity of remaining charges must increase, causing charges to exit across S_1 and S_2 with increased velocity. In order to

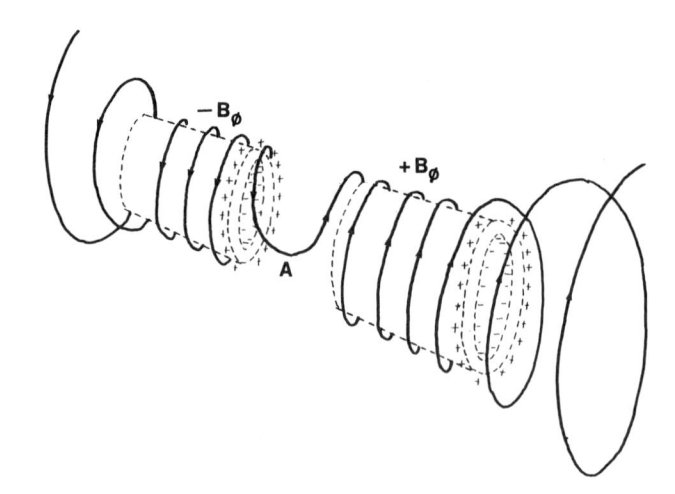

Figure 1. Stretching and winding of a magnetic field line by differential rotation of type $\omega(r)$ in a magnetic vortex tube. Note reversal of B_φ to either side of the apex A of the loop, and the consequent reversal of $J_{\hat z}$.

maintain J constant at S_1 and S_2, charges must enter region with reduced velocity. Increased exit rate and reduced entry rate will in time denude the region $\mathfrak{R}$. If the charges are electrons and positrons, the above argument is straightforward, because both carriers have equal drift velocities. If we replace the positrons with protons, there will arise space charge whose field retards electrons and accelerates protons in order to ensure that the difference between exit and entry rates is the same for electrons and ions in the steady state.

Denudation of region $\mathfrak{R}$ causes conduction current in $\mathfrak{R}$ to be replaced by displacement current $(S/4\pi)\delta E_{\mathfrak{z}}/\delta t$, where $S = 2\pi R \Delta R$. The axial electrostatic field $E_{\mathfrak{z}}(t)$ increases until all of the energy in the field $B_\varphi(t)$ has been transferred into $E_{\mathfrak{z}}(t)$ as previously. Then the energy oscillates between the fields, until dissipated. The equivalent circuit for this situation is shown in Figure 2b. We represent the increasing denudation of the region $\mathfrak{R}$ by a parallel plate capacitor whose plates are gradually separated from a touching initial position. In most laboratory situations the drift velocity instability is too slow to be of interest, but an exception may be acceleration in the magnetic pinch of the "plasma focus" device. Drift velocities are much higher in astrophysical situations.

In the case of denudation, $E_{\mathfrak{z}}(t)$ oscillates between a maximum value $E_{\mathfrak{z}}$ and zero as extrema. Because there is no reversal of polarity, successive bunches of charges ejected from one end of $\mathfrak{R}$ have the same sign, resulting in a beam which is macroscopically charged (Figure 3b). The beam from the other end is oppositely charged.

Integration of the energy density in $B_\varphi(r)$ yields

$$\mathcal{E}_B = \int_R^{r_1} \frac{B_\varphi^2}{8\pi} Z 2\pi r \, dr = k_1 Z \left(\frac{J}{c}\right)^2 = \frac{k_1 R^2 B_\varphi^2(R) Z}{4} \tag{10}$$

where $k_1 = \ln(r_1/R)$, r_1 being the maximum radial extent of the field as determined by return current.

Integration of the energy density in the electric field, assumed uniform over the volume $S\Delta z$, yields

$$\mathcal{E}_E = \int_0^{\Delta z} \frac{E_{\mathfrak{z}}^2}{8\pi} S \, dz = \frac{2\pi Q^2 \Delta z}{S} = \frac{k_2 R^2 E_{\mathfrak{z}}^2 \Delta z}{4} \tag{11}$$

where $\pm Q$ is the accumulated charge at either end of the break and $k_2 = \Delta R/R$.

The field $E_{\mathfrak{z}}(t)$ reaches maximum strength when it has acquired all of the energy in $B_\varphi(t)$. From $\mathcal{E}_E = \mathcal{E}_B$ we obtain amplitude

$$E_{\mathfrak{z}} = K B_\varphi(R) \tag{12}$$

where

$$K = \left(\frac{k_1 Z}{k_2 \Delta z}\right)^{1/2} = \left(\frac{k_1 R Z}{\Delta R \Delta z}\right)^{1/2} \tag{13}$$

The maximum energy W gained by charge e in the field $E_{\mathfrak{z}}(t)$ is

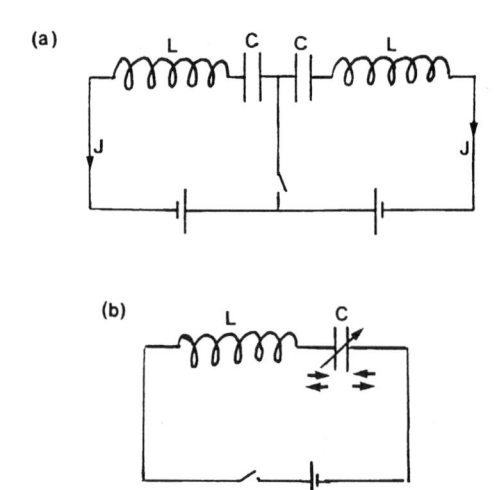

Figure 2. Equivalent circuits for two situtions in which energy oscillates between an azimuthal magnetic field B_φ and an axial electrostatic field $E_{\hat{z}}$. In case (a) there are two acceleration regions with oppositely directed fields $\pm E_{\hat{z}}$, and each field oscillates with reversal of polarity. In case (b) there is one acceleration region, and now $E_{\hat{z}}$ oscillates without reversal of polarity between zero and a maximum value.

$$W = eE_{\hat{z}}\Delta z = KeB_\varphi(R)\Delta z \tag{14}$$

A more useful form of (14) is

$$W = K'eB_\varphi(R)R = \frac{2K'eJ}{c}$$
$$K' = \left(\frac{k_1 Z\Delta z}{R\Delta R}\right)^{1/2} \simeq 3 \tag{15}$$

The numerical value is based on assumptions $k_1 \simeq 9$ and $Z\Delta_{\hat{z}} \simeq R\Delta R$.

A self-inductance L and capacitance C of the equivalent circuit can be defined from the energies in the fields by

$$\mathcal{E}_B = \frac{LJ^2}{2}, \qquad L = \frac{2k_1 Z}{c^2}$$
$$\mathcal{E}_E = \frac{CV^2}{2}, \qquad C = \frac{k_2 R^2}{2\Delta z} \tag{16}$$

where $V = E_{\hat{z}}\Delta z$. The period of the electromagnetic "ringing" oscillation is

$$P = 2\pi(LC)^{1/2} = \frac{2\pi K'' R}{c}$$
$$K'' = \left(\frac{k_1 Z\Delta R}{R\Delta z}\right)^{1/2} \tag{17}$$

A circuit with capacitance C, self-inductance L and resistance R_Ω in series has charge $Q(t)$ on the capacitor at time t, where $L\ddot{Q} - R_\Omega \dot{Q} + (Q/C) = 0$. The time scale for transfer of energy from electrostatic to magnetic fields (reactive time scale) is $t_R = (LC)^{1/2} = P/2\pi$. Ohmic dissipation rate is $R_\Omega \dot{Q}^2$, and magnetic energy is $L\dot{Q}^2/2$, so dissipation time is $L/2R_\Omega$.

The above mechanism has important implications for reconnection of magnetic field lines (Gekelman and Pfister 1988, Cowley 1981, Baum and Bratenahl 1980). Field lines can reconnect only in the time scale for dissipation of the induced currents which oppose any change of magnetic field, and this time has usually been assumed to be that for Ohmic dissipation, which is uncomfortably long (e.g. Wild et al. 1981). If we recognize that magnetic field energy can be converted into electrostatic field energy in the reactive time scale (17), and then dissipated by charges which are accelerated to velocities above escape velocity, the difficulty disappears. It follows that acceleration of charges will always accompany reconnection of magnetic field lines, the particles being accelerated in the direction normal both to the approaching opposite field lines and to the velocity with which they approach. One might regard electrons with collimated motions as a hallmark of field-line reconnection, a conclusion not at present recognized by laboratory plasma physicists.

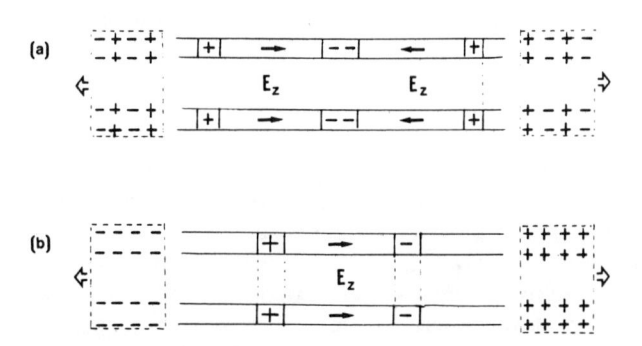

Figure 3. In situations where energy oscillates between an azimuthal magnetic field B_φ and an axial electrostatic field E_z, bunches of charged particles are accelerated in E_z, emerging as beams which are macroscopically neutral when E_z reverses polarity (case a) and are macroscopically charged when E_z does not reverse polarity (case b).

3. Cosmic Rays

(i) Origin of High Energy Cosmic Rays

Acceleration mechanisms hitherto considered fail to produce particles with the highest cosmic ray energies $\simeq 10^{21}$ eV. The mechanism herein proposed succeeds in this objective. The mechanism operates at a pinched region of a magnetic vortex tube (MVT) which enters and leaves the polar caps of a star or galaxy. In both the stellar and galactic cases, the maximum cosmic ray energies can be achieved.

In the stellar case, the acceleration occurs in MVTs associated with highly degenerate stars, specifically some X-ray pulsars (Joss and Rappaport 1984) identified with white dwarfs, and most γ-ray burst sources (Mazets et al. 1981) here identified with neutron stars. Photons with energies in the range 2–20 PeV (1 PeV = 10^{15} eV) from some X-ray pulsars, notably Cyg X-3, Her X-1, LMC X-4 and Vela X-1, have been detected by air-shower techniques, identification being made by the periods characteristic of the sources (see review by Hillas 1984). There can be no doubt that particles are accelerated to energies $> 10^{16}$ eV in these sources. Cyclotron lines from the sources confirm the presence of magnetic fields in the terragauss range (1 TG $\equiv 10^{12}$ G). Cyclotron lines in the range 20–70 keV are seen also in many γ-ray burst sources, demonstrating the presence of fields in the 1.7–6 TG range.

(ii) Cosmic Rays from Stellar MVTs

X-ray bursters are white dwarf stars which exhibit bursts of X-rays with high luminosities ($10^{36} - 3 \times 10^{38}$ erg s^{-1}) and short durations (10–1000 s), repeating with irregular intervals of 10^4–10^5 s. The spectrum is that of a black body of changing temperature and sometimes changing radius. At peak a typical temperature is $kT \simeq 3$ keV. The model which best explains such bursts has been mentioned briefly previously (Browne 1986, 1990, 1991), and will be presented in more detail elsewhere; it differs radically from the conventional model of accretion from a companion star in a binary system. The new model attributes the X-rays to thermal emission from the hot isothermal interior of a degenerate star which is exposed transiently to view through an MVT "window". The transient window is created when gas in the core of a polar MVT is expelled by radiation pressure, creating a low density path to the degenerate interior which cannot refill by lateral inflow due to the strong magnetic field at the MVT wall. The reason for the "blow out" is an imbalance between internal power generation P and surface luminosity L, causing a slow increase of radiation pressure at the base of the surface layer of normal gas. Note that isothermal degenerate gas cannot expand in response to a pressure gradient, as would happen in a normal star. The surface layer "punctures" where a polar MVT enters the star, because here the threshold pressure for "lift off" is least. After release of excess energy the outflow ceases and the window closes. The process now repeats because the imbalance $P > L$ is reestablished. The star responds to the imbalance by release of excess energy in a quasi-periodic sequence of outbursts. The precise flux profile of the X-ray burst depends on the degree to which the expelled gas front is transparent to the X-rays from beneath. As expelled gas is first compressed the optical thickness of the front can increase before it decreases, producing a double-peak flux profile. The values $L = 3 \times 10^{38}$ s^{-1} and $kT = 3$ keV imply emitting area πR^2, where $R = 10$ km, the "blackbody radius" of the source. This radius now is that of the MVT where it enters the star.

The same model can be applied to γ-ray bursts, but with the more extreme parameters of neutron stars. The sources of γ-ray bursts are unknown, because usually no identifiable source is found in the direction from which the burst comes. In general the bursts appear not to repeat, but there are a few exceptions, the so-called "repeaters". The repeaters are distinguished in another respect; their flux is considerably softer than that of the majority of bursts. Previously (Browne 1990, 1991) I have suggested that the softening of the flux (from $kT \simeq 300$ keV to $kT \simeq 30$ keV) is due to reprocessing in the escaping gas in analogy to reprocessing in the X-ray bursts, specifically to Compton scattering in the outflowing plasma. Why the majority of bursts appear not to repeat can now be understood; their emission is directed into a small solid angle Ω about a direction which varies from burst to burst, so that the chance of the Earth intercepting successive bursts is small. In the case of the repeaters, the Compton scattering enlarges Ω sufficiently for successive bursts to be detected.

One burst, GBS 05236 occurring on 5 March 1979, came from a direction which was located to within 0.1 arcmin^2 by nine widely separated spacecraft (Cline et al. 1982). In this direction only, the supernova remnant N49 in the Large Magellanic Cloud is within the error box. This would place the source at the distance 55 kpc, in which case its luminosity is 7×10^{44} $(\Omega/4\pi)$ erg s^{-1}, where Ω is the solid angle into which the emission is directed. The γ-ray rays had energies extending above 300 keV during the initial spike, but for the lower intensity flux following the initial spike the spectrum was considerably softer with $kT \approx 30$ keV (Golenetskii et al. 1984), suggesting reprocessing of the primary flux. Evidently the outflow becomes optically thick to Compton scattering, which requires a column density as high as $\approx 10^{24}$ cm^{-2}, but this is to be expected for gas escaping from a neutron star. Confirmation of the reprocessing comes from the observation of repeat bursts (Rothschild and Lingenfelder 1984). The blackbody temperature $kT \approx 300$ keV is typical for primary escaping flux. Values between 90 keV and 870 keV are inferred by Mazets et al. (1981) from a survey of spectra. Choosing arbitrarily $\Omega = 0.1$ sterad, we find $L \approx 5.6 \times 10^{43}$ erg s^{-1}, and then $kT = 300$ keV yields $R = 0.46$ km.

For $kT = 300$ keV the degenerate plasma of the neutron star is relativistic, because $kT/mc^2 \approx 3/5$. There will be a high rate of pair creations, and for equilibrium density of electrons and positrons there will be a high density of positronium, which annihilates into two 511 keV photons. A broad 511 keV line is seen in most γ-ray bursts, but with a substantial redshift. In GBS 05236 the line is seen at 430 keV, implying redshift $z = 511/430 = 1.19$. The redshift can be attributed to the transverse (second order) Doppler effect, $z = (1 - v^2/c^2)^{-1/2} \equiv \gamma$, where v is thermal velocity of positronium. Then from $3kT/2 = 2\gamma mc^2$ we obtain $kT = 128$ keV. This temperature may refer to cooler outflowing plasma.

The γ-ray burst sources exhibit in addition to the positronium line also a cyclotron line, usually in absorption (Mazets et al. 1981). The cyclotron line is seen in the range 20–70 keV, which implies the presence of a magnetic field in the range 1.7–6.0 TG. If one substitutes into formula (15) the values $B \approx 6$ G and $R \approx 0.5$ km, then with $K' = 3$ we obtain $W = 3 \times 10^{20}$ eV, in good agreement with the highest cosmic ray energy.

The isotropic distribution of γ-ray bursts has interesting implications for the number of undetected neutron stars surrounding the Earth, and possibly for the undetected dark matter of the universe.

(iii) Cosmic Rays from Galactic MVTs

Turning now to galactic MVTs, one might adopt $B_\varphi \approx 10$ mG and $R \approx 10^{13}$ cm. The value of R comes from the shortest time scales for variability of X-ray flux from AGN, and independent justification is given in section 7 (iii). The value for B_φ comes from the observed synchrotron spectra and flux densities (section 4i). With these values, formula (15) yields $\gamma = 2 \times 10^8$ or $W \approx 10^{14}$ eV.

The highest energy synchrotron emission from a quasar is probably that of PKS 0208-512 which extends from 30 MeV to 4 GeV with power-law flux spectrum, where $\alpha = 0.69$ (Bertsch et al. 1993). Synchrotron emission at 4 GeV in a magnetic field of 10 mG requires that $\gamma \approx 6 \times 10^9$ according to (18), which is electron energy 3×10^{15} eV.

Whether the value of B_φ at a magnetic pinch can be great enough to provide ions of energy 10^{21} eV remains an open question. The threshold condition for the Cerenkov continuum (section 7vi) requires that $\gamma^2 n_H > 2 \times 10^{23}$ cm^{-3}, so that if $n_H \approx 2000$ cm^{-3} we require $\gamma > 10^{10}$ for ions. Such a proton would have energy $> 10^{19}$ eV, which is approaching the maximum cosmic ray energy.

4. Synchrotron Radiation

(i) Non-uniform Field Model

Active galactic nuclei (AGNs), which embrace quasi-stellar objects (QSOs), BL Lac objects, and Seyfert 1 nuclei, radiate a power-law continuum spectrum extending from radio to X-ray frequencies. Flux density $\Im_\nu \delta\nu$ in spectral range $\nu \to \nu + \delta\nu$ has the form $\Im_\nu \propto \nu^{-\alpha}$ over the frequency range $10^8 < \nu < 10^{18}$ Hz for a large number of sources. In general, $\alpha \approx 1$ (Neugebauer et al. 1979, Richstone and Schmidt 1980, Moore and Stockman 1981, Cruz-Gonzalez and Huchra 1984, Ledden and O'Dell 1985).

Electrons which have been accelerated in the axial field $E_{\hat{z}}$ in a denuded region of current channel (section 2) emerge into a non-uniform magnetic field $B_\varphi(r) = 2J/rc$. Synchrotron power radiated by an electron with Lorentz factor γ in a transverse magnetic field B_φ occurs at frequencies

$$\nu \approx \frac{\gamma^2 e B_\varphi}{2\pi mc} = 2.8 \times 10^6 \gamma^2 B_\varphi \qquad (18)$$

Traditionally, one obtains the spectrum for a distribution of electron energies in a uniform magnetic field. Here we shall obtain the spectrum for monoenergetic electrons (of Lorentz factor γ) in a nonuniform magnetic field ($B_\varphi = 2J/rc$). The density of electrons $n_e(r)$ also may depend on r. The task is to calculate $\Im_\nu \delta\nu$.

Noting that

$$\nu \propto \gamma^2 B_\varphi(r) \propto r^{-1} \qquad (19)$$

the emission in frequency band $\nu \to \nu + \delta\nu$ comes from electrons in radius band $r \to r + \delta r$, where $\delta\nu/\nu = -\delta r/r$. An electron radiates synchrotron power,

$$P_s = \frac{2}{3} d^2 c \gamma^2 B_\varphi^2(r) \qquad (20)$$

where $d = e^2/mc^2$. Electrons in the cylindrical annulus $r \to r + \delta r$ of length Z generate synchrotron luminosity

$$L_\nu \delta\nu = 2\pi r \delta r Z n_e(r) P_s \frac{\delta\nu}{\nu} \qquad (21)$$

Noting that $\delta v/\delta r = -v/r$ and using $B_\varphi(r) = 2J/rc$, we obtain

$$vL_v = \frac{16\pi}{3c} d^2 Z n_e(r) \gamma^2 J^2 \tag{22}$$

Writing $n_e(r) \propto r^{-\delta} \propto v^\delta$, one obtains the power-law spectrum

$$L_v \propto v^{-\alpha}, \quad \alpha = 1 - \delta \tag{23}$$

Most observed spectra fit this law with values of α between 1 and 0. The value $\alpha = 1$ implies $\delta = 0$, and hence uniform electron density n_e. The value $\alpha = 0$ implies $n_e(r) \propto r^{-1}$. The BL Lac source AO 235+164 shows a flat spectrum over five orders of magnitude in an unresolved core of < 0.1 mas (Jones et al. 1984). The variation $n_e(r) \propto r^{-1}$ seems to hold close to the acceleration site.

The values $R = 3 \times 10^{12}$ cm and $B_\varphi(R) = 10$ mG (appropriate for the pinch) in the acceleration formula (15) yield $W = 10^{14}$ eV, corresponding to electron Lorentz factor $\gamma = 2 \times 10^8$. Then $J/c = RB_\varphi(r)/2 = 1.5 \times 10^{10}$ emu. Then, taking the length of the radiating column to be $Z = 4 \times 10^{15}$ cm, and the electron density to be 3×10^6 cm^{-3} (column density $Zn_e = 1.2 \times 10^{21}$ cm^{-2}) one obtains from (22) $vL_v = 4.3 \times 10^{45}$ erg s^{-1}, which agrees with observation (Moore and Stockman 1981).

Superimposed on the v^{-1} spectrum of QSOs is a high energy tail in some sources. Centaurus A has a power-law intensity spectrum with index $\alpha = 0.59$ between 70 keV and 8 MeV (Gehrels et al. 1984). The Seyfert galaxy NGC 4151 has a power-law spectrum with $\alpha = 0.6$ in the range 2 keV–2 MeV (Baity et al. 1984), but spectral indices $\alpha = 0.1, -0.1, 0.43$ and 0.53 also have been observed (see Ubertini et al. 1984). The QSO 3C 273 shows a 30–100 keV power-law spectrum with $\alpha = 0.3$ (Bezler et al. 1984). The QSO PKS 0208-512 has γ-ray flux from 30 MeV to 4 GeV with spectral index $\alpha = 0.69$ (Bertsch et al. 1993)..

One expects that synchrotron radiation is beamed into a cone of solid angle $\Omega(\lambda)$ which depends on wavelength λ. The forward-beaming of radiation from an individual electron with Lorentz factor γ is into a cone of semi-angle γ^{-1}, which is exceedingly small for $\gamma \approx 10^8$, so that overall beaming is determined by the spread of electron trajectories. The trajectories will be most collimated close to the MVT wall of radius R where X-rays are produced, and will be most spread at large values of r where radio emission is generated.

The limited radiative lifetimes of electrons in hot-spots of extended radio-jets (Kapahi and Schilizzi (1979) and optical emissions from jets or knots (Butcherer et al. 1980, Meisenheimer and Roser 1986) both are arguments for local acceleration of electrons. According to Meisenheimer and Heavens (1986) shock waves can account for acceleration in the hot spot of the jet of 3C 273.

(ii) Variability

Because the X-rays arise close to the MVT core of radius R, the time scale for variations of flux will be shortest for X-rays and longest for radio flux. This is observed. For the BL Lac source A0235+164. Jones et al. (1984) write (omitting their references):

Perhaps the best known property of AO 0235+164 is its variability, which is dramatic at both radio and optical wavelengths. The time scale of this variability ranges from months at low frequencies to weeks at frequencies ≈ 10 GHz to days at optical frequencies.

QSO 4C 29.45 has shown optical variability on time scale 12 min (Grauer 1984). X-ray flux varies on the shortest time scales. Matilsky et al. (1982) report 200 s variability of X-rays in QSO 1525+227. Marshall et al. (1983) report 70% increase of X-ray flux in 150 s for Seyfert galaxy NGC 4051. Tennant et al. (1981) have reported 100 s X-ray fluctuations in Seyfert galaxy NGC 6814, and Kuneida et al. (1990) found a factor of two change of 2–20 keV flux in 50 s and a correlated change in a 6.35 keV line with delay less than 256 s. A dimension of only 1.5×10^{12} cm is implied for NGC 6814.

At radio frequencies polarization position angle χ may rotate through more than a revolution. During the 1975 outburst in AO 0235+164, χ at centimeter wavelengths rotated over nearly 180° in an apparently linear manner (Ledden and Aller 1979). Similar rotations were found in three other sources, 0607-157, 0727-115, and BL Lac (Aller et al. 1981). In the latter two sources rotations exceeding 360° have been observed. In regard to 0727-115, Aller et al. write:

Although the first year of data is consistent with a steady rate of rotation, the remainder of the changes of χ appear to occur in a series of jumps.

Jumps of between 55° and 82° occurred in single months. In BL Lac, total rotation range was 440° in a 38 day period. Large changes of χ coincide with outbursts of flux. After an outburst, χ at 8.0 GHz and 14.5 GHz differed by 360° from χ at 4.8 GHz, which had not changed. Infrared polarization behaves similarly (Holmes et al. 1984a, Holmes et al 1984b).

The model which best accounts for the above behaviour envisages a bunch of relativistic electrons with collimated motions propagating toward the observer through the core of an MVT which points toward the observer. The core of the MVT can be field-free, or may have a weak purely axial field $B_{\mathcal{g}}$, but the strong helical magnetic field at the wall traps the electrons. Such a field configuration can be seen for transverse viewing in the case of 3C 66B, mapped by the Hubble Space telescope (Macchetto et al. 1991). Each time the electron is deflected toward the axis by the wall magnetic field, its acceleration is toward the axis in the direction specified by the azimuthal coordinate ϕ of the reflection point so that $\chi = \phi$. Successive reflections cause a series of jumps of χ. The jump is 180 for an electron whose trajectory lies on a plane containing the axis of the MVT. More generally, the electron follows the path of a light ray in an off-axis mode of an optical fibre. If electrons follow a helical trajectory and are deflected by a helical field, the polarization position angle χ may rotate smoothly with time.

In B2 1308+326 optical polarization has position angle χ which varies with wavelength λ as well as with time t or axial distance z (Sitko et al. 1984). On March 16, 1983, $\Delta\chi/\Delta t$ was $-8°$ per day, and $\Delta\chi/\Delta\lambda$ was $+6°$ per 0.5 μm. On June 8, 1983, both quantities had reversed sign, $\Delta\chi/\Delta t$ becoming $+5.5°$ per day and $\Delta\chi/\Delta\lambda$ becoming $-6°$ per 0.5 μm. Since $dx/d\lambda = (dx/dz)(dz/d\lambda)$, it follows that $d\lambda/dz$ does not reverse sign, and hence dB_φ/dz does not reverse sign, because for synchrotron radiation $\nu \propto B_\varphi$ according to (25). It follows, therefore that dB_φ/dz remains positive, as would happen if electrons were approaching a magnetic mirror or were entering a magnetic pinch without passing through it. The field configuration shown in Figure 1 also is excluded.

With regard to polarization, we expect bimodal behaviour. (1) Large rotations of χ are observed only when the line of sight is nearly parallel to the MVT axis, which sources will tend to be brightest. (2) Off-axis viewing selects emission from those electrons which have been deflected into the line of sight, so that acceleration is in the plane containing the line of sight and the MVT axis, causing a preferred χ and small percentage polarization for the majority of less bright sources. Exactly this behaviour has been observed. To quote from Angel and Stockman (1980), one luminous group (with 9 quoted examples) "shows no tendency for a preferred angle, and data span all points of the compass", whilst in the case of another less luminous group (with 12 quoted examples) "repeated measurements show a restricted range of angles".

Another finding from VLBI measurements is a "strong trend for the radio E vector to be normal to, and for the optical E vector to be parallel to, the VLBI structural axis" (Rusk and Seaquist 1985). The radio emission must come from regions where $B_{\mathfrak{z}}$ dominates, and the optical emission from regions where B_φ dominates, assuming transverse viewing in each case. When B_φ dominates, the jet edges should be circularly polarized and the center linearly polarized for acceleration parallel to the axis.

(iii) Superluminal Expansion

The phenomenon of superluminal expansion refers to the expansion of radio knots at angular rates which imply, at the distance of the source, superluminal velocities. The phenomenon can be explained by assuming that the source approaches the observer with ultrarelativistic motion in a direction making angle θ with the line of sight where $0 < \theta < \pi/2$. Resolving the velocity βc of the source into components ($\beta c \cos\theta$, $\beta c \sin\theta$) parallel and transverse to the line of sight, successive light signals which are emitted with time separation Δt are received with time separation, $\Delta t' = (1 - \beta \cos\theta)\Delta t$, due to Doppler effect. This shortening makes the transverse displacement of the source $\beta c(\sin\theta)\Delta t$ seem to occur in time $\Delta t'$, which represents transverse velocity

$$v_\perp = \beta c \sin\theta \frac{\Delta t}{\Delta t'} = \frac{\beta c \sin\theta}{1 - \beta \cos\theta} \tag{24}$$

If the moving radio component is a bunch of relativistic electrons, a circumstance that might now be predicted, then $\beta = 1$, so that

$$v_\perp \simeq c \cot\left(\frac{\theta}{2}\right), \quad (\beta \simeq 1) \tag{25}$$

If $\theta < \pi/2$, it follows that $v_\perp > c$. By measuring $v_\perp$, we can infer θ. For example, a single radio knot expands from the nucleus of 3C 273 with angular velocity 0.76 milliarcsecond per year. From the redshift $z = 0.158$, we assign distance 1.47×10^{27} cm to 3C 273. Then $v_\perp = 9.6c$ (Pearson et al. 1981). Substituting this value into (25), it follows that $\theta = 12°$.

Emitted knots may change direction, so that $v_\perp$ changes along their trajectory, as has been shown for 3C 345 (Biretta et al. 1983, Moore et al. 1983). Presumably the electron bunch is guided by being confined to the core of a curved MVT. The distinction between the moving knots, which are electron bunches with superluminal apparent motions, and stationary knots should be noted. For example, synchrotron emission with a power-law spectrum from radio to X-rays comes from stationary knots in Centaurus A (M87) at distance 5 Mpc (Burns et al. 1983). The stationary knots presumably are regions where $B_\varphi > B_{\bar{g}}$, perhaps magnetic pinches.

Relationship (25) would provide an absolute measurement of $v_\perp$ if θ could be measured independently of this relationship. Lynden-Bell (1977) postulated that separating superluminal knots seen in 3C 120 were emitted in opposite directions. This enabled him to write two equations

$$v_{\perp 1} = c \cot\left(\frac{\theta_1}{2}\right), \quad v_{\perp 2} = c \cot\left(\frac{\theta_2}{2}\right) \qquad (26)$$

where $\theta_1 + \theta_2 = \pi$. Thus $v_{\perp 2} = c \tan(\theta_1/2)$ so that $v_{\perp 1} v_{\perp 2} = c^2$. Then, if one measures angular expansion rates $\alpha_1 = v_{\perp 1}/d$ and $\alpha_2 = v_{\perp 2}/d$, where d is the distance of the source, we obtain $d = c(\alpha_1 \alpha_2)^{-\frac{1}{2}}$. However, it is unlikely that the assumption $\theta_1 + \theta_2 = \pi$ is valid for 3C 120 (Cohen et al. 1979). What is noteworthy is that any relationship between θ_1 and θ_2 yields a reliable measurement of the Hubble constant.

5. Galactic Magnetic Vortex Tubes

(i) Trailing Magnetic Vortex Tubes

As discussed elsewhere (Browne 1985, 1993) an axial vortex forms during early evolution of both stars and galaxies at a stage when both still are surrounded by discs. In the stellar case, bipolar outflows occur orthogonally to the plane of the disc. In the galactic case, bipolar outflows from QSOs are indicated by the Doppler shifts of the narrow absorption lines which are consistent with flows driven by resonance-line radiation pressure, in this case a Cerenkov field. The broad emission lines originate in the same outflow in response to the Cerenkov resonance line fields, the broadening being attributed to saturation pumping (Browne 1983, see section 7). Cloud models of these sources are not required, and are totally misleading in regard to the picture they evoke.

One usually observes a pair of synchrotron jets extending in opposite directions along the rotation axis of the galaxy. A synchrotron jet will be assumed to delineate a magnetic vortex tube, which will distort because vortex lines comove with the fluid medium. Often the jets trail as the galaxy moves through a material medium. The various morphologies of jet-counterjet systems can be understood as trailing of different degrees viewed from different perspectives (Browne 1985).

The synchrotron emission from jets is anisotropic, being beamed along the jet axis. Such sources are therefore brighter when the MVT is viewed "end-on". A classic example is Cygnus A. Assuming Cygnus A has motion along the line of sight, the trailing jets are seen transversely close to the galaxy but axially at the extremities where jets are tangent to the line of sight. The transition from ordered filaments in the jet to disordered filaments in the lobes (Perley et al. 1984) is understandable if the latter are merely end-on projections of the former. The source will be brightest at the extremities, explaining the symmetric radio lobes. The radio lobes will be equidistant from the galaxy and collinear with the galaxy, all three being coplanar. The observation of a bow shock wave outside the trailing arms (Carilli et al. 1988) also can be expected.

The filamentary character of synchrotron jets in general is a consequence of each MVT being a bundle of sub-MVTs. The range of jet scales, from < 1 pc to > 1 Mpc, can be expected for the same reason, as can their co-alignment (Bridle and Perley 1984).

(ii) Precessing Magnetic Vortex Tubes

Magnetic vortex tubes have a tendency to precess. The reason is a change in the axis of resultant angular momentum of fluid being expelled, which is the cause of precession of the vortex at the exit of a bathtub. The jets of galaxies in some cases show clear evidence of precession in the form of periodic wiggles (Icke 1981) or circular arcs (Dreher and Feigelson 1984, Hunstead et al. 1984). Luminous arcs on the scale of clusters of galaxies, with center of curvature toward a cD galaxy at the center of the cluster (Paczynski 1987, Lynds and Petrosian 1989), may indicate precession of an MVT of scale comparable to the cluster.

(iii) Pairs of Magnetic Vortex Tubes

There is observational evidence for pairing of QSOs. The classic double quasar 0957+561 has components with identical spectra separated by 6 arcsec (Walsh et al. 1979). The hypothesis of a double image of a single QSO stimulated a search for further examples. The idea is that the bending of light in the gravitational field of an intervening galaxy is comparable to the bending in a Fresnel biprism or a Billet split lens. There followed the discovery of PG1115+080, a triple source with member separations < 2.7 arcsec (Weymann et al. 1980), 2345+007 (Weedman et al. 1982), 2016+112 (Lawrence et al. 1984), 1635+267 (Djorgovski and Spinrad 1984) and 2237+030 (Huchra et al. 1985). A pair 1146+111 B,C with large separation of 157 arcsec, showed spectral similarities (Turner et al. 1986), but later differences were found in the red (Shaver and Cristiani 1986) and in the ultraviolet (Huchra 1986). Djorgovski et al. (1987) have noted a pair of sources QQ1145-071 which would be classified as a double image on the basis of optical data alone, yet cannot be such when radio data are considered. The existence of slight differences between members of a pair which have strong similarities makes the double-image hypothesis unlikely for any pairs. There exist also close pairs ($<1'-2'$) whose members have quite different properties; for example, in 1038+528 (33") the member redshifts are 0.678 and 2.296 (Owen et al. 1980). A common absorption system has been found for the pair Q0307-195 A,B (58") (Shaver and Robertson 1983).

If the MVT emanating from either side of a galaxy trails after the galaxy which recedes along the line of sight, so that the line of sight is tangent to the MVTs (as happens in Cyg A) the beamed emission from the MVT and counter MVT could produce a pair of QSOs. It is possible for a pair of MVTs to emanate from the same side of a galactic disc, an example being 3C 75 (Owen et al. 1985). It is required, of course, that acceleration in both MVTs produces identical spectra, both in continuum and line emissions. Certainly, if one does not believe the double image hypothesis, the pairing of QSOs contains an important clue as to their character.

The distribution of QSOs in the sky can show other types of order. For example, the linear alignment of eight QSOs across the galaxy NHC 3384 (Arp et al. 1979) might indicate a spiral MVT, with a QSO being seen at points where the line of sight is tangent to the spiral. Arp and Hazard (1980) found six QSOs in a region of dimensions 15'×15' with redshifts 2.22, 2.12, 1.10, 1.01, 1.01 and 0.86, and also three nearby QSOs with redshifts of 1.93, 1.89 and 1.67.

6. Cerenkov Resonance-Line Radiation

(i) Resonance Line Cerenkov Field

The rate at which an ultrarelativistic particle of charge q moving with velocity βc along the z-axis loses energy to a polarizable medium with dielectric constant ϵ underlies the calculation of Cerenkov radiation (Jackson 1975). An atom at position (0, b, 0) experiences a pulse of electric field (E_x, E_y, 0) and of magnetic field (0, 0, B_z) given by

$$E_z = -\epsilon^{-1} q\gamma \beta c t \left(b^2 + \gamma^2 \beta^2 c^2 t^2\right)^{-3/2}$$
$$E_y = -\epsilon^{-1} q\gamma b \left(b^2 + \gamma^2 \beta^2 c^2 t^2\right)^{-3/2} \tag{27}$$
$$B_x = -\epsilon \beta E_y$$

where $\gamma = \left(1 - \epsilon \beta^2\right)^{-1/2}$ with $\epsilon = \epsilon(\omega)$. We obtain the Fourier transforms

$$E_y(\omega) = (2\pi)^{-1/2} \int_{-\infty}^{\infty} E_y(t) \exp(i\omega t) dt$$

$$= (2\pi)^{-1/2} \int_{-\infty}^{\infty} \epsilon^{-1} q b \gamma \left(b^2 + \gamma^2 \beta^2 c^2 t^2\right)^{-3/2} \exp(i\omega t) dt \tag{28a}$$

$$= (2\pi)^{-1/2} \frac{2q\omega}{\epsilon \gamma \beta^2 c^2} K_1\left(\frac{\omega b}{\gamma \beta c}\right)$$

$$E_z(\omega) = (2\pi)^{-1/2} \int_{-\infty}^{\infty} E_z(t) \exp(i\omega t) dt$$

$$= (2\pi)^{-1/2} \int_{-\infty}^{\infty} \epsilon^{-1} q\gamma \beta c t \left(b^2 + \gamma^2 \beta^2 c^2 t^2\right)^{-3/2} \exp(i\omega t) dt \tag{28b}$$

$$= -i(2\pi)^{-1/2} \frac{2q\omega}{\gamma^2 \beta^2 c^2} K_0\left(\frac{\omega b}{\gamma \beta c}\right)$$

where $K_1(x)$ and $K_0(x)$ are modified Bessel functions of order unity and zero respectively. Defining

$$\lambda = \frac{\omega}{\gamma \beta c} = \frac{\omega}{\beta c}\left(1 - \epsilon \beta^2\right)^{1/2} \tag{29}$$

and using the asymptotic forms

$$x \gg 1: \quad K_\nu(x) \to \left(\frac{\pi}{2x}\right)^{1/2} \exp(-x) \tag{30}$$

we obtain

$$E_{\hat{z}}(\omega) \to \frac{iq\omega}{c^2}\left(1 - \frac{1}{\epsilon\beta^2}\right)(\lambda b)^{-1/2} \exp(-\lambda b)$$

$$E_y(\omega) \to \frac{q}{\epsilon\beta c}\left(\frac{\lambda}{b}\right)^{1/2} \exp(-\lambda b) \tag{31}$$

The flow of energy across a cylinder of radius b around the trajectory of the particle is given by the integral of the Poynting vector across this cylinder:

$$\begin{aligned}P_C &= \int_{-\infty}^{\infty} \frac{cE_{\hat{z}}B_x}{4\pi} 2\pi b\, dx = \tfrac{1}{2}\beta c^2 b \int_{-\infty}^{\infty} E_{\hat{z}}(t)B_x(t)dt \\ &= -\beta c^2 b \int_0^{\infty} B_x^*(\omega)E_{\hat{z}}(\omega)d\omega \\ &= -i\frac{q^2\beta}{c}\int_0^{\infty}\left(\frac{\lambda^*}{\lambda}\right)^{1/2}\omega\left(1-\epsilon^{-1}\beta^{-2}\right)\exp\left[-(\lambda^* + \lambda)b\right]d\omega\end{aligned} \tag{32}$$

where we use $dx = \beta c\, dt$ to convert the integral from dx to dt and then we convert from dt to $d\omega$ using

$$(2\pi)^{-1}\int_{-\infty}^{\infty} \exp[i(k-k')x]dx = \delta(k-k') \tag{33}$$

If P_C is not to decrease exponentially with b, we require that $\lambda + \lambda^* = 0$, which is satisfied by purely imaginary λ, requiring that $\epsilon\beta^2 > 1$. Then

$$P_C = \frac{q^2\beta}{c}\int\left(1-\epsilon^{-1}\beta^{-2}\right)\omega\, d\omega \tag{34}$$

Introducing index of refraction μ by $\mu^2 = \epsilon$, the condition for imaginary λ can be expressed by

$$\mu\beta > 1 \quad \text{or} \quad \beta^2\gamma^2(\epsilon - 1) > 1 \tag{35}$$

It is useful to compare the Cerenkov power per unit bandwidth $dP_C/d\omega$ with the synchrotron power per unit bandwidth obtained from (20) and (18). With $q = e$,

$$\frac{dP_C}{d\omega} = \frac{e^2\beta}{c}\left(1-\epsilon^{-1}\beta^{-2}\right)\omega$$

$$\frac{P_s}{\omega} = \frac{2}{3}deB = \frac{2e^2\omega}{3c\gamma^2} \qquad (36)$$

For the ratio we find

$$q \equiv \frac{dP_C}{d\omega}\cdot\frac{\omega}{P_s} = \frac{3\beta\gamma^2}{2}\left(1-\epsilon^{-1}\beta^{-2}\right) \qquad (37)$$

If we write $\epsilon\beta^2 = 1+\delta$, so that $\delta>0$ is the condition for Cerenkov emission, then $q = 3\gamma^2\delta/2\epsilon\beta \simeq (3/2)\gamma^2\delta$. Regarding the relationship between synchrotron and Cerenkov radiations, see Schwinger et al. (1976).

For a gas with n_i atoms per unit volume in state i one has

$$\epsilon - 1 = 4\pi dc^2 \sum_i \sum_j \frac{n_i f_{ij}}{\omega_{ij}^2 - \omega^2 + i\gamma_{ij}\omega} \qquad (38)$$

where f_{ij} is oscillator strength and γ_{ij} damping constant for transition $i \to j$ of frequency ω_{ij}. Near the resonance $\omega \simeq \omega_{ij}$, the summations can be reduced to two terms only. Noting that $f_{ji} = -f_{ij}$, and writing $\epsilon = \epsilon' + i\epsilon''$, one finds

$$\epsilon' - 1 = \frac{4\pi dc^2(n_i - n_j)f_{ij}(\omega_{ij}^2 - \omega^2)}{(\omega_{ij}^2 - \omega^2)^2 + \gamma_{ij}^2\omega^2} \qquad (39)$$

The condition (35) is most easily satisfied at resonance. However (39) assumes that all atoms are tuned within a natural linewidth γ_{ij} of ω_{ij}, whereas usually one the fraction $\gamma_{ij}/\delta\omega$ are so tuned, where $\delta\omega$ is the thermal broadening. Understanding $n_i - n_j$ to be the total population difference we must include a factor $\gamma_{ij}/\delta\omega$ in (39). Then, noting that $\omega_{ij}^2 - \omega^2 = \omega\gamma_{ij}$ we obtain at resonance the threshold condition

$$\epsilon' - 1 \simeq 4\pi dc^2 \frac{(n_i - n_j)f_{ij}}{\omega_{ij}\delta\omega} > (\beta\gamma)^{-2} \qquad (40)$$

This condition has been obtained previously (Browne 1983), but its implications have not been fully analysed.

Dielectric constant ϵ, refractive index μ, and electric susceptibility χ are related by $\mu^2 = \epsilon = 1 + 4\pi\chi$. Writing $\mu = \mu' + i\mu''$ and $\chi = \chi' + i\chi''$, and noting that $\mu' \simeq 1$ and $\mu'' \ll 1$, we obtain $\mu' - 1 = 2\pi\chi'$ and $\mu'' \simeq 2\pi\chi''$. Because the amplitude of an electromagnetic wave propagating along the x-axis is proportional to $\exp(i\mu\omega x/c) = \exp(-\mu''\omega x/c)\exp(i\mu'\omega x/c)$, and because intensity varies as the square of amplitude, the absorption coefficient is $\kappa = 2\mu''\omega/c = 4\pi\chi''\omega/c$.

(ii) Stimulated Raman Emission

In addition to elastic scattering due to the two-photon transition $i \to j \to k$ there is inelastic scattering due to the two-photon transition $i \to j \to k$ (stimulated

Raman scattering), where state k has energy between the energies of j and i and where the direct transition i → k is forbidden. In the elastic process a photon $\hbar\omega$ ($\simeq \hbar\omega_{ij}$) is absorbed and re-emitted. In the inelastic process a photon $\hbar\omega$ ($\simeq \hbar\omega_{ij}$) is absorbed and a photon $\hbar\omega'$ ($\simeq \hbar\omega_{jk}$) is emitted. The cross sections are comparable; specifically, the inelastic cross section is less than the elastic cross section by the factor $\gamma_{ij}\gamma_{jk}/\gamma^2$ where $\gamma = \sum_k \gamma_{jk}$ with summation being over all states between j and i (Loudon 1983). The resonance $\omega \simeq \omega_{ij}$ is not essential, but it greatly increases the cross section. The stimulated Raman scattering is proportional to the intensity of radiation fields at both frequencies ω and ω' (Yariv 1989).

The stimulated Raman transition has been used to obtain a submillimeter laser using a rotational transition of a molecule such as CH_3F in an excited vibrational state pumped by 10.6 μm radiation from a CO_2 laser (*e.g.* Levy 1981, Tobin 1985).

Coherence of the wavefunction is preserved during the 2-photon transition i → j → k. There is also a competing 2-transition process i → j followed by j → k in which coherence is not preserved. Competition between the two processes was observed in experiments of Wiggins *et al.* (1978) at different gas pressures.

The behaviour of an atom with natural frequencies $\omega_{ij} = (E_j - E_i)/\hbar$ which is exposed to strong radiation fields at frequencies ω and ω' has been treated by several authors (Javan 1957, Panock and Temkin 1977, Chang 1977). The atom is perturbed by the combined field,

$$E(t) = E_o \exp(i\omega t) + E'_o \exp(i\omega' t) \qquad (41)$$

where E_o and E'_o include phase factors $\exp(i\phi)$ and $\exp(i\phi')$. Associated with the transition dipole moments p_{ij} and p_{jk} are interaction energies and frequencies,

$$\Omega_{ij} = \frac{p_{ij}E_o}{2\hbar}, \quad \Omega_{jk} = \frac{p_{jk}E'_o}{2\hbar} \qquad (42)$$

Conventional time-dependent perturbation theory cannot be used when E_o and E'_o are strong. We require the polarization

$$P = [\chi(\omega)E_o \exp(i\omega t) + \chi(\omega')E'_o \exp(i\omega' t)] \qquad (43)$$

where $\chi(\omega)$ is complex electric susceptibility. Panock and Temkin (1978) use a density matrix approach to find

$$\chi(\omega) = \frac{p_{ij}^2(n_{ij} - \Omega_{ij}^2 RK)}{\hbar(i\gamma - \delta)}, \quad \chi(\omega') = \frac{p_{jk}^2(n_{jk} + \Omega_{jk}^2 RK)}{\hbar(i\gamma + \delta')} \qquad (44)$$

where

$$\delta = \omega - \omega_{ij} \qquad \delta' = \omega' - \omega_{jk}$$

$$K = n_{ij}(\delta - i\gamma)^{-1} + n_{jk}(\delta' + i\gamma)^{-1}$$

$$\delta' = \delta: \quad R = (\gamma^2 + \delta^2)\left[\delta(\Omega_{jk}^2 - \Omega_{ij}^2) + i\gamma(\gamma^2 + \delta^2 + \Omega_{ij}^2 + \Omega_{jk}^2)\right]^{-1}$$

The spontaneous decay time on switching off $E(t)$ is γ.

For saturation pumping, when populations n_j and n_i are comparable, both the resonance line absorption and the stimulated Raman emission transitions are greatly broadened by AC Stark effect. Moreover the line profile develops two peaks separated by the Rabi frequency $2\Omega_{ij} = p_{ij}E_o/\hbar$, where E_o is the amplitude of the pump field at frequency ω_{ij}. The blueward peak may be suppressed when a Cerenkov field (as opposed to a laser field) provides the pumping for the following imply asymmetry of the emission line. The stimulated Raman emission lines due to Cerenkov pumping will therefore have an intrinsic redshift of magnitude $\simeq \Delta\omega_{ij}$.

The Cerenkov field produces off-resonance pumping ($\delta > \gamma$) because it is confined to the redward wing of the resonance line. The various cases have been treated by Panock and Temkin (1977). When emission intensity is weak, so that $\Omega_{jk} < \gamma$, the emission profile shows two peaks of equal heights displaced by

$$\delta'_\pm = \frac{\delta}{2} \pm \left(\Omega_{ij}^2 + \delta^2/4\right)^{1/2} \simeq \pm\Omega_{ij} \qquad (45)$$

When emission intensity is strong, so that $\Omega_{jk} > \gamma$, the profile continues to have two peaks, but the redward peak at δ'_- increases whilst the blueward peak at δ'_+ decreases, and disappears when $\Omega_{jk} \gg \gamma$.

7. Cerenkov Effects

(i) Lα Forest

In the case of the Lyman α we have λ_{ij} = 1215 Å, v_{ij} = 2.47×10^{15} Hz, f_{ij} = 0.832 and γ_{ij} = 3.76×10^9 Hz. Then, assuming thermal velocity 150 km s^{-1}, condition (40) for Cerenkov emission yields $\gamma^2 n_H > 4.5\times 10^{19}$ cm^{-3}. Taking as previously $\gamma = 2\times 10^8$, the condition is $n_H > 1000$ cm^{-3}.

The strong Lα Cerenkov radiation will exert strong outward pressure on H atoms. Of course only photons which are scattered incoherently exert radiation pressure; the coherent component of the resonance scattering does not affect the motion of the scatterer. The result will be a high-velocity outflow channeled through the core of the MVT, rather like the bipolar outflows from protostars. The radiation pressure will provide continued acceleration, so one may expect the highest velocities to occur farthest from the source of the particle beam. Moreover, resonance-line radiation pressure on ions drives an axial current if the drag forces on ions and electrons differ, which is a radiation battery. Such a force due to resonance line radiation pressure segregates chemical elements in Ap stars (Browne 1968, 1993). The electrons are co-accelerated in a space charge field, so that a current depends on different drag forces for ions and electrons.

A characteristic of outflows driven by resonance-line radiation pressure is a tendency to form shells with optical depth unity, each shell being driven by radiation in a discrete band of frequencies $\delta\omega$ determined by velocity spread δv within the shell. Within an individual shell the atoms at the front are screened by the atoms at the rear, causing a circulation of atoms from rear to front and back again.

At resonance the cross section for scattering of a photon is $\sigma = \lambda_{ij}^2/2\pi$, where $\lambda_{ij} = 2\pi c/\omega_{ij}$ (Loudon 1983). If n_H is the density of H atoms, the fraction which at any instant are tuned to within γ_{ij} of radiation of frequency ω in thermal bandwidth $\delta\omega$ is $(\gamma_{ij}/\delta\omega)n_H$. Consequently, the shell thickness Δz required for optical depth unity at resonance is given by

$$\left(\lambda_{ij}^2/2\pi\right)\left(n_H\gamma_{ij}/\delta\omega\right)\Delta z = 1 \qquad (46)$$

For Lα we have $\lambda_{ij} = 1.215 \times 10^{15}$ cm, $\gamma_{ij} = 6.2 \times 10^8$ s^{-1} and $\delta\omega/\omega_{ij} = \delta v/c$ where $\delta v \approx 15$ km s^{-1} (Boksenberg 1978). Thus $\delta\omega/\gamma_{ij} \approx 200$, and (46) yields $n_H \Delta z = 8.8 \times 10^{12}$ cm^{-2}. If $n_H \approx 3 \times 10^6$ cm^{-3} then $\Delta z \approx 30$ km.

Different shells are accelerated by different bands of radiation, and they may attain different velocities accounting for the multiplicity of narrow Lα absorption lines. Because acceleration is continual, the largest velocities will be attained at the largest distances. Material at different distances may show quite different Doppler blueshifts.

The observations, reviewed by Strittmatter and Williams (1976), Perry et al. (1978) and Weymann et al. (1981), confirm these predictions in detail. The absorption line spectrum blueward of Lα. QSO PHL 957 shows well over 200 narrow lines (Lα forest). A separation of 149 km s^{-1} occurred at least seven times between Lα forest lines in PKS 0237-23 (Boksenberg and Sargent 1975). Perry et al. (1978) write:

> The extreme narrowness of the lines implies very small internal velocity dispersions in the absorbing gas at a few 10^4 °K. At very high resolution ($\Delta\lambda \leq 1$ Å) components as narrow as 10–20 km s^{-1} are reported in the optical data (Boksenberg 1978) and as narrow as 4 km s^{-1} in the 21-cm absorption in 3C 286 (Wolfe et al. 1976). If the gas is intrinsic to the QSO, this implies that there is no dispersion in velocity relative to the central source except for random thermal motions. Several of the narrow lines are essentially black, implying total obscuration of the central source.

The narrow absorption lines exhibit redshifts z_{abs} substantially less than the emission lines z_{em}. The difference is particularly large in PHL 938 ($z_{em} = 1.95$, $z_{abs} = 0.61$), where it implies relative outflow velocity $\approx 0.5 c$ (Burbidge et al. 1968).

In a few QSOs direct evidence for outflow comes from broad absorption troughs instead of a multitude of discrete lines. The instability leading to shells with discrete velocities evidently is not fully developed in these sources. The broad absorption troughs in PHL 5200 and RS 23 are adjacent to the emissions, but are separated from the emissions in Q1246-057 and in the latter source structure is developing at the bottom of the troughs (Perry et al. 1978). The latter source has evidently evolved further toward the Lα forest. No Lα emission usually is seen at the absorption redshifts (Elston et al. 1991).

The rich absorption line spectrum of QSO OQ 172 reveals three redshift systems, $z_{abs} = 3.092$, 2.651 and 3.066. The first two redshifts include over 20 absorption lines in the range $\lambda = 3325$–4530 Å attributed to the Lyman and Werner bands of the H$_2$ molecule, and for each redshift system column density $N(H_2) \approx 4 \times 10^{19}$ cm^{-2} is found (Levshakov and Varshalovich 1979). Total column density is proportional to the number of contributing shells, and to the random velocity spread δv within each shell. The value $\approx 10^{13}$ cm^{-2} inferred from (47) was for a shell with very small thermal velocity spread δv. Total column densities for HI range up to 10^{21} cm^{-2} in PHL 957, 1331+170 and 0528-250 (Perry et al. 1978). In Mk 231 a new absorption redshift has appeared at 4660 km s^{-1}, which is less than the systematic velocity by 8240 km s^{-1}. (Boroson et al. 1991).

Weymann et al. (1981) summarize the observations of redshift systems as follows (with reversal of their sign convention for outflow velocity v):

> (a) A narrow peak at $z_{abs} \approx z_{em}$ with systems extending to about -3000 km s^{-1}; (b) A nearly uniform distribution from $v \approx 0$ out to very large ($\approx -60{,}000$ km s^{-1}) velocities; (c) An excess of systems from the velocity of the peak at ≈ -3000 km s^{-1} to $\approx -20{,}000$ km s^{-1}.

(ii) Stimulated Raman Emission Lines

Cerenkov fields at the frequencies of $L\beta$, $L\gamma$, etc. of H will excite emission lines due to transitions $j \to k$ originating from the upper resonance level j. Specifically, in addition to the resonance elastic scattering $i \to j \to i$ there occurs resonance inelastic scattering $i \to j \to k$, where k is a level above the ground level i which cannot be populated directly because $i \to k$ is forbidden (section 5 ii). The transitions $i \to j$ followed by $j \to k$ may occur independently, with absorption of a photon $\hbar\omega \simeq \hbar\omega_{ij}$ and emission of a photon $\hbar\omega' \simeq \hbar\omega_{jk}$, or there may occur a two-photon single transition $i \to j \to k$ ("stimulated Raman scattering"). The probability of the latter is proportional to the product of the intensities of the two radiation fields $I(\omega)I'(\omega')$, so that at high intensities it is dominant. Resonance $\omega \simeq \omega_{ij}$ is not essential for the process, but it enhances the cross section greatly.

If the pumping field (usually a laser field, but here a Cerenkov field) at frequency ω is strong enough to saturate the $i \to j$ transition (equalize the populations in levels i and j), then the level j is much broadened by AC Stark effect. Both the $i \to j$ and $j \to k$ transitions are broadened. Moreover the broadened line profile develops two peaks separated by the Rabi frequency $\Omega_{ij} = p_{ij}E_o/2\hbar$, where p_{ij} is the transition electric dipole moment and E_o is the amplitude of the pump field. It is proposed to explain the broad emission lines of AGNs (with widths ten times the narrow emission line widths) as stimulated Raman transitions for a Cerenkov pumping field at resonance-line frequencies.

Because of anomalous dispersion the condition for a Cerenkov field $\beta\mu > 1$ can be satisfied only for the redward wing of a resonance line where $\mu > 1$ (Jackson 1975). Thus the pump field will be off-resonance with redward displacement $\simeq \Omega_{ij}$. From (45) the stimulated emission field also will be off-resonance with an approximately equal redward displacement. In other words, the broad emission lines have an intrinsic redshift Ω_{ij} proportional to the amplitude E_o of the pump field. Since E_o varies from resonance line to resonance line the intrinsic redshift will vary from line to line. In particular, E_o will be weakest for high ionization lines owing to low density of exciting charges. It is also possible for E_o to vary from region to region of the outflow.

There is no stimulated Raman scattering for $L\alpha$ pumping, but $L\beta$ pumping gives $H\alpha$ emission. The threshold condition for the $L\beta$ Cerenkov field is slightly more stringent than for the $L\alpha$ field. The values to be substituted into (40) now are $\omega_{ij} = 1.83 \times 10^{16}$ rad s^{-1}, $g_i f_{ij} = 0.158$, giving for thermal velocity 150 km s^{-1}, $\gamma^2 n_H > 3.3 \times 10^{20}$ cm^{-3}.

The existence of a threshold condition (40) will mean that certain sources exhibit broad lines whilst other sources, similar in other respects, do not. The classic example is the complete absence of emission lines in BL Lacs in contrast to QSOs, despite their otherwise similar properties. The sudden appearance of broad emission lines in BL Lac source PKS 0521-36 (Ulrich 1981) suggests that the threshold for Cerenkov emission, not normally met in BL Lacs, was suddenly exceeded in this source. The distinction between Seyfert 1 and Seyfert 2 nuclei may be another case. Without Cerenkov radiation no explanation is apparent. It is possible for the threshold to be exceeded in the core of an outflow but not in the periphery, in which case broad and narrow emission lines will coexist, as in Seyfert 1.5 galaxies.

(iii) Intrinsic Redshifts

The narrow-line regions show line asymmetry with excess blueward flux, whereas the broad lines show line asymmetry with usually excess redward flux. The excess blueward flux is due to Doppler effect for outflowing gas approaching the observer (section 5). The excess redward flux is due to superposition of an intrinsic redshift for the broad emission lines, usually but not always large enough to dominate the blueshift caused by outflow, because both sets of lines now originate in the same outflow.

Sulentic (1989) has made a general study for classification purposes of broad-line profiles in AGNs, for a total sample of 61 objects. He writes:

> Quasars and Seyfert galaxies (and also broad-line radio galaxies) exhibit more complex emission-line spectra that show both "narrow" (200–800 km s^{-1} FWHM) and broad (2500–6000 km s^{-1} FWHM) emission features. The fact that the narrow emission lines arise from forbidden transitions suggest that these lines arise from gas that is spatially (as well as kinematically) distinct from the broad line emitting gas. Thus the concept of two distinct zones in AGNs has arisen. Actually, recent observations suggest that, in many cases, no well defined boundary exists between these two regions; however, the broad (BLR) versus narrow (NLR) distinction is still clearly fundamental in some way.

These well-judged remarks are precisely what the present theory predicts. Both narrow and broad lines originate in the same region, and there is a fundamental distinction between them, but of physical emission process rather than region of origin or kinematics. The Cerenkov excitation process requires that the lines originate from levels connected to the ground level by permitted transitions.

Sulentic goes on to remark (omitting references):

> Both NLR and BLR profiles are complicated in at least two other ways: (1) they are frequently observed to be asymmetric and (2) redshifts determined from the NLR and BLR often differ significantly. The line shifts and asymmetries are now observed both toward the red and the blue with displacements as large as 5000 km s^{-1}. While existing models can deal with systematic red or blueward effects the presence of both greatly complicates the problem. There also appears to be a systematic blueshift of the high ionization relative to the low ionization, broad lines in these objects.

Again these observations are in complete accord with what we would now expect. The broad emission lines originate in outflowing gas because the Cerenkov pumping field exerts strong radiation pressure. Their intrinsic redshift therefore will be superimposed on a Doppler blue shift, and because in general the redshift outweighs the blueshift the asymmetry of the broad lines is toward the red most often. The lower redshift for higher ionization lines is predicted above (section (i)). The narrow emission lines are from atoms not subject to resonance-line radiation pressure, so that such atoms fall inwards if they are not entrained in the outflow.

The statistics for 61 sources are summarized by Sulentic as follows:

> (a) Only 9 objects show symmetric and unshifted profiles; (b) a large number (15) of displaced profiles are symmetric, with redshifts twice as common as blueshifts; (c) 35 objects show a measurable wavelength shift relative to the narrow-line region [O III] (from 120 to 1900 km s^{-1}), with redshifts (20) slightly more common than blueshifts; (d) a shifted profile is very unlikely to show a blue asymmetry (red asymmetries are 8 times more common); and (e) 37 objects show an asymmetrical profile with red (27) ones almost three times as common as blue.

(iv) Anomalous Balmer Decrements

If the broad lines are stimulated Raman scattering for a Cerenkov resonance-line field then levels connected to the ground level by allowed transitions will be heavily overpopulated relative to levels populated by recombinations. For recombinations in hydrogen, which is optically thick only in the Lyman lines (case B), the expected level populations yield line intensity ratios $L\alpha/H\alpha = 12$, $H\alpha/H\beta = 2.9$ and $P\alpha/H\beta = 0.3$, and hence $L\alpha/H\beta = 35$. The intensity ratios observed are $L\alpha/H\alpha \approx$ 1–2, $H\alpha/H\beta \approx 3$–5, and $P\alpha/H\beta \approx 0.3$ for QSOs (Soifer et al. 1981), and $L\alpha/H\beta \approx 22$ for Seyfert 1 nuclei (Clarke et al. 1986). In order to account for the anomalous Balmer decrements it has been necessary to consider multiple optically thick clouds occupying a small fraction of the volume (e.g. Canfield and Puetter 1981).

On recognizing the Cerenkov excitation mechanism, the problem disappears. The level populations will depend on the intensities of the resonance lines $L\beta$, $L\gamma$, etc. Thus $H\alpha$ is excited by the $L\beta$ Cerenkov field, whilst both $H\beta$ and $P\alpha$ are excited by the $L\gamma$ field. The intensity ratio $H\alpha/H\beta$ can therefore be highly anomalous, whilst the intensity ratio $P\alpha/H\beta$ must remain normal since both $H\beta$ and $P\alpha$ originate from the same level. This is observed.

The threshold condition (40) is most easily satisfied for H lines because of the abundance of H. Less abundant species will require higher values of γ to compensate for their reduced abundance, and since there will be less high γ exciting charges their Cerenkov fields will be weaker if they exist at all. One notes the role of thermal broadening in the threshold condition (40); the less is thermal broadening (for example within velocity-selected shells) the lower the threshold condition.

For 5 Seyfert 1 galaxies Clavel and Joly (1984) found a UV spectrum (1200–3200 Å) rich in emission lines, and in particular FeII emission from levels connected to the ground level by 5–9 eV resonance lines. The four main emission lines have different widths and profiles, which they ascribed to various sub-regions with different physical conditions. However, it is difficult to envisage different kinematic conditions specific to different spectral lines. The difficulty now disappears, because the broadening is not kinematical but proportional to the amplitude E_o of the Cerenkov pump field at the particular resonance line.

It should be noted that the stimulated Raman emission will populate fine structure levels of the ground electronic configuration. Populated fine structure levels have been observed, although collisions are held responsible. To quote from Perry et al. (1978):

> Of the ions usually observed in absorption, only C II, Si II and Fe II have ground-state fine structures; thus it is only in systems where these lines are seen that such arguments (density limits for collisional excitation) can be used. Despite the oft repeated statement that excited fine structure is almost never present and that therefore upper limits on the density $\sim 10^2$ cm^{-3} are general, when one examines the list of ~230 absorption systems one finds that actually 40% of them contain identifications of any C II, Si II, or Fe II lines and about 20% of that those that do, show fine structure lines.

(v) Broad-Line Variability

Variable line intensity may be expected for Cerenkov excited lines. Pressure waves which affect gas density will influence the strength of the Cerenkov pumping field, and hence the emission line strengths. Variability is a feature of the broad emission lines (Peterson *et al.* 1985, Zheng 1988). The continuum and Hβ usually co-vary, but not always. Time scales are a few months to a few years. One assumes that the broad emission lines originate from gas which is outflowing within the core of an MVT, the lines being pumped by the Cerenkov field generated by ultrarelativistic electrons also moving outward within the MVT core. Variation in the density of the outflowing gas will cause the Cerenkov pumped stimulated Raman emission lines to vary. The synchrotron radiation from the electrons also may vary because the B_φ/B_z ratio is affected. The velocity of the density wave is unknown but should be comparable to general outflow velocity, perhaps 100 km s^{-1} (see Sulentic 1989). Then the distance traveled in one year is 3×10^{14} cm.

(vi) Cerenkov Continuum

An interesting problem is posed by the so-called "blue-UV" bump which is superimposed on the power law continuum spectrum of QSOs and Seyfert I galaxies (Grandi and Phillips 1980, Malkan and Sargent 1982, Oke *et al.* 1984, Bechtold *et al.* 1987). This additional continuum tends to co-vary with the broad emission lines, an example being NGC 4151 (Antonucci and Cohen 1983). Thus there is reason to suppose that the blue-UV bump may represent Cerenkov continuum.

The threshold for Cerenkov continuum is considerably more stringent than for resonance-line fields. For the Hα line condition (4) is $\gamma^2 n_H > 2.3\times10^{17}(\delta\omega/\gamma_{ij})$ cm^{-3}, where $\delta\omega/\gamma_{ij}$ is the thermal broadening factor. It is possible for $\delta\omega/\gamma_{ij}$ to be as small as 200. For continuum emission $\delta\omega/\gamma_{ij}$ is replaced by $\omega_{ij}/\gamma_{ij} \approx 4\times10^6$. Thus the continuum threshold is a factor 2×10^4 above minimum threshold. Nevertheless, this higher threshold may well be reached.

(vii) Cerenkov Polarization

In the case of Cerenkov radiation one selects a direction of polarization by direction of viewing relative to the direction of motion of the ultrarelativistic exciting charges because of forward beaming of the emission. The preferred direction of acceleration, and hence of the E vector, lies in the plane containing the line of sight and the source axis. This should apply also to Cerenkov stimulated Raman transitions. Osterbrock and Matthews (1986) write:

> The emission lines in many AGNs are fairly strongly polarized, and in several Seyfert 1s the broad lines and narrow lines show different degrees of polarization (Schmidt and Miller 1985).

They suggest electron scattering, but now the explanation is different. The situation is somewhat analogous to that discussed in section 3(ii) for the polarization of the synchrotron emission of QSOs.

(viii) Dimensions of AGNs from AC Stark Effect

The observed value for Ω_{ij} yields a direct value for the amplitude E_o of the Cerenkov resonance-line field within the source. If this can be related to the amplitude of the same field at Earth then a value for R/d can be inferred, where R is source radius and d source distance. From (42) we have

$$2\hbar\Omega_{ij} = p_{ij}E_o \qquad (47)$$

where p_{ij} and E_o can be eliminated in favour of $g_i f_{ij}$ and $I(\omega_{ij})$ by

$$p_{ij} = \frac{3\hbar g_i f_{ij} e^2}{2m\omega}, \quad I(\omega_{ij}) = \frac{cE_o^2}{8\pi} \qquad (48)$$

resulting in

$$I(\omega_{ij}) = \frac{137}{3\pi} m\omega_{ij} \cdot \Omega_{ij}(g_i f_{ij})^{-1} \qquad (49)$$

Then, by relating pump intensity $I(\omega_{ij})$ at source to flux density $\mathfrak{J}(\omega_{ij})$ at Earth a value for the ratio of source radius R to source distance d is obtained.

For 3C 273 the equivalent width of $H\alpha$ is 330 Å (Oke 1965). We adopt half this value for Ω_{jk}, so that $\Omega_{jk}/\omega_{jk} = 165/6562 = 2.5 \times 10^{-2}$. For $L\beta$ we have $\omega_{ij} = 2\pi c/\lambda_{ij} = 1.84 \times 10^{16}$ rad s^{-1} and $g_i f_{ij} = 0.158$. Then (49) yields $I(\omega_{ij}) = 8 \times 10^{18}$ erg s^{-1} cm^{-2}. At Earth one observes $\mathfrak{J}(\omega_{ij}) = 3.8 \times 10^{-2}$ erg s^{-1} cm^{-2} for $L\beta$ (Boggess et al. 1979). Hence $R/d = 6.9 \times 10^{-16}$. But for 3C 273 the redshift $z = 0.158$ implies a distance of $d = 1.46 \times 10^{27}$ cm, for $H = 100$ km s^{-1} Mpc^{-1}. Hence $R = 1.4 \times 10^{12}$ cm.

The result is in agreement with source dimensions inferred from the time scale of most rapid X-ray variability. Kuneida et al. (1990) inferred a dimension 1.5×10^{12} cm for Seyfert galaxy NGC 6814 from a factor of two change in 2–20 keV X-ray flux in 50 s. If a reliable independent method of measuring R were available, it would be possible to use the foregoing argument to obtain an absolute measurement for d, and hence for the Hubble constant.

8. Summary

It is shown that magnetic vortex tubes (MVTs), which are essentially vortex tubes with a magnetic field twisted into helical form and concentrated toward the wall of the vortex tube (*i.e.* the core of radius R), plays a basic role in the synchrotron jets which emanate from active galactic nuclei (AGNs), including quasi-stellar objects (QSOs), BL Lac sources and Seyfert 1 nuclei. The MVTs channel outflows of gas, rather like the bipolar outflows from stars at an early phase of their evolution, when they are surrounded by discs. The mechanism of the outflow is exactly the same mechanism which drives the updraught at the centre of a terrestrial tornado (Browne 1991, 1993). The outer lobes of radio galaxies are supplied by gas in the same manner, but because the jets feeding the lobes are viewed transversely to their axes, the beamed emission in the jets does not intercept the Earth.

In order to account for the synchrotron radiation from these jets, one requires a mechanism for acceleration of charged particles to $>10^{15}$ eV, which is in the upper cosmic ray range, and the mechanism must provide collimated beams of charges in order to account for the beaming of the synchrotron radiation generated by the charges. This problem was solved previously (Browne 1985b, 1986, 1988a, b, c). This mechanism has been reviewed in section 2 of this paper because the production of charges with energies in the upper cosmic ray range is essential for the

Cerenkov broad line emissions from AGNs, was well as for their synchrotron radiation.

It is argued, again following previous work (Browne 1983), that the broad emission lines of AGNs are excited by a strong Cerenkov radiation field at resonance line frequencies. The Cerenkov field populates the upper level j from which the emission line transition j→k arises. The resonance transition i→j and the emission transition j→k are not sequential at strong pumping fields, but occur as a single transition involving two photons, the absorption of a photon $\hbar\omega(\approx \hbar\omega_{ij})$ and the emission of a photon $\hbar\omega'(\approx \hbar\omega_{ij})$. This process, stimulated Raman scattering, has a cross section comparable with resonance fluorescence i→j→i. It has been thoroughly investigated in quantum electronics in relation to far infrared lasers, although apparently unknown to astrophysicists. The broadening of the lines is not of Doppler origin, as commonly assumed, but is due to AC Stark effect associated with saturation pumping. Due to anomalous dispersion, a Cerenkov field can develop only for the redward wing of the resonance line, the Cerenkov threshold condition not being satisfied for the blueward wing, so that the broad emission lines always have an intrinsic redshift comparable to their broadening. However, this redshift is superimposed on a Doppler blueshift due to the outflow of the gas toward the observer. Numerous papers have been published attempting to fit the spectral observations to multiple cloud models, but, however contrived the envisaged physical scene, they have not been fully satisfactory. This problem now disappears. So do many other problems, such as the anomalous Balmer decrements.

The outflow is driven by resonance-line radiation pressure, which can persist for a very long distance from the source. The complexity of the absorption line blueshifts seen in QSO spectra is attributed to structure within the outflow. Because of self-screening of the resonance line flux, there is a tendency for the outflowing gas to split into numerous shells, each with optical thickness about unity to resonance radiation and each driven by radiation in a separate frequency band. The existence of different redshift systems in the rich absorption spectra is attributed to gas at different distance which has different ouflow velocities due to the continued acceleration. Thus, these absorption spectra are all associated with the source, and are not intergalactic, as often supposed.

It is shown that a dimension for the AGN jets at source can be inferred from the AC Stark effect, because the Rabi peaks in the line profile are displaced by an amount which depends on the intensity of line radiation at the source. Comparing this intensity with that at Earth, and knowing the distance of the source from the Hubble redshift, one obtains a redshift for the emitting area. A value of 1.4×10^{12} cm is obtained for 3C 273. An upper limit to source dimension of 1.5×10^{12} cm has been inferred for Seyfert galaxy NGC 6814 from the time-scale of X-ray variability (Kuneida et al. 1990).

In regard to the problem of intrinsic redshifts, one such effect has been identified as being superimposed on the Hubble redshift, but it is no greater than the maximum broadening of the lines. The agreement between the source dimension inferred from the AC Stark effect and X-ray variability tends to confirm that the sources really are at the distances indicated by the Hubble interpretations of their redshifts. The problems of large discrepancies between emission and absorption line redshifts has been resolved by recognizing that the outflow velocities can be relativistic. Nevertheless, there remains one other type of intrinsic redshift for QSOs, which can be comparable to the Hubble redshift, and which will be discussed in a subsequent paper.

References

Angel, J.R.P. and Stockman, H.S., 1980, *Ann. Rev. Astron. Astrophys.* 18:321.
Aller, H.D., Hodge, P.E. and Aller M.F., 1981, *Ap. J. (Letters)* 248:L5.
Antonucci, R.R.J. and Cohen, R.D., 1983, *Ap. J.* 271:564.
Arp, H., 1987, *Quasars, Redshifts and Controversies*, Interstellar Media.
Arp, H. and Hazard, C., 1980, *Ap. J.* 240:726.
Arp, H., Sulentic, J.W. and di Tullio, G., 1979, *Ap. J.* 229:489.
Baity, W.A., et al., 1984, *Ap. J.* 279:555.
Baum, P.J. and Bratenahl, A., 1980, *Adv. Electronics Electron Phys.* 54:1.
Bechtold, J., Czerny, B., Elvis, M., Fabbiano, G. and Green, R.F., 1987, *Ap. J.* 314:699.
Bertsch, D.L. et al., 1993, *Ap. J. (Letters)* 405:L21.
Bezler, M. et al., 1984, *Astron. Astrophys.* 136:351.
Biretta, J.A., Cohen, M.H., Unwin, S.C. and Pauliny-Toth, I.I.K., 1983, *Nature* 306:42.
Boggess, A. et al., 1979, *Ap. J. (Letters)* 230:L131.
Boksenberg, A., 1978, *Phys. Scripta* 17:205.
Boksenberg, A. and Sargent, W.L.W., 1975, *Ap. J.* 198:31.
Boroson, T.A., Meyers, K.A., Morris, S.L. and Persson, S.E., 1991, *Ap. J. (Letters)* 370:L19.
Browne, P.F., 1983, *Phys. Lett.* 99A:196.
Browne, P.F., 1985a, *Astron. Astrophys.* 144:298.
Browne, P.F., 1985b, in: ed. J.E. Dyson, *Active Galactic Nuclei*, Manchester Univ. Press, p. 365.
Browne, P.F., 1986, *IEEE Trans. Plasma Sci.* PS-14:718.
Browne, P.F., 1987, in: R. Beck and R. Grave (eds.), *Interstellar Magnetic Fields*, Springer-Verlag Publ., p.211.
Browne, P.F., 1988a, *Astron. Astrophys.* 193:334.
Browne, P.F., 1988b, *J. Phys. D: Appl. Phys.* 21:596.
Browne, P.F., 1988c, *Laser and Particle Beams* 6:409.
Browne, P.F., 1990, in: R. Beck et al., (eds.), *Galactic and Intergalactic Magnetic Fields*, IAU Sympos. No. 140, p. 136.
Browne, P.F., 1992, in: L. Vittone and L. Errico (eds.), *Stellar Jets and Bipolar Outflows* (Kluwer Publ.).
Browne, P.F., 1993, *Ap. J.* (submitted).
Burbidge, E.M., Lynds, C.R. and Stockton, A.N., 1968, *Ap. J.* 152:1077.
Burns, J.O., Feigelson, E.D. and Schreier, E.J., 1983, *Ap. J.* 273:128.
Butcher, H.R., van Breugel, Wh. and Miley, G.K., 1980, *Ap. J.* 235:749.
Canfield, R.C. and Puetter, R.C., 1981, *Ap. J.* 243:390.
Carilli, C.L., Perley, R.A. and Dreher, J.H., 1988, *Ap. J. (Letters)* 334:L73.
Chang, T.Y., 1977, *IEEE J. Quant. Electronics* QE-13:937.
Clarke, J.T., Bowyer, S. and Grewing, M., 1986, *Ap. J.* 305:167.
Clavel, J. and Joly, M., 1984, *Astron. Astrophys.* 131:87.
Cline, T.L. et al., 1982, *Ap. J.* 255:L45.
Cohen, M.H. et al., 1979, *Ap. J.* 231:293.
Cowley, S.W.H., 1981, *Nature* 292:191.
Cruz-Gonzalez, I. and Huchra, J.P., 1984, *Ap. J.* 89:441.
Dreher, J.W. and Feigelson, E.D., 1984, *Nature* 308:43.
Djorgovski, S. and Spinrad, H., 1984, *Ap. J. (Letters)* 282:L1.
Djorgorvski, S. Perley, R., Meylan, G. and McCarthy, P., 1987, *Ap. J. (Letters)* 321:L17.
Elston, R., Bechtold, J., Lowenthal, J. and Rieke, M., 1991, *Ap. J. (Letters)* 373:L39.
Gehrels, N. et al., 1984, *Ap. J.* 278:112.
Gekelman, W. and Pfister, H., 1988, *Phys. Fluids* 31:2017.
Golenetskii, S.V., Ilyinskii, V.N. and Mazets, E.P., 1984, *Nature* 307:41.
Grandi, S.A. and Phillips, M.M., 1980, *Ap. J.* 239:475.

Grauer, A.D.: 1984, *Ap. J.* 277:77.
Holmes, P.A., Brand, P.W.J.L., Impey, C.D. and Williams, P.M., 1984, *M.N.R.A.S.* 210:961.
Holmes, P.A., et al., 1984, *M.N.R.A.S.* 211:947.
Huchra, J.P., 1986, *Nature* 323:784.
Huchra, J. et al., 1985, A. J. 90:691.
Hunstead, R.W., Murdoch, H.S., Condon, J.J. and Phillips, M.M., 1984, *M.N.R.A.S.* 207:55.
Icke, V., 1981, Ap. J. (Letters) 246:L65.
Jackson, J.D., 1975, *Classical Electrodynamics*, 3rd edit., pp. 554, 625, 635, 638-9.
Javan, A., 1957, Phys. Rev. 107:1579.
Jones, D.J., Baath, L.B., Davis, M.M. and Unwin, S.C., 1984, *Ap. J.* 284:60.
Joss, P.C. and Rappaport, S.A., 1984, *Ann. Rev. Astron. Ap.* 22:537.
Kapahi. V.K. and Schilizzi, R.T., 1979, *Nature* 277:610.
Kunieda, H. et al 1990, *Nature* 345:786.
Lawrence, C. et al., 1984, *Science* 223:46.
Ledden, J.E. and Aller, H.D., 1979, *Ap. J. (Letters)* 229:L1.
Ledden, J.E. and O'Dell, S.L., 1985, *Ap. J.* 298:630.
Lerner, E.J., 1986, *Laser Part. Beams* 4:193 and 215.
Levshakov, S.A. and Varshalovich, D.A., 1979, *Ap. Lett.* 20:67.
Levy, G.F.D., 1981, *Opt. Commun.* 38:143.
Loudon, R., 1983, *The Quantum Theory of Light*, 2nd edit., Clarendon Press, pp. 320-326.
Lynds, R. and Petrosian, V., 1989, *Ap. J.* 336:1.
Lynden-Bell, D., 1977, *Nature* 270:396.
Macchetto, F. et al., 1991, *Ap. J. (Letters)* 373:L55.
Madejski, G.M. and Schwartz, D.A., 1983, *Ap. J.* 275:467.
Malkan, M.A. and Sargent, W.L.W., 1982, *Ap. J.* 254:22.
Marshall, F.E., Holt, S.S. and Mushotzky, R.F., 1983, *Ap. J. (Letters)* 269:L31.
Matilsky, T., Shrader, C. and Tananbaum, H., 1982, *Ap. J. (Letters)* 258:L1.
Mazets, E.P. et al., 1981, *Nature* 290:378.
Meisenheimer, K. and Heavens, A.F., 1986, *Nature* 323:419.
Meisenheimer, K. and Roser, H.-J., 1986, *Nature* 319:459.
Moore, R.L. and Stockman, H.S., 1981, *Ap. J.* 243:60.
Moore, R.L., Readhead, A.C.S. and Baath, L., 1983, *Nature* 306:44.
Nardi, V., 1983, in: V. Nardi, H. Sahlin and W.H. Bostick (eds) *Energy Storage, Compression and Switching*, Vol. 2, Plenum Press.
Nardi, V., Bostick, W.H., Feugeas, J. and Prior, W., 1980, *Phys. Rev. A* 22:2211.
Neugebauer, G., Oke, J.B., Becklin, E.E. and Matthews, K., 1979, *Ap. J.* 230:79.
Oke, J.B., 1965, *Ap. L.* 141:6.
Oke, J.B., Shields, G.A. and Koreansky, D.G., 1984, *Ap. J.* 277:64.
Osterbrock, D.E., 1991, *Rep. Prog. Phys.* 54:579.
Osterbrock, D.E. and Mathews, W.G., 1986, *Ann. Rev. Astron. Astrophys.* 24:171.
Owen, F.N., Wills, B.J. and Wills, D., 1980, *Ap. J. (Letters)* 235:L57.
Owen, F.N., O'Dea, C. P., Inoue, M. and Eilek, J.A., 1985, *Ap. J. (Letters)* 294:L85.
Paczynski, B., 1987, *Nature* 325:572.
Panock, R.L. and Temkin, R.J., 1977, *IEEE J. Quant. Electronics* QE-13:425.
Pearson, T.J. et al., 1981, *Nature* 290:365.
Perley, R.A., Dreher, J.W. and Cowan, J.L., 1984, *Ap. J.* 285:L35.
Perry, J.J., Burbidge, E.M. and Burbidge, G.R., 1978, *PASP* 90:337.
Peterson, B.M., Crenshaw, D.M. and Meyers, K.A., 1985, *Ap. J.* 298:283.
Richstone, D.O. and Schmidt, M., 1980, *Ap. J.* 235:361.
Rothschild, R.E. and Lingenfelter, R.E., 1984, *Nature* 312:737.
Rusk, R. and Seaquist, E.R., 1985, A. J. 90:30.
Schmidt, G.D. and Miller, J.S., 1985, *Ap. J.* 290:517.
Shaver, P.A. and Robertson, J.G., 1983, *Ap. J. (Letters)* 268:L57.
Shaver, P.A. and Cristiani, S., 1986, *Nature* 321:585.
Sitko, M.L., Stein, W.A. and Schmidt, G.D., 1984, *Ap. J.* 282:29.
Soifer, B.T., Neugebauer, G., Oke, J.B. and Mathews, K., 1981, *Ap. J.* 243:369.
Strittmatter, P.A. and Williams, R.E., 1976, *Ann. Rev. Astr. Ap.* 14:307.

Sulentic, J.W., 1989, *Ap. J.* 343:54.
Swinger, J., Tsai, W.-Y. and Erber, T., 1976, *Ann. Phys.* 96:303.
Syrovatskii, S.I., 1966, *Sov. Phys. JETP* 23:754.
Tennant, A.F., Mushotzky, R.F., Boldt, E.A. and Swank, J.A., 1981, *Ap. J.* 251:15.
Tobin, M.S., 1985, *Proc. IEEE* 73:61.
Turner, E.L. *et al.*, 1986, *Nature* 321:142.
Ubertini, P. *et al.*, 1984, *Ap. J.* 284:54.
Ulrich, M.H., 1981, *Astron. Astrophys. (Letters)* 103:L1.
Walsh, D., Carswell, R.F. and Weymann, R.J., 1979, *Nature* 279:381.
Weedman, D.W., Weymann, R.J., Green, R.F. and Heckman, T.M., 1982, *Ap. J. (Letters)* 255:L5.
Weymann, R.J. *et al.*, 1980, *Nature* 2856:641.
Weymann, R.J., Carswell, R.F. and Smith, M.G., 1981, *Ann. Rev. Astr. Ap.* 18:41.
Wiggins, J.D., Drozdovicz, Z. and Temkin, R.J., 1978, *IEEE J. Quant. Electronics* QE-14:23.
Wild, N., Gekelman, W. and Stenzel, R.L., 1981, *Phys. Rev. Lett.* 46:339.
Wolfe, A.M., Broderick, J.J., Condon, J.J. and Johnston, K., 1976, *Ap. J. (Letters)* 208:L47.
Yariv, A., 1989, *Quantum Electronics*, 3rd edit. (New York: Wiley), pp. 469-473.
Zheng, W., Burbidge, E.M., Smith, H.E., Cohen, R.D. and Bradley, S.E., 1987, *Ap. J.* 322:164.

Computer Simulations of Galaxies

W. Peter and E. Griv

Department of Physics
Ben-Gurion University
Beersheva, Israel

A. L. Peratt

Los Alamos National Laboratory
Los Alamos, NM 87545
USA

This paper is dedicated to the memory of Oscar Buneman

Gravitational N-body experiments can be used to simulate the stability of model galaxies, the evolution of disk-shaped galaxies, and spiral structure. However, they usually suffer from overheating (*i.e.*, large velocity dispersions of the interacting stars), and non-flat rotational curves. For the N-body simulations to agree with observation of actual normal galaxies, it is necessary to assume the presence of a large spherical slowly rotating subsystem (a halo) with the bulk of the galactic mass. An Alfvén model based on the interaction of two galactic-sized plasma filaments (or equivalently, two inviscid fluid vortices) may explain interesting features of galactic morphology including double radio lobes and the presence of extragalactic jets from classic double radio sources.

1. Introduction

One of the most important problems in stellar dynamics is to construct models of stable flattened galaxies which have properties consistent with observed ones in actual collisionless galaxies of stars. Gravitational N-body experiments, in spite of shortcomings due to the finiteness of the number of particles, yield interesting information on the stability of model galaxies (see Bardeen 1975, Bertin 1980, and Athanassoula 1984 for reviews). Galactic simulations can also be used to trace the evolution of disk-shaped galaxies and the development of spiral structure.

Three methods are currently employed to simulate evolution of collisionless disk galaxies by N-body experiments. Miller *et al.* (1970) and Quirk (1971) proposed a model that is doubly periodic in which the forces, star positions, and velocities are allowed to attain only discrete values; stars jump between integer values of position and velocity. A typical disk contains about 10^5 total particles. In the experiments of Hohl (1971a,b; 1972), the disk plane was covered with a difference grid. The stars within the grid cell did not interact gravitationally; the potential at the center of each cell was then computed, and the particles moved under the action of a force equal to the gradient at the centers of cells. These simulations utilized a relatively large number of particles (typically of the order of 10^4–10^5) to represent the 10^{11} stars in an actual galaxy; each particle is then considered to be a cloud of stars of 10^6 to 10^7 solar masses. Miller et al. and Hohl used a fast Fourier transform technique to solve for the gravitational potential. Ostriker and Peebles (1973), Morozov (1981a) and Grivnev (1985) investigated the evolution and final stationary states of a series of models for an isolated disk of stars by direct integration over a time span of equations of motion

$$\frac{d^2\vec{r}_i}{dt^2} = \frac{1}{N}\sum_{j\neq i}^{N}\frac{\vec{r}_j - \vec{r}_i}{\left[(\vec{r}_j - \vec{r}_i)^2 + r_c^2\right]^{3/2}} \qquad (1)$$

The cutoff radius r_c of the Newtonian potential is introduced in order to eliminate collisions between the model particles. It is so specified to ensure that the system remains collisionless throughout the time interval covered by the calculation. This "softening" parameter reduces the interaction at short ranges and puts a lower limit on the "size" of the particles, i.e., the stars in the system can no longer be considered as point-masses.

Initially, the evolution of infinitesimally thin systems in cartesian coordinates was simulated. Hockney and Brownrigg (1974), Hohl (1978), Morozov (1981b) and Miller (1985) included both the z-motions of stars and a spherical nonrotating subsystem. Miller (1976) has simulated disks by performing the calculations in polar coordinates.

2. Numerical Experiments: Initial Conditions

Linear stability analysis demonstrates that an infinitely thin disk will be locally Jeans-stable against small-scale axisymmetric perturbations of gravitational potential and density only if the radial-velocity dispersion of random velocities of stars exceeds Toomre's analytical limit (Toomre 1964; see also Bardeen 1975, Bertin 1980, Athanassoula 1984 and Fridman and Polyachenko 1984),

$$\sigma_T = 3.36 \frac{Gn_0}{\chi} \qquad (2)$$

where $n_0(r)$ is the local surface density, $\chi = 2\Omega[1 + (r/2\Omega)(d\Omega/dr)]^{1/2}$ is the local epicyclic frequency, and $\Omega(r)$ is the angular velocity of disk rotation. Toomre's criterion (2) is the local stability criterion which applies to axisymmetric ring or tightly wound spiral modes. It does not take into account the effects of tangential forces. In order to suppress the axisymmetric instability that would cause the disk to break up, numerical experiments superpose random radial and azimuthal velocities, with dispersions equal to or a little more than the one from (2), on the particle circular velocities $v_0 = r\Omega$. The model disks initially conform to several different surface-density distributions $n(r) \sim (1 - r^2/L^2)^{1/2}$: a nearly homogeneous disk; a power law disk with $n(r) \sim r^{-1}$; and, according to observations, a series of exponential disks $n(r) \sim \exp(-r/L)$. For actual galaxies that are seen in the sky, collisional relaxation times due to binary star-star encounters are several hundred rotation periods of the system (Chandrasekhar 1942). For the parameters used in the simulation, the model can also be considered collisionless for tens or hundreds of rotations (Hohl 1973).

3. Numerical Experiments: Principal Results

As all computer simulations show, the model evolution, irrespective of the initial distribution, entails primarily a reorganization of the system toward a state where in the final quasi-steady state the surface density approaches the sum of two exponentials (Hohl 1971a, b 1972). One exponential describes the central core component and the other the extended disk (Bardeen 1975). The original axisymmetric shape of the disk becomes distorted into a barlike structure similar to elongated elliptical galaxies or to the bars of Type SB galaxies. As the model evolves during 2–3 rotations the spiral structure will develop. The spiral arms are evidently density waves and not material arms, since test particles pass right through them (Miller 1971, Grivnev 1985). In many cases, initially the normal spiral modes of two tightly wound arms grow exponentially; the growth then saturates and the unstable spiral modes are converted into bars. Bar formation is a partial gravitational collapse (Jeans instability); angular momentum must be lost in order that contraction can occur (Lynden-Bell and Kalnajs 1972). Two unexpected results are obtained. (1) The system heats up very quickly. The stars quickly acquire large random velocities ~ 2–3 times more than Toomre's criterion, equation (2), predicts. (2) In a steady state the velocity dispersion at the edge of the disk is equal to half the circular velocity. In real normal galaxies such large velocity dispersions are not observed. For example, according to Hohl (1972) and Miller (1978), the model steady-state velocity dispersions scale ~ 150 km s^{-1} in the center of the galaxy and ~ 60 km s^{-1} at the edge. This is in contrast to ~ 30 km s^{-1} observed in the neighborhood of the sun in our own Galaxy.

With regard to the first difficulty, the explanations are speculative (see next section); with regard to the second, there must be something present to stabilize a model galaxy (Miller 1976). The principal stabilizing factor would evidently be a spherical component (a central bulge and/or an extended halo subsystem) which can be called the halo (Ostriker and Peebles 1973, Morozov 1981b). According to calculations of Ostriker and Peebles, a disk is stable against bar modes only if a substantial part of the mass of the system is in the form of the halo. The computer simulations agree with the observed value for ratio of the velocity dispersion to the circular velocity in the solar vicinity in the event that $M_H/M_D \simeq 2$ where M_H and M_D are the masses of halo and disk. A halo strongly concentrated toward the center will exert the best stabilizing effect (Morozov 1981b). The addition of a nonrotating massive core/halo component shows that multiarmed tightly-wound spiral structure develops and persists for many rotations in an evolving manner (Hockney and Brownrigg 1974, Hohl 1978).

4. Stability of Stellar Disks

To describe the collisionless star disks of galaxies, the Boltzmann kinetic and Poisson equations will be used. Let $\delta\Phi$ and δf be the amplitudes of small perturbations of the gravitational potential and distribution function, respectively. The linearized kinetic equation becomes (Shu 1970, Bertin 1980, Morozov 1981b)

$$\delta f = \frac{\partial f_o}{\partial E}\delta\Phi[r(t)] + i\left[(\omega - m\Omega)\frac{\partial f_o}{\partial E} + \frac{2\Omega}{\chi^2}\frac{m}{r}\frac{\partial f_o}{\partial r}\right]\int_{-\infty}^{0} dt'\,\delta\Phi[r(t')]e^{i\omega t' - m\theta(t')} \qquad (3)$$

where $E = \frac{1}{2}(u^2 + v^2)$, and u and v are the radial and azimuthal components of the peculiar velocity, respectively. Also, ω is the frequency of the excited waves, and f_o is the steady-state distribution function. Equation (3) and Poisson's equation give a complete statement of the problem of the modes of a disk. These modes are either normal spiral modes or barred and spiral modes in transition, depending on the properties of the equilibrium model (see Lin and Bertin 1984 and papers cited therein). The role of spiral modes is to carry angular momentum from the inner parts to the outer parts of the disk (Lynden-Bell and Kalnajs 1972). Numerical experiments have confirmed this prediction (Sellwood and James 1979). The Jeans instability, by increasing the random velocities of the particles, brings the disk to the limit of gravitational stability. Therefore, the spiral waves in computer-generated galaxies are short-lived, and dissipate after 2–3 rotations of the system.

Thermodynamically, under the action of growing spiral waves, the energy of the star's random motions will increase. That is, the velocity dispersion will increase and the system will approach a steady state (Lynden-Bell and Kalnajs 1972). Carlberg and Sellwood (1985) investigated with a computer model the process of the increase of random velocities by the interaction of stars and unstable spiral waves.

To second order in the asymptotic theory of Lin et al.(1969), Equation(3) and the Poisson equation give the following relation for small arbitrary nonaxisymmetric perturbations (Bertin 1980, Griv 1992a)

$$\epsilon^2 \frac{d^2 \delta \Phi}{dt^2} + Q(\omega, k, r)\delta \Phi = 0 \qquad (4)$$

where $\epsilon^2 \ll 1$ and k is the wave number. The complex eigenvalues ω of the discrete spiral modes are determined from equation(4) by means of the "quasi-classical quantization rules" of Bohr-Sommerfeld. This has been done in fluid dynamics and in stellar dynamics (Bertin 1980). For weakly inhomogeneous stellar systems of galaxies a discrete spectrum will differ little from a continuous one, and in the *zeroth* approximation may be regarded as continuous. In this approximation, omitting the first term in equation (4), we obtain the generalized dispersion relation. This relation for axisymmetric perturbations coincides with the well-known Lin-Yuan-Shu dispersion equation (Lin et al. 1969), and in the simplest case describes two ordinary gravitational (Jeans) branches of oscillations (Morozov 1981b, Grivnev 1988). Using the dispersion relation, we can obtain to second order in asymptotic theory a generalized stability criterion: In order for arbitrary nonaxisymmetric perturbations to be stable, the value of the stellar radial-velocity dispersion $\sigma(r)$ should satisfy the inequality (Morozov 1981b)

$$\sigma \geq \sigma_T \left\{ \left[1 + \left|\frac{\lambda}{1+\lambda}\right| \tan^2 \varphi \right]^{1/2} + \left|(1+\zeta-1.1\eta)\frac{\sin \varphi}{k_T L}\right|^{2/3} \right\} \qquad (5)$$

Here σ_T is Toomre's limit, equation (2), and $\lambda = r(\partial \Omega/\partial r)/4\Omega$; φ is the pitch angle of spiral arms, $k_T = \chi^2/2\pi G n_o$; $L^{-1} = \partial \ln n_o/\partial r$, $\zeta = \partial \ln(2\Omega/\chi)/\partial \ln n_o$; $\eta = \partial \ln \sigma/\partial \ln n_o$. Griv (1992a) has obtained the value of critical dispersion in the next order of asymptotic theory. The stabilizing effect of the disk's thickness was determined by Toomre (1964) and Shu (1970). Grivnev (1988) has considered the effect of stellar drift motion on the stability of the system. The oscillation and stability of a stellar and molecular cloud disk of our Galaxy with account taken of binary collisions is considered by Griv (1992b).

It is clear from equation (5) that stability of nonaxisymmetric perturbations ($\sin\varphi \neq 0$) in a differentially rotating disk ($\lambda \neq 0$) requires a larger velocity dispersion of stars than Toomre's critical value σ_T. The effects of tangential forces were considered in the gaseous approach also (Lau and Bertin 1978, Lin and Lau 1979). So Toomre's initial radial-velocity dispersion, equation (2), in accordance with the results of numerical experiments, should stabilize all small-scale perturbations. The disk however should still be unstable against slowly growing nonaxisymmetric perturbations. In the final (steady) state the radial-velocity dispersion of disk particles should be equal to the critical expression from equation (5). It has been confirmed by N-body simulations the theoretical criterion (5) predicts a value of σ which now is consistent with the result of numerical experiments (Morozov 1981a, b, Grivnev 1985).

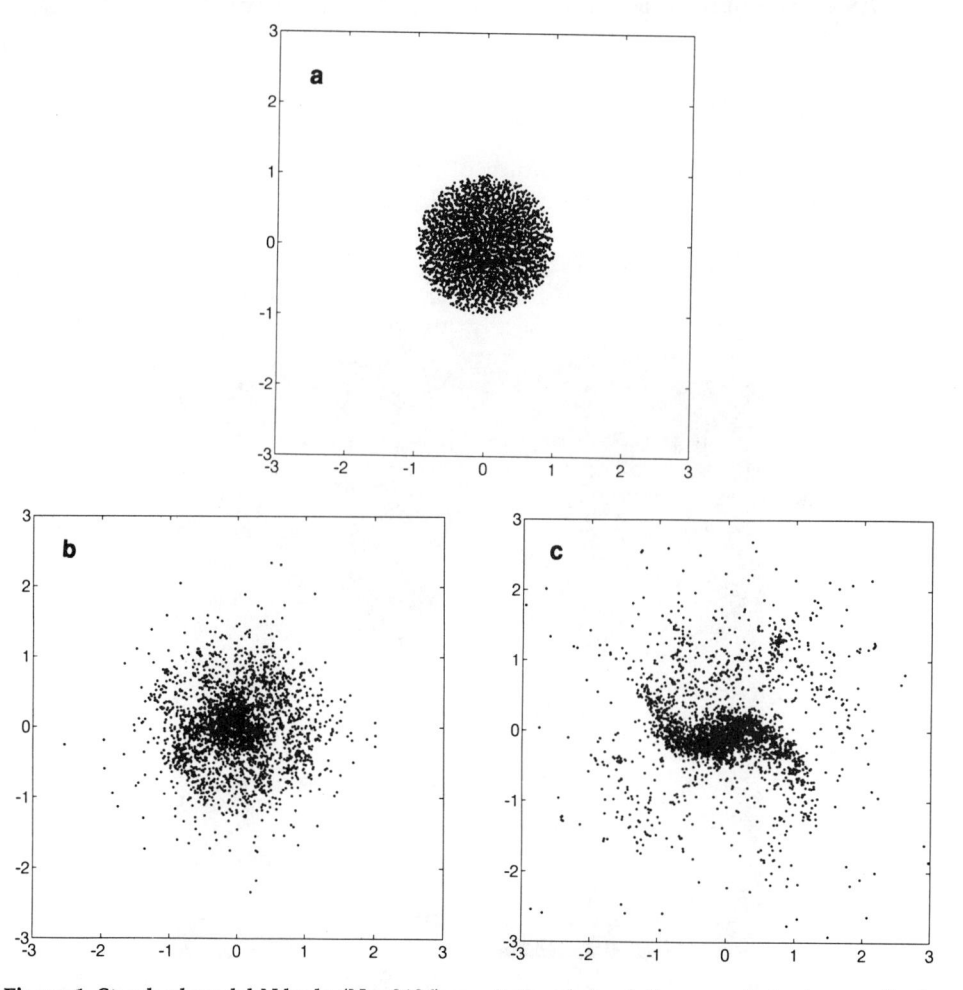

Figure 1. Standard model N-body ($N = 3186$) gravitational simulation snapshots at normalized times (a) $t = 0$, (b) $t = 400$, and (c) $t = 1000$. The time is normalized so that the time $t = 1000$ corresponds to a single revolution of the initial disk. The central dense core and underdense 'corona' is due to the Jeans instability. Note the large velocity dispersions of the particles at the end of the calculation.

In Figure 1 we show a series of three snapshots from a typical three-dimensional, N-body simulation ($N = 3186$) run on a Cray XMP supercomputer. The simulation initializes the particles on a set of 40 circular rings with a circular velocity in the xy-plane. For the initial disk, solid body rotation and a nearly homogeneous distribution of surface density was assumed. The particle positions and velocities were then randomized (making sure to keep the energy of each ring constant so as to ensure the equilibrium between the centrifugal and gravitational forces). Note that Maxwellian-distributed random velocities with radial and azimuthal dispersions according to Toomre [equation (2)] were added to the initial velocities of the particles. The sense of disk rotation was taken to be counterclockwise. It is seen from Figure 1 that the effects of the Jeans instability appear quickly in the simulation. At first, a tightly-wound spiral structure appears, but this gives way to a more open spiral structure with a larger pitch angle. In the computer simulations, the bar is seen to rotate about the center of the system slower than the stars. At the end of the calculation the criterion (5) is satisfied.

5. Plasma Simulations of Interacting Galactic-Sized Birkeland Filaments

Because of the long-range nature of the gravitational (or electromagnetic) force, N-body simulations are computationally intensive. Each body interacts with

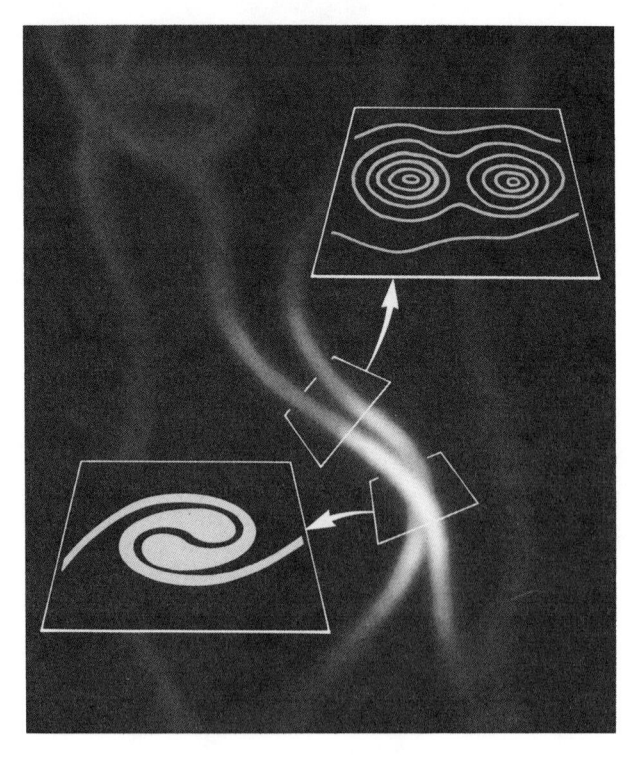

Figure 2. Schematic drawing of the Alfvén model: the intergallactic medium is made up of large-scale, field-aligned plasma filaments which interact to form double radio lobes and spiral galaxies.

the N–1 other bodies, and the interactions must be carried out for each of the N bodies. Hence the number of operations per time step $\propto N^2$, which for a typical run of 10^5 particles and 10^4 timesteps may take as long as a month to run. By introducing a grid into the calculation, each particle's charge is accumulated on the grid points and the field equations are solved. Forces are then interpolated onto the particle position, and the run time is considerably reduced. This is the *particle-in-cell* approach.

Particle-in-cell methods have been employed successfully in gravitational N-body experiments, but their application to plasma physics problems is more developed. With suitable particle shaping, charge and current allocation, and field solving algorithms, particle-in-cell codes can push a large number (> 10^5) of particles a significant amount of timesteps (> 10^4). Consider, for example, three-dimensional particle-in-cell simulations of interacting galactic-sized plasma filaments with magnetic-field-aligned electric currents (Peratt 1992). This describes a nonstandard astrophysical (but standard plasma physics) Alfvén model of a plasma universe threaded by field-aligned plasma currents and filaments (Figure 2). The simulations show the attraction and eventual coalescence of two field-aligned current filaments into an object whose cross-section is strikingly similar to a spiral galaxy. Some interesting features of the simulations are: (1) no center massive galactic halos are required to keep the galactic system stable, (2) results of synchrotron emission suggest that the isophotal contours may be identified as double radio lobes, and (3) the rotational velocities of the galaxy are *flat*, i.e., consistent with present-day observation without requiring the presence of dark matter (Figure 3).

The isophotal contours of the synchrotron emission from a typical simulation are plotted in Figure 4 as an increasing function of time, spanning 10.7 to 58.7 Myr after the start of the filament interaction. Initially, the current is uniformly distributed over each pinch, but, after a few Myr, the current density hollows out. The most intense current is then at the outer boundaries.

The monochromatic power of quasars and double radio galaxies spans a range of about 10^{33} to 10^{39} W. For example, a 'prototype' double radio galaxy Cygnus A has an estimated radio luminosity of 10^{37} W. Together with the power calculated from the simulations, the simulation isophotes are in good agreement with those observed from this object. Figure 2 may suggest that double radio lobes in galaxies are not unrelated, and may belong to the same species of object seen at different times in its evolution.

The simulations of the two interacting Birkeland currents discussed here aresimilar to the interaction of two magnetically confined electron columns in the laboratory (Driscoll and Fine 1990). This phenomenon, in turn, is known to be an excellent experimental manifestation of the two vortex instability in a constant density, inviscid fluid. This instability accounts for the pairing (and merger) of two isolated vortices into a larger structure with well-developed spiral arms. It is known that the two-dimensional drift-Poisson equations governing a magnetized electron column are isomorphic to the Euler equations governing a constant density inviscid fluid (Levy 1965, 1968). Surface charge perturbations on the electron column (diocotron modes) are equivalent to surface ripples on extended vortices; unstable diocotron modes on hollow electron columns are examples of the Kelvin-Helmholtz instability. Simulations of two interacting field-aligned current filaments (Peratt 1992) would also describe the interaction between two vortices in a constant-density, inviscid fluid.

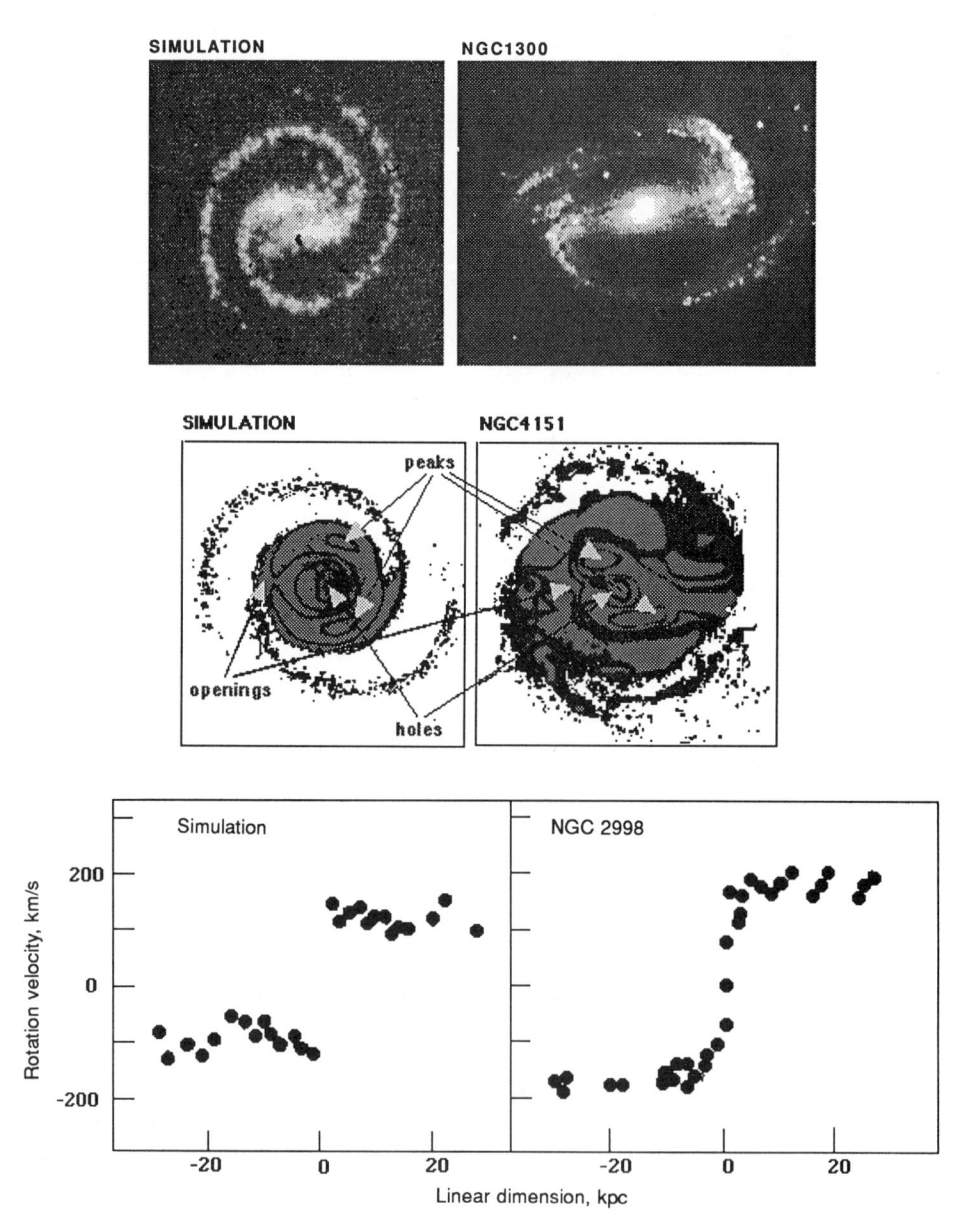

Figure 3. Alfvén model plasma simulations. (top) barred spiral galaxy simulation at t = 1749. The galaxy NGC 1300 is shown for comparison. (middle) normal spiral galaxy simulation. Overlayed on this simulation are contours of neutral hydrogen attached to magnetic field lines. For comparison an overlay of HI contours on the galaxy NGC 4151 is shown. Note that both figures show a double HI structure centered about a void, and both display a cusp opening towards a spiral arm. (bottom) Rotation velocity versus distance curve obtained from the simulation (top, left). For comparision the rotation characteristic of NGC 2998 is shown. Note that both simulation and galaxy curves have a *flat* profile modulated by a diocotron instability.

a OBSERVATION **b** SIMULATION

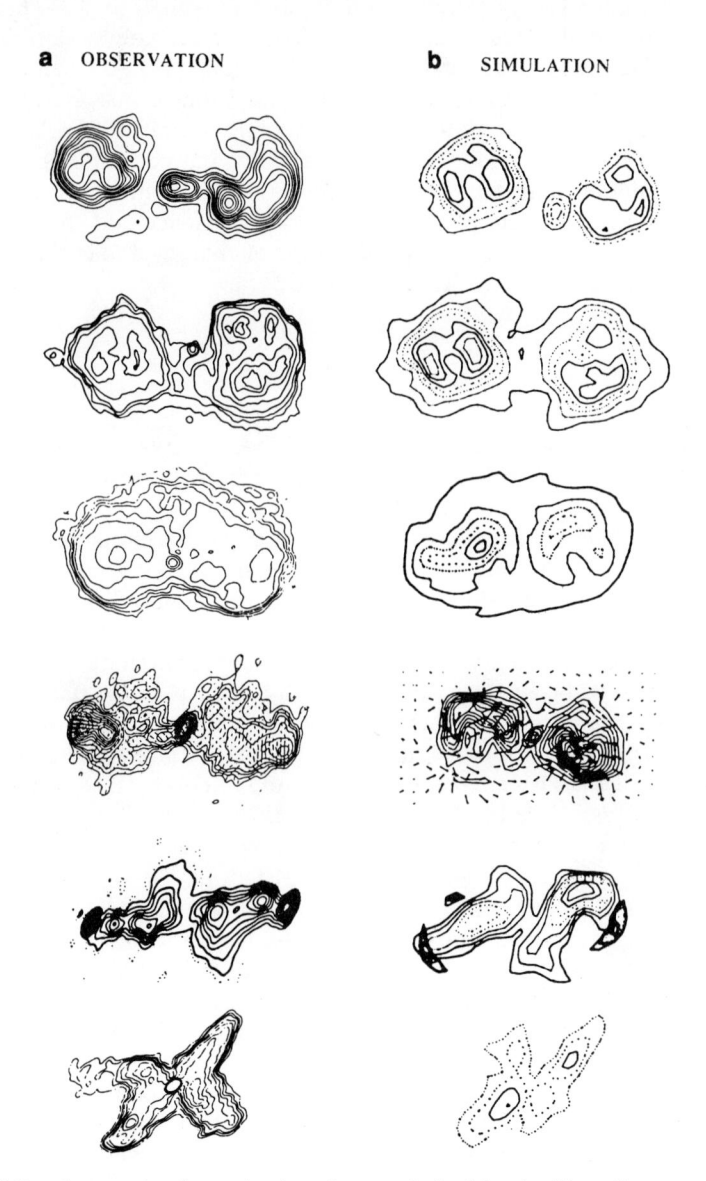

Figure 4. (a) Synchrotron isophotes (various frequencies) of the double radio sources 0844+319, Fornax A, 3C310, 2355+490, 3C192, and 3C315 (left-hand column). (b) Three-dimensional simulation analogs at times 10.4 Myr to 58.7 Myr. Time increases from top to bottom.

An interesting feature of double radio sources is the presence of jets, or collimated supersonic outflows of plasma (*e.g.*, Burns *et al.* 1991). Class II sources usually have only a single jet present, and this jet typically points towards one of the lobes. One explanation for the absence of a second jet has been a relativistic beaming effect. In the filamentary plasma model, jets are likened to sheet beams which filament into current bundles because of the diocotron (or Kelvin-Helmholtz) instability. In this scheme, the absence of a second jet follows from the fact that synchrotron radiation from the channel plasma in the simulations is possible only when the polarity of the induction field in the channel is correct (Peratt 1986). An explanation for jet production in this model (or, equivalently, the constant density inviscid Eulerian fluid model) may be inferred from the experimentally observed interaction between charged vortices produced by nucleation in a cloud chamber (Wong *et al.* 1991). Such a system is known to produce jets of charges due to the conservation of flux and the conservation of angular momentum.

6. Discussion

The circumstance between criterion (2) and the radial-velocity dispersion which was obtained from numerical experiments results from Toomre's neglect of *nonaxisymmetric* perturbation instabilities. The tendency of numerical N-body systems to get hot (typically, the velocity dispersion at the edge of the disk is about half the circular velocity) may be due to the neglect of stabilizing factors [*i.e.*, a central massive bulge or extended halo]. The way to construct stellar disks which are stable against bar modes and have properties consistent with observed ones is to put a bulk of the galactic mass in a spherical slow-rotating subsystem.

Gravitational N-body simulations require a halo mass in flat galaxies to be somewhat larger than the mass of the disk; in our Galaxy the spherical component mass is about twice that of the disk mass. Up to now, *no direct observational evidence for such massive halos has been found*. Marochnik and Suchkov (1969) investigated the spiral density waves in a model galaxy consisting of rotating and nonrotating subsystems. It has been shown that the Cherenkov instability (*i.e.*, inverse Landau damping) in such a galaxy develops just like it would, for example, in a plasma. This instability is a kinetic one and is not associated with the gravitational Jeans instability. It arises from the interaction between resonant stars of the nonrotating subsystem with the wave field. In principle, this resonant instability may explain the appearance of long-lived sprial structure in a disk with a massive halo seen in the computer experiments of Hockney and Brownrigg (1974). This question has not been studied in detail eitherin computer simulations or in theory.

A direct numerical integration of the collisionless Boltzmann and Poisson equations, which are free from the shortcomings of numerical experiments (the finiteness of the number of particles) and of analytical methods (the difficult task in the nonlinear regime), has been proposed by Fujiwara (1981) as another approach to investigate the stability of galaxies. However, the results of a Boltzmann simulation are almost the same as an N-body calculation (Nishida *et al.* 1984). Note that the gaseous component may play an important role in the development and stability of spiral structures (Fridman and Polyachenko 1984).

Gravitational N-body calculations assume that gravitational forces are the overriding influence on galactic morphology. Large-scale magnetic fields (of order of tens of μG) have now been observed and carefully measured in spiral galaxies (Beck 1986). The fact that these fields *tend to follow the orientation of the optical spiral arms* in the form of large-scale (~ tens of kpc) filaments may suggest that magnetic fields may play a more important role in spiral galaxies than is believed today. For instance, based on observations, Arp (1986) argues for a mass ejection model from central regions as the explanation for spiral galaxies. This concept (Bucerius 1938) postulates that spiral arms represent material streaming out of two diametrically opposed points on a rotating disk. The gaseous matter (which should at least be partially ionized) might be stabilized in these arms by the large-scale magnetic fields.

It is interesting to note that the ejection model for spiral arms may be consistent with the plasma simulations (Peratt 1992) if we adopt the mathematically equivalent scenario that the simulations describe two interacting fluid vortices (Section 5). In this case, the interactions of the two vortices would cause material outflow out of the system in the form of spiral arms even without the presence of a magnetic field. The interaction of two nonneutral electron columns in the laboratory (Driscoll and Fine 1990) can be used to model the interaction of two vortices in a fluid without, for example, the disturbing effects of a boundary layer. The experimentally observed morphology of the system during the merging of the two vortices agrees very well with the simulations of Peratt (1992), and demonstrates the formation of long-lived spiral arms.

Acknowledgments

The authors would like to thank C. R. Keys for his encouragement. This work is partially sponsored by the Israeli Ministry of Science, the Israeli Ministry of Absorption, and the U.S.–Israel Binational Science Foundation. Supercomputer calculations were supported by a grant from the San Diego Supercomputer Center which is funded by the National Science Foundation.

References

Arp, H., 1986, *IEEE Trans. Plasma. Sci.* PS-14:748.
Athanassoula, E., 1984, *Phys. Rep.* 114:319.
Bardeen, J.M., 1975, in: *Dynamics of Stellar Systems*, IAU Symposium No. 69, A. Hayli, Ed. (D. Reidel, Dordrecht, p. 297).
Beck, R., 1986, *IEEE Trans. Plasma. Sci.* PS-14:740.
Bertin, G., 1980, *Phys. Rep.* 61:1.
Bucerius, H., 1938, *Astron. Nachr.* 267:93.
Burns, J. O., Norman, M. L., and Clarke, D. A., 1991, *Science* 253:522.
Carlberg, R. G., and Sellwood, J. A. ,1985, *Astrophys. J.* 292:79.
Chandrasekhar, S., 1942, *Principles of Stellar Dynamics*, Univ. of Chicago Press (Reissued by Dover, 1960).
Driscoll, C. F., and Fine K. S., 1990, *Phys. Fluids* 2:1359.
Fridman, A. M., and Polyachenko, V. L., 1984, *Physics of Gravitating Systems*, Springer-Verlag, Berlin-New York.
Fujiwara, T., 1984, *Publ. Astron. Soc. Japan* 33:531.
Griv, E., 1992a, in: *Proceedings of the ESO/EIPS Workshop on Structure, Dynamics, and Chemical Evolution of Early-Type Galaxies*, Marciana Marina, Italy (to be published).

Griv, E., 1992b, in: *Proceedings of the Third Annual October Conference: Back to the Galaxy*, University of Maryland (to be published).
Grivnev, E. M., 1985, *Sov. Astron.* 29:400.
Grivnev, E. M., 1988, *Sov. Astron.* 32:139.
Hockney, R. W., and Brownrigg, D. R. K., 1974, *Mon. Not. RAS* 167:351.
Hohl, F., 1971a, *Astrophys. J.* 168:343.
Hohl, F., 1971b, *Astrophys. Space Sci.* 14:91.
Hohl, F., 1972, *J. Comput. Phys.* 9:10.
Hohl, F., 1973, *Astrophys. J.* 184:153.
Hohl, F., 1978, *Astron. J.* 83:768.
Levy, R. H., 1965, *Phys. Fluids* 8:1288.
Levy, R. H., 1968, *Phys. Fluids* 11:920.
Lau, Y. Y. and Bertin, G., 1978, *Astrophys. J.* 226:508.
Lin, C. C. and Bertin, G., 1984, *Advances in Appl. Mech.* 24:155.
Lin, C. C. and Lau, Y. Y., 1979, *Studies in Applied Math.* 60:97.
Lin, C. C. Yuan, C. and Shu, F. H., 1969, *Astrophys. J.* 155:721.
Lynden-Bell, D. and Kalnajs, A. J., 1972, *Mon. Not. RAS* 157:1.
Marochnik, L. S. and Suchkov, A. A., 1969, *Sov. Astron.* 13:411.
Miller, R. H. Prendergast, K. H. and Quirk, W., 1969, *Astrophys. J.* 161:903.
Miller, R. H., 1971, *Astrophys. Space Sci.* 14:73.
Miller, R. H., 1976, *J. Comput. Phys.* 21:400.
Miller, R. H., 1985, *Celestial Mechanics* 37:307.
Morozov, A. G., 1981a, *Sov. Astron.* 25:19.
Morozov, A. G., 1981b, *Sov. Astron.* 25:421.
Nishida, M.T., Watanabe, Y., Fujiwara, T. and Kato, S., 1984, *Publ. Astron. Soc. Japan* 36:27.
Ostriker, J. P. and Peebles, P. J. E., 1973, *Astrophys. J.* 186:467.
Peratt, A. L., 1986, *IEEE Trans. Plasma Sci.* PS-14:639.
Peratt, A. L., 1992, *Physics of the Plasma Universe*, Springer-Verlag, New York.
Quirk, W. J., 1971, *Astrophys. J.* 167:7.
Sellwood, J. A. and James, R. A., 1979, *Mon. Not. RAS* 187:483.
Shu, F. H., 1970, *Astrophys. J.* 160:99.
Toomre, A., 1964, *Astrophys. J.* 139:1217.
Wong, A. Y. Bonar, D. and Wuerker, R.F., 1991, *Bull. of Amer. Phys. Soc.* 36:2303.

Are There Gamma-Ray Burst Sources at Cosmological Distances?

W. Tkaczyk

Institute of Physics, University of Łódź
ul. Pomorska 149/153, 90-236 Łódź, Poland

The experimental data for gamma-ray bursts are reviewed and discussed from the point of view the localization of sources. The spectral features in the 60 keV and 400 keV energy ranges, as well the time history (0.2 ms time variations) favour a Galactic source distribution. The dipole and quadrupole moments show isotropic distribution inconsistent with the disk neutron star population. The BATSE test results indicate an inhomogeneous cosmological source distribution. The difficulty with the conventional cosmological model is that the extremely luminous objects must be very compact. No known extragalactic objects have the spatial distribution observed for gamma-ray burst sources. This may point toward unconventional cosmological models.

1. Introduction

Gamma-ray bursts (GRBs) are short flashes of hard photons with energies of 10 keV to few tens of MeV, and thus have the hardest spectra of any known class of astrophysical objects. Gamma-ray bursts have remained an enigma for two decades. In the last five years, a distribution hypothesis favouring an extragalactic location of GRB sources has gained a large number of followers (Paczyński 1991). Gamma-ray bursts were discovered in 1969 by the Vela satellites, originally designed to detect GRBs from nuclear explosions above the atmosphere. The discovery paper by Klebesadel *et al.* (1973) was published four years later, when the authors were finally convinced that the bursts they saw were from cosmic sources. In the twenty years since the discovery of gamma-ray bursts was reported, a rich collection of data on several hundred events has been accumulated from the KONUS experiment (Mazets *et al.* 1981, 1988). The spectroscopic study of gamma-ray burst spectra on the Solar Maximum Mission (SMM) satellite (Nolan *et al.* 1984, 1983) has given renewed impetus to the quest to understand the origin of these events. For detailed discussions of the experimental data on gamma-ray bursts, see review articles by Mazets and Golenetskii (1988), Nolan *et al.* (1984), Teagarden (1984) and Harding (1991).

The most powerful laboratory providing recent measurements of GRBs is BATSE (Burst And Transient Source Experiment), launched in April 1991. BATSE consists of eight uncollimated detector modules arranged on the corners of the Compton Gamma-ray Observatory to provide the maximum unobstructed view of the celestial sphere. In full operation, the experiment detects about one GRB per day. The number of GRBs from BATSE has exceeded the number of events detected by any other experiment to date.

Measuring the energy spectrum from bursts provides additional information about their origin. In the burst, both the continuum emission and the spectral features are characterized by rapid variability. The spectroscopic study of gamma-ray spectra from the KONUS experiment (Mazets *et al.* 1981), GINGA, SIGNE and from the Gamma-ray Spectrometer on the Solar Maximum Mission (Nolan *et al.* 1983) have provided rich material on their characteristic features. The spectra of many bursts contain absorption ($E < 100$ keV) and emission ($E \approx 350$–450 keV) features which have been interpreted respectively as broad cyclotron scattering and gravitationally redshifted annihilation lines (Mazets *et al.* 1985). This favours a neutron star as the source of gamma bursts. One of the most important results from the both KONUS and SMM experiments is the fact that some classes of bursts have hard tails in the energy spectrum. The hard spectrum begins near the energy range 400–500 keV.

Before the BATSE era, a consensus had developed that GRBs originate on or near (magnetic) neutron stars. This conclusion was based largely on the $\gg 8$ sec

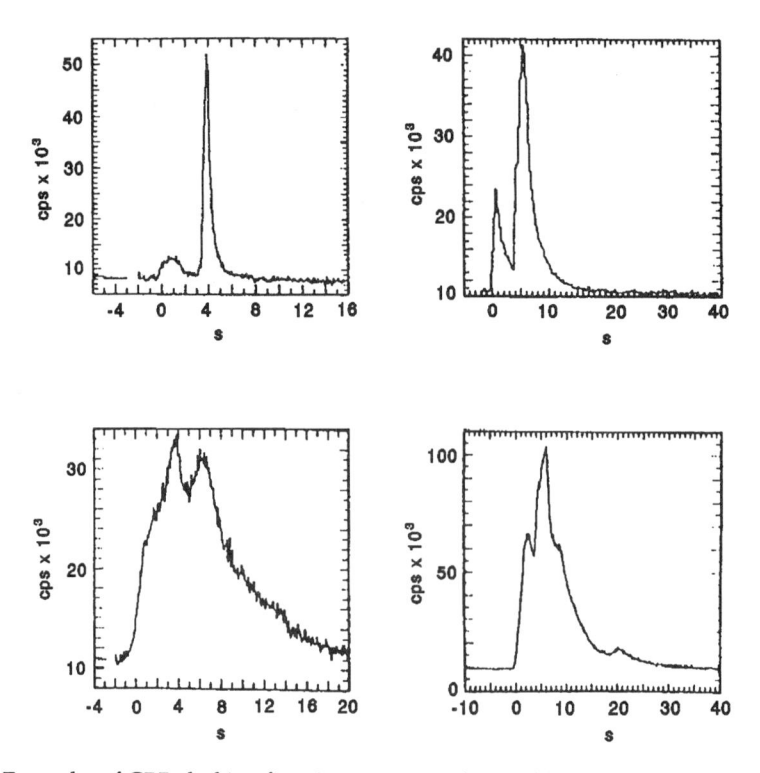

Figure 1. Examples of GRBs lacking fine time structure observed by BATSE (Fishman *et al.* 1993)

oscillations seen in the tail of the March 5, 1979 event and the rapid intensity variation seen in many other events. The low-energy features (≈ 50 keV) reported in 30% of events and interpreted as absorption or emission cyclotron lines, and the high energy (400 keV) features reported in about 7% events and interpreted as a redshifted pair annihilation line, also support neutron stars as a source of GRBs.

The data from GINGA and BATSE experiments, which registered GRBs with higher sensitivity, show that the sources are isotropically distributed on the sky. These results support either a close local or a distant cosmological distribution.

The new characteristics discovered by activated experiments were followed by a wide variety of theoretical ideas to explain the origin of the bursts.

The primary reason for the persistent aura of mystery surrounding GRBs is that their distances are not known. The one important question which GRB models have, as yet, not answered is: What are the sources of gamma-rays bursts? It is quite easy to propose a model in which part of the energy is emitted in this energy range, but relatively difficult to say why almost all the energy output is in gamma-rays. The gamma-ray production process requires a relativistic plasma in the source, and this plasma should also produce substantial luminosity in X-ray or optical wavelengths.

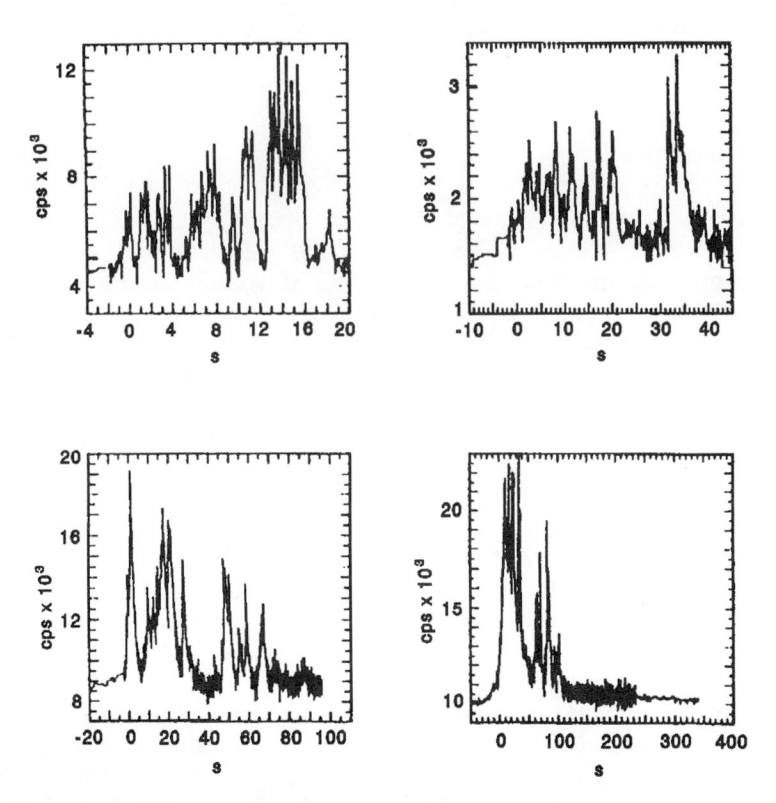

Figure 2. Examples of GRBs with complex time profiles observed by BATSE in energy range 60 keV to 300 keV (Fishman et al. 1993)

2. Time History

The time history of bursts has a complicated structure, and can be extended from several milliseconds up to a few hundred seconds. The time history or light curve of typical GRBs is characterized by a decay phase which may be quite complex and multipeaked. The duration is measured from the beginning of the rise to the time when the decay phase reaches noise level. The observed time profiles of gamma bursts show an enormous diversity in overall shape, pulse structure and duration. The duration of a typical GRB is in the range from a few ms to few hundreds of seconds, with a mean duration 15 s. A few examples of the variety of GRBs observed by BATSE (Fishman et al. 1993) are shown in Figures 1 and 2.

Recently, an extremely short rise time ~ 0.2 msec was recorded (Cline et al. 1980) in the time history of GB790305. This burst is unique in that it was the first event observed from the source, which is positionally coincident with the radio source N49, a supernova remnant in the Large Magellanic Cloud. The decay phase of this burst also showed at least 22 cycles of strong 8 s periodicity (Mazets et al. 1979, Barat et al. 1979). This last feature can be interpreted as the precession or rotation of a neutron star.

BATSE experiments have found some evidence for a sub-millisecond structure in GRBs (Bhat et al. 1992). The extremely short time-scale variability (~0.2) msec has been detected in GB910711, in the absence of relativistic beaming. Such a fast temporal structure limits the emission region to ≈ 60 km, suggesting a compact object (neutron star).

3. Spectral Features

Observations of the spectra of GRBs are usually performed with satellite-borne scintillators. The response functions of these devices are complicated. The observed photon count-rate spectrum is really a convolution of the true GRB spectrum with the instrument response function. Thus, the estimate of the incident spectrum from the data and the detector response function is nontrivial. As a result of these problems, the shape of GRB continua remains highly controversial, and still not confirmed by all experiments.

The spectra of GRBs are remarkable in that nearly all of the emission is in gamma-ray energies. The earliest observed GRB continuum spectra appeared to be thermal, with temperatures of a few hundred keV derived from fits with thermal *bremsstrahlung* models (Mazets et al. 1981), and are in general variable over the burst duration. Golenetskii et al. (1984), based on an analysis of GRBs in the KONUS catalogue, suggested that there is a correlation between temporal variability of luminosity and temperature in many bursts. This has been confirmed by data from SIGNE and PROGNOZ 9.

The gamma-ray spectrometer on SMM confirmed early indications that a power-law component was required to fit some spectra above 400 keV (Matz et al. 1985). From the SMM data, it was determined that most bursts have this high energy emission, and that most of the burst energy is above 1 MeV (hard tail). Continuum spectra of GRBs seem to be variable on time scales as short as the detector resolution time. In general, for the wide energy range, the spectra are fitted by a broken inverse power-law model with X-ray photon number index $\gamma_X = 1$ and a gamma-ray spectral index γ_γ varying from 1.5–3.

The transition between the X-ray and gamma regimes occurs between 100 keV and 1 MeV, suggesting a link between the shape of GRB spectra and the rest mass-energy of the electron. This is a strong indication that annihilation processes can essentially contribute photons to GRB spectra. The most convincing evidence for an annihilation line in GRB spectra was obtained by KONUS for burst GB 790305. The line shows asymmetric features at 430 ± 30 keV which clearly stand above the continuum and have a relatively high statistical significance $\sim 5\sigma$. The SIGNE team data provided confirmation of annihilation lines in this burst (Cline 1984). The emission features appear in spectra of 10–20% of GRBs at energies of 350–400 keV. Only the SMM detector did not register emission line features in any of its GRB spectra. The annihilation features appear to be variable from burst to burst, and, in particular, in bursts on time scales that are short compared to the total burst duration. In some cases the annihilation lines are so narrow that required pair temperatures are much lower than the continuum temperatures. In other cases, the features can be fit with broad line profiles (hard tail) which may contribute significantly to the continuum above the line. In a paper by Tkaczyk and Karakula (1985), we proposed annihilation of unthermalized plasma as the mechanism producing GRB spectra with line and hard tail features. Recently, we have found the peak energy

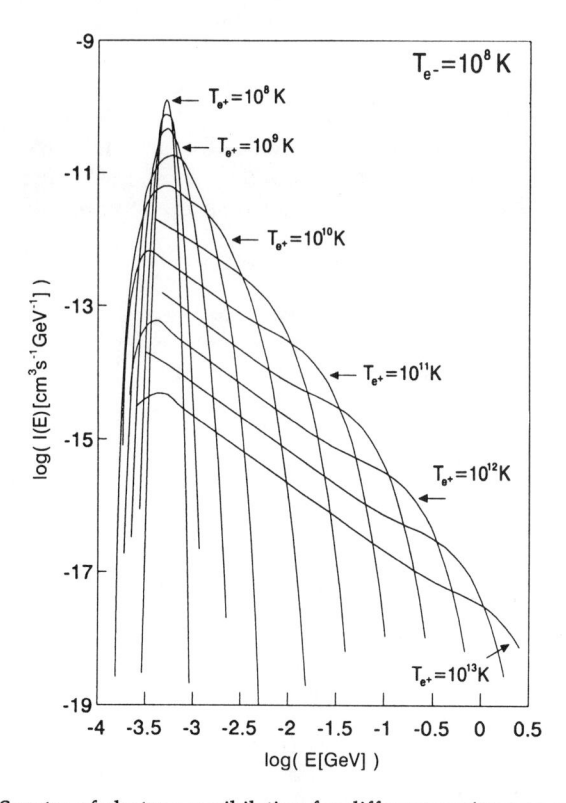

Figure 3. Spectra of electron annihilation for different positron temperatures.

and widths of spectra which are significantly different from annihilation of unthermalized ($T_{e+} \neq T_{e-}$) electrons and positrons (Tkaczyk and Karakula 1985, 1987), in comparison to the case when plasma is thermalized ($T_{e+} = T_{e-}$). Figure 3 shows, as an example, the spectra of electron annihilation at a temperature $T_{e-} = 10^{8}$ ° K for different temperatures of positrons. We have shown that the nine GRB spectra with hard tails measured by KONUS, SMM and HEAO A-4 experiments have good fits with thermal *bremsstrahlung* in soft energy regions, and annihilation lines of nonthermalized plasma in hard energy regions. We argue that conditions for the existence of nonthermalized plasma occur close to the surfaces of neutron stars. There are two possible cases where positrons will not thermalize with electrons: with injection by turbulent motion of hot positrons or electrons into a colder plasma region, the positrons are additionally heated (accelerated) by an electric potential caused by the charge separation between light electrons and heavy protons in the stationary Eddington limited matter accreting onto a neutron star. In my opinion, the fact that some of the BATSE spectra cannot be fitted by a single power-law (Schaefer et al. 1992) is a weak indication that these can also be fitted by our model. Figure 4 shows, as an example, KONUS GB191178 and the best fit spectrum by our model.

In the KONUS catalogue (Mazets et al. 1983, 1981), 30% of the GRBs show single-absorption features at energies between 20 and 60 keV. The data from the GINGA spectrometer have confirmed these features by detecting double-absorption lines in three GRBs (Murakami et al. 1988). Absorption features also have been detected in GB780325 by HEAO-1 (Heuter 1988). The interpretation of these features as cyclotron absorption or emission lines implies strong magnetic fields in the range 2–5×10^{12} G and a magnetic neutron star origin for them. There are no known astronomical objects except neutron stars having such strong magnetic fields. Although this result leads us to conclude that GRBs originate from neutron stars in our Galaxy, the distance to the objects is still unknown. In strong magnetic fields the effective temperature of positrons is higher than the temperature of electron pairs (Tkaczyk and Karakula 1992).

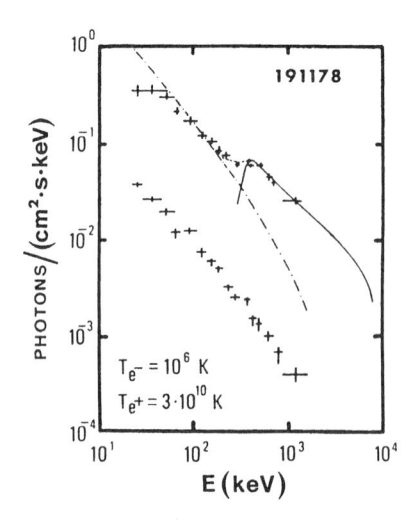

Figure 4. KONUS GB191178 and the best fit photon spectrum from the present model (full line from annihilation process and broken line from *bremsstrahlung*)

4. Spatial Distribution

One simple method to estimate the distance to the sources of GRBs is to plot their distribution on the sky. If we find an anisotropy in their distribution similar to the clustering in the Galaxy or Local Group of galaxies, we can estimate the distance scale. Three previous sky distributions have been obtained from the KONUS, interplanetary network on the PVO, IC3 Helios and VENERA satellites and BATSE for 153 bursts (Meegan et al. 1992). The plots did not show any anisotropy and were consistent with a uniform sky distribution (Klebesadel et al. 1982, Atteia 1987, Golenetskii 1988). Constraints on the source distribution can also be obtained from measurements of the dipole and quadrupole moments of the source distributions. The dipole moment is proportional to $\cos\theta$, where θ is the angle of the source with respect to the Galactic Centre. In the case of a uniform distribution of detectable sources, it should be $\cos\theta = 0$. The quadrupole moment is proportional to $\sin^2 b - 1/3$, where b is the galactic latitude, and in a uniform distribution of sources, the mean $\sin^2 b$ is equal to $1/3$. For BATSE bursts (Meegan et al. 1992), $\cos\theta = -0.002 \pm 0.006$, as compared with 0 ± 0.046 for an anisotropic distribution and $\sin^2 b = 0.31 \pm 0.006$, compared with 0.333 ± 0.023 for an isotropic distribution. Another method widely used for estimating the distance scale is a $\log N(>S)$–$\log S$ plot of GRBs, where S is the luminosity of the GRBs and N is the number of detected events from the whole sky above a certain level, S, of intensity. The intensity of sources can be expressed by their luminosity L and distance r

$$S = \frac{L}{4\pi r^2} \tag{1}$$

The number of detected events $N(>S)$ can be expressed for a spherical distribution of sources with density ρ by

$$N(>S) = \rho \frac{4}{3}\pi \left(\frac{L}{4\pi}\right)^{3/2} S^{-3/2} \tag{2}$$

For sources distributed in a disc with thickness h, it is

$$N(>S) = \pi r^2 h \rho = \rho h \frac{L}{4S} \tag{3}$$

In most cases of balloon-born experiments, this type of analysis has shown a break in the shape the $\log N$-$\log S$ relation. But this can be caused by limitations on the detection response and sensitivity. To overcome the difficulties between the sky distribution and the $\log N$-$\log S$ relation, a test called V/V_{max} is applied. The definition of V/V_{max} is as follows: the counting rate in maximum light curve from a source with luminosity L and at distance r is:

$$C_{max} = \eta \frac{L}{4\pi \cdot r^2} \Leftrightarrow r^2 = \frac{\eta \cdot L}{4\pi \cdot C_{max}}, \tag{4}$$

Minimum counting rates are from sources at r_{max}, so the number of GRBs relative to the threshold of the detector is given by equation (7),

$$C_{min} = \frac{\eta \cdot L}{4\pi \cdot r_{max}^2}, \tag{5}$$

The distance to the GRB source with a counting rate at maximum light curve C_{max} is

$$r = r_{max} \cdot \sqrt{\frac{C_{min}}{C_{max}}}, \tag{6}$$

$$N = \frac{4}{3} \cdot \pi \cdot \rho \cdot r_{max}^3 \cdot \left(\frac{C_{max}}{C_{min}}\right)^{-3/2} \tag{7}$$

The V/V_{max} expressed in terms of radius and counting rate is

$$\left(\frac{C_{max}}{C_{min}}\right)^{-3/2} \cong \left(\frac{r}{r_{max}}\right)^3 \cong \frac{V}{V_{max}}, \tag{8}$$

Although instrumental sensitivity affects the values of C_{min} and C_{max}, the value V/V_{max} does not depend on the sensitivity of each detector because variations in C_{min} from burst to burst are due mainly to the incident angle, and changes in the background are taken into account, so the value V/V_{max} does not depend on changes in background and sensitivity threshold. If the distribution of GRB sources is uniform in a volume, the values of V/V_{max} should be uniformly distributed between 0 and 1. Thus the mean V/V_{max} is expected to be 0.5.

5. Discussion and Conclusions

The results of the V/V_{max} test are summarized in Table 1. It is clear that all data are significantly different from the 0.5 value predicted by homogeneously distributed sources of gamma-ray bursts. From BATSE, the power index of the plot $\log N$–$\log C_{max}/C_{min}$ is –0.8 (Meegan 1992), and does not follow the –3/2 power law expected for a spherical extended homogeneous distribution of sources. There is a statistically noticeable deficit of weak bursts.

The mean value V/V_{max} indicated in Table 1 from this experiment is not consistent with a homogeneous source distribution. Bearing in mind measured

Table 1. V/V_{max} test results from different experiments

Experiment	V/V_{max}
KONUS	0.43±0.024
Solar Maximum Mission	0.40±0.03
Pioneer Venus Orbiter (PVO)	0.46±0.024
Single, Apex and Lilas	0.43±0.024
GINGA	0.35±0.035
BATSE	0.348±0.042
	0.33±0.02

dipole and quadruple moments, which are consistent with isotropic distribution, this is a strong indication that GRB sources are at cosmological distances. Detailed comparisons show that no known galactic objects have a spatial distribution that satisfies the BATSE data. Sources distributed in the disk would produce unacceptable quadrupole moments for distant objects or unacceptable V/V_{max} for nearby displacements (Paczyñski 1990). Nearby extragalactic models have difficulty with the lack of correlation with objects in the Local Group and Virgo cluster. Local Galactic models require energy output of 10^{37} erg, and assume that sources are neutron stars. This provides a very good explanation of the spectral features and time variation. The cosmological models need more luminous sources ~10^{51} erg. No known extragalactic object is so compact. If the BATSE data are not an apparatus effect, the results indicate that GRB sources are an unknown class of object. Their existence may suggest a non conventional model of the Universe.

6. Acknowledgments

This work was supported in part by grant UL 505.

7. References

Atteia, J.L. et al., 1987, *Astrophys. J. Suppl.* 61:305.
Bhat, P.N. et al., 1992, *Nature* 359:217.
Barat, C. et al., 1979, *Astron. Astrophys.* 79:L24.
Cline, T.L. et al., 1980, *Astrophys. J.* 237:L1.
Cline, T.L., 1984, in: *High Energy Transient in Astrophysics* (ed. Woosley, S.E.) p.333.
Fishman, G.J. et al., 1993, *Astron. Astrophys. Suppl. Ser.* 97:17.
Golenetskii, S., 1988, *Adv. Space Res.* 8:653.
Golenetskii, S.V. et al., 1984, *Nature* 307:41.
Harding, A.K., 1991, *Phys. Rep.*, 206 No. 6:327.
Heiter, G.J., 1988, in: *AIP Conf. Proc. Nuclear Spectroscopy of Astrophysical Sources* (eds. Gehrels, N. et al.) 285.
Klebesadel, R.W., Strong, I.B. and Olson, R.A., 1973, *Astrophys. J.* 182:L85.
Klebesadel, R.W. et al., 1982, in: *AIP Conf. Proc. Gamma-Ray Transients and Related Astrophysical Phenomena* (eds. Lingenfelter R.E. et al.).
Matz, S.M. et al., 1985, *Astrophys. J. Lett.* 282:L37.
Mazets, E.P. and Golenetskii, S.V., 1988, *Sov. Sci. Rev. E. Astrophys. Space. Phys.* 6:281.
Mazets, E.P. et al., 1981, *Nature* 290:378.
Mazets, E.P. et al., 1985, in: *Proceedings of the 19th ICRC* 9:415.
Mazets, E.P. et al., 1979, *Nature* 282:587.
Meegan, C.A., 1992, *Nature* 355:143.
Nolan, P.L. et al., 1983, in: *AIP Conf. Proc.*, ed. Burns et al., 101:59.
Nolan, P.L. et al., 1984, in: *AIP Conf. Proc.*, ed. Woosley, S.E., 115:399.
Paczyñski, B., 1990, *Astrophys. J.* 348:485.
Paczyñski, B., 1991, *Acta. Astr.* 41:157.
Schaefer, B.E. et al., 1992, *Astrophys. J.* 393:L51.
Teegarden, B.J., 1984, in: *AIP Conf. Proc.* 115:45 (ed. Woosley, S.E.).
Tkaczyk, W. and Karakula, S., 1985, in: *UH-UHE Behaviour of Accreting X-Ray Sources*, eds. Giovannelli, F. and Mannocchi, G, p.177.
Tkaczyk, W. and Karakula, S., 1987, in: *Multifrequency Behaviour of Galactic Accreting Sources*, ed. Giovannelli, F., p.254.
Tkaczyk, W. and Karakula, S., 1992, in: *NASA Conf. Pub. 3137 The Compton Observatory Science Workshop*, ed. Shrader C.R..

Gamma-Ray Emission Regions in AGNs

Svetlana Triphonova and Anatoly Lagutin

Institute of Nuclear Physics
Moscow State University
Moscow, Russia

In the cascade model for the production of gamma-radiation from AGNs one would expect gamma-ray flux as well as radiation at other wavelengths to vary. Recent observations provide strong evidence for variability of high-energy gamma-ray flux. In this paper we suggest a technique for calculating gamma-radiation variations in the cascade model. The results obtained are in agreement with available data. By considering mean gamma-ray spectra and their variations together, one can obtain more detailed information about the structure of gamma-ray emission regions in AGNs.
Key words: gamma-rays — AGN: NGC 4151, 3C 273, 3C 279 — cascade model

1. Introduction

Since Seyfert galaxies were recognized as galaxies having unusually luminous nuclei with strong broad emission lines, and quasars were identified as distant extragalactic radio sources, with similar strong emission lines, all the evidence has tended to indicate these are not different kinds of objects, but physically similar entities. Both observation and theory increasingly show the continuity of physical appearance from the least luminous Seyfert 1 galaxies to the rarest but most luminous types of AGNs—quasars and QSOs. Recent observations using high-quality digital detectors and subtraction techniques have enabled a number of authors to detect faint galaxy images surrounding AGNs previously described as quasars or QSOs (Osterbrock 1991). What were previously regarded as seperate phenomena have been shown to be different manifestations of the same underlying physical process, distinguished by luminosity, observer distance, orientation or physical environment.

Attempts have been made to incorporate all observed nuclear activity within a unified framework (Blandford 1990). In the general theory of AGNs, one of the most important parameters controling AGN properties is the rate of accretion onto a black hole. The black hole hypothesis is supported by much of the evidence, but it lacks definitive proof. The nature of the source of power that produces a continuum (from radio to gamma) nonthermal spectrum with a highly variable flux in many wavelength ranges is the main astrophysical problem posed by AGNs. Gamma-quanta of ultra-high energies have been detected from AGNs: up to 5 GeV from quasar 3C 279 (Hartman et al. 1992), up to 1 TeV from BL Lac object Mkn 421 (Punch et al. 1992). The radio emission consists of synchrotron radiation produced by relativistic particles in a strong magnetic field. This—along with the apparent superluminal expansion and relativistic jets observed in some blazars and quasars—provides strong evidence for an acceleration of particles up to relativistic velocities in AGNs.

Obtaining observations of the gamma-ray radiation and interpreting the observations are both problematic. The sources are not resolved in gamma-rays, and the region where radiation is generated is unknown. Numerous mechanisms have been suggested for producing gamma-radiation in association with active galactic nuclei and quasars; a sample of these is discussed, for example by Hartman et al. (1992). The most widely studied mechanism is the so called synchrotron self-Compton (SSC) model (Grindlay 1975). According to the model, gamma-quanta are produced in inverse Compton upscattering of relativistic electrons off their own synchrotron photons. High-energy gamma-rays might be expected in significant quantities from second-order Compton scattering, but simple SSC models have trouble not only producing gamma-ray luminosity as high as that observed and explaining the mismatch between spectral indices of the radio and hard X-rays (Dean et al. 1990), but also interpreting the observed lack of correlation of variations between different wavelengths (Bednarek and Calvani 1991). The most important point is that steep high-energy gamma-ray spectra from most AGNs suggest that photon-photon absorption is actually taking place in the objects.

This process is taken into account in the currently very popular cascade models. The high luminosity and rapid variability of the low-energy radiation from AGNs suggests that a compact emission region of dense soft photons exists in the source. If relativistic particles propagate through this region, electrons lose their energy through inverse Compton scattering off the soft photons, and gamma-quanta are absorbed, producing electron-positron pairs on the soft photons. The inverse Compton gamma-quanta can produce further pairs; the pairs in turn lose their energy through Compton scattering. Estimates of optical depths based on the observed high luminosity and rapid variability of the soft radiation show that relativistic electron-photon cascades (EPC) may well develop in such objects. According to the cascade models, an essential part of the gamma-radiation observed from AGNs is of cascade origin. Using these models, it is possible to describe the mean observed gamma-ray spectra and explain their characteristic properties (Aharonian et al. 1984, Fabian et al. 1986, Lightman and Zdziarski 1987, Svensson 1987, Ivanenko et al. 1991a).

The variability of low-energy radiation from AGNs is a well-established fact. Recently, variations of high-energy gamma-radiation have been observed (Kanbach 1992). In the cascade model, we should naturally expect the gamma-radiation to vary, since luminosity in soft photons is variable. In fact, changes in the soft photon spectra (the medium of the cascade development) should produce definite characteristic variations in gamma-ray spectra (resulting from the development of electron-photon cascades).

In this paper, we consider Seyfert galaxy NGC 4151 and radio loud quasar 3C 273 and OVV quasar 3C 279—objects each a different distance, luminosity, spectrum—in the framework of the one and the same cascade model. By bringing together results on mean gamma-ray spectra and variations, it is possible to draw some conclusions as to the region of the gamma-ray spectra production.

2. Mean Gamma-Ray Spectra and Variations

Gamma-quanta with energies E in the interval from 1 MeV to 100 MeV are most effectively absorbed in collisions with the X-ray photons, because the cross section of e^+e^--pair production is maximum at $3m^2c^4/E$ (Aharonian et al. 1985).

The energy output of the objects under consideration either is dominated by hard X/low energy gamma-ray emission (NGC 4151, 3C 273) or is larger in the high-energy gamma-ray band than in any other frequency band (3C 279). Therefore, we suppose that the photon energy density dominates the energy density of magnetic field, and that the latter can be neglected when considering the production of the gamma-ray radiation.

We suppose that high-energy electrons ($\alpha = e$) and/or gamma-quanta ($\alpha = \gamma$) with differential energetic spectrum $S_\alpha(E)$ generated in the core propagate through the region of dimension R isotropically filled with X-ray photons with spectral density $n(\omega) = n_o/\omega^a$ in the interval $\omega_1 < \omega < \omega_2$. Allowance is made for e^+e^--pair production in photon-photon collisions and inverse Compton scattering of electrons off X-ray photons. We are interested in the cascade spectrum of the gamma-quanta escaping the X-ray emission region.

If the dimension of the X-ray emission region R is measured in cascade units $t_o = (4\pi r_o^2 n_o)^{-1}$, then cascade gamma-ray spectra depend on the value $t = R/t_o$ (see equation (2) below). We now introduce differential energetic spectra $I_\alpha(t, E^*)$, $\alpha = e, \gamma$, such that $I_\alpha(t, E^*)dE^*$ is the mean number of the gamma-quanta with energies between E^* and $E^* + dE^*$, escaping a region of thickness t. $I_\alpha(t, E^*)$ may be written as

$$I_\alpha(t, E^*) = \int S_\alpha(E) q_\alpha(t, E; E^*) dE \quad (1)$$

where $q_\alpha(t, E; E^*) \equiv q_\alpha(t, E)$ is the so-called importance—in other words, the contribution from an individual cascade initiated by a single primary particle of type α with energy E. For $q_a(t, E)$, the following adjoint cascade equations are valid (Ivanenko et al. 1991b)

$$\frac{\partial}{\partial t} q_e(t, E) + \sigma_e(E) q_e(t, E) - \int_{E^*}^{E_\gamma} dE' W_{e\gamma}(E, E') q_\gamma(t, E') -$$

$$\int_{\max\{E^*, E-E_e\}}^{E} dE' W_{e\gamma}(E, E-E') q_e(t, E) = 0 \quad (2)$$

$$\frac{\partial}{\partial t} q_\gamma(t, E) + \sigma_\gamma(E) q_\gamma(t, E) - 2 \int_{\max\{E^*, E_{\gamma 1}\}}^{E_{\gamma 2}} dE' W_{\gamma e}(E, E') q_e(t, E') = \delta(t) \delta(E - E^*)$$

Here $W_{e\gamma}$ and $W_{\gamma e}$ are the differential cross-sections of the inverse Compton scattering and pair production, σ_e and σ_γ — the corresponding total absorption cross-sections

$$\sigma_e(E) = \int_0^{E_e} W_{e\gamma}(E,E')dE' \qquad \sigma_\gamma = \int_{E_{\gamma 1}}^{E_{\gamma 2}} W_{\gamma e}(E,E')dE'$$

$$E_e = \frac{E}{1+ E_{th}/4E}, \qquad E_{\gamma 1,2} = 0.5E\left(1 \pm \sqrt{1 - E_{th}/E}\right)$$

and $E_{th} = m^2 c^4 / \omega_2$ is the energy threshold of pair production in a given photon field.

The cross-sections $W_{\alpha\beta}$ are obtained by integrating the cross-sections $W_{\alpha\beta}(E,E';\omega)$ of the interactions in the unit density field of monoenergetic photons with energy ω (cf. Aharonian et al. 1985) over the X-ray photon spectral density $n(\omega)$

$$W_{\alpha\beta}(E,E') = \int W_{\alpha\beta}(E,E';\omega)n(\omega)d\omega \qquad (3)$$

It follows from formulae (1–3) that $I_\alpha(t,E^*)$ is a functional of the functions $n(\omega)$ and $S_\alpha(E)$, so variations of the X-ray photon spectrum as well as variations of the primary cascade-initiating particle spectrum cause changes in the spectrum of escaping gamma-quanta $I_\alpha(t,E^*)$.

To obtain the functional variations, we use the sensitivity technique (Oblow 1973, 1976). First, one finds the mean value of the functional with realistic values of functions. Next, the sensitivity coefficients are calculated. They represent differential rates of change of the functional with respect to the differential changes in the functions used to calculate the mean value of the functional.

Thus, the functional derivative $\delta I_\alpha(t,E^*)/(\delta S_\alpha(E)dE)$ gives the change of $I_\alpha(t,E^*)$ corresponding to the change per unit of primary particle number inside unit energy range in the vicinity of the point E. Similarly, the functional derivative $\delta I_\alpha(t,E^*)/(\delta n(\omega)d\omega)$ gives (in the linear approximation) the variation $\delta I_\alpha(t,E^*)$ as a result of a unit change of $n(\omega)$ in unit energy interval near the point ω. Then the variation $\delta I_\alpha(t,E^*)$ caused by certain variations $\delta n(\omega)$ and δS_α is written as (taking into account equation (1))

$$\delta I_\alpha(t,E^*) = \int q_\alpha(t,E)\delta S_\alpha(E)dE + \int dE S_\alpha(E) \int \frac{\delta q_\alpha(t,E)}{\delta n(\omega)d\omega} \delta n(\omega)d\omega$$

Thus, it is sufficient to know the importances $q_\alpha(t,E)$ and sensitivity coefficients $\delta q_\alpha(t,E)/(\delta n(\omega)d\omega)$ to calculate the mean gamma-quanta spectra and their variations in the cascade model.

The equations for the sensitivity coefficients of the importance, i.e. $\delta q_\alpha(t,E)/(\delta n(\omega)d\omega)$, may be derived from equations (2) by applying the functional derivative $\delta/(\delta n(\omega)d\omega)$ to the left-hand and right-hand sides of the equations; these differ from equations (2) on their right-hand sides. The equations are then solved numerically (Pljasheshnikov et al., 1979). By making test calculations, it has been found that computations of relative variations of the importance $\delta q_\alpha(t,E)/q_\alpha(t,E)$ are accurate to within a few percent (Ivanenko et al. 1992).

The mean X-ray spectra from AGNs are accurately described by a single power law with photon spectral indexes α_x around a "universal" value of 1.7 (Turner et al. 1991). We suppose that the primary cascade-initiating particle spectra are power-laws $S(E) \sim E^{-\beta}$ with indexes $\beta \approx 2$, as predicted by shock acceleration models. By comparing the calculated spectra with the observed ones, the value of t

is obtained. Using the known expression for the X-ray photon density (Herterich 1974)

$$n(\omega) = \frac{F_x(\omega)d^2}{R^2 c}$$

we obtain

$$t = R/t_o = 4\pi r_o^2 \frac{F_x d^2}{Rc} \omega^\alpha \qquad (4)$$

Here d is the distance to the source, R–the dimension of the X-ray emission region of the AGN and $F_x(\omega)$–the observed flux of X-rays with energy ω. The values of d and F_x are assumed to be known; thus, the value of R can be derived.

3. Results

NGC 4151

Figure 1 shows the spectrum of Seyfert galaxy NGC 4151. The mean X-ray spectrum of this galaxy is represented by a single power law spectrum of index $\alpha_x = 1.6$ over the energy range 1–150 keV. The calculated mean gamma-ray spec-

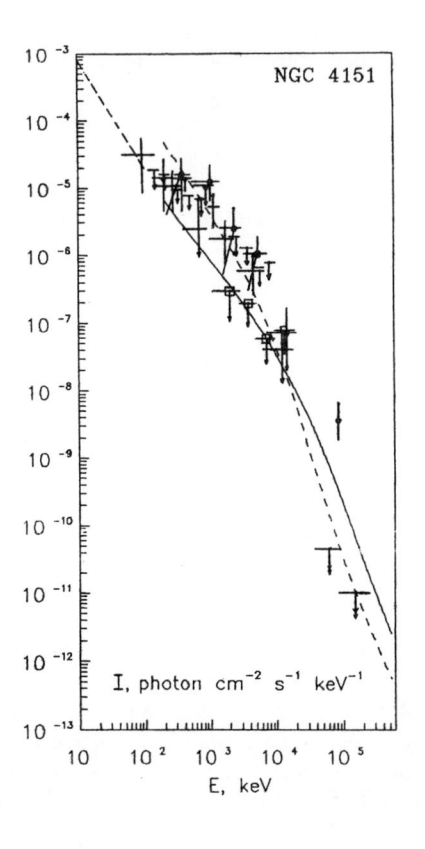

Figure 1. The spectrum of Seyfert galaxy NGC 4151; observational data: ✝ are from Galper et al. (1973), the other data from Perotti et al. (1981).

trum (dashed curve) well describes the experimental data when the primary electron spectrum has an index $\beta_e = 2.2$ in the interval 50–10^6 MeV and the radius of the cascade development region is on the order of $R = 5 \cdot 10^{13}$ cm. This value is obtained provided $d = 20$ Mpc and the 10 keV X-ray photons flux is $F_x = 8 \cdot 10^{-4}$ (cm^2 s keV)$^{-1}$.

The X-ray variability of NGC 4151 is now a well- established fact. Variations of 2-10 kev photon flux by a factor of 2–10 on various time scales have been measured (see refs in Perotti et al. 1991). The minimum X-ray flux-doubling time scale of a few times 10^4 s (Yaqoob and Warwick, 1991) agrees with the value ~10^{14} cm obtained in the cascade model. Hard X- ray photon flux variability by a factor of 4 on a time scale of several months can be assumed from observations. Low-energy gamma-ray emission up to few MeV shows evidence for intensity variability by a factor of 4 on the same time scale (Perotti et al. 1991). Over the set of Ginga observations, the 2–10 keV flux from the source varied by a factor of 4 (Yaqoob and Warwick 1989).

We have considered as perturbed the state of the X-ray photon field in which the photon density decreased to one fourth of the mean density. The corresponding gamma-ray spectrum is shown in Figure 1 (solid curve).

It is evident that there exists a characteristic anti-correlation between variations of the gamma-quanta fluxes with energies below and above 10 Mev: as X-ray luminosity decreases, soft gamma-quanta flux also decreases, while the hard gamma-ray flux increases. The calculated variations are greatest (by a factor of 7) in the energy range up to 1 Mev and around 100 MeV. The observational data are the most dispersed in the same regions—by a factor of up to 10. When the X-ray flux is varied by a factor greater than 4 (see above), we can easily obtain gamma-ray flux variations by a factor of 10.

If a simultaneous reduction of the cascade-initiating electron spectrum is assumed, the variations in the 0.5–5 MeV region should increase, while they decrease at 100 MeV due to the above noted anti-correlation. However, before making any inferences about changes in the primary electron spectrum, allowance must be made for a softening of the X-ray spectrum as the source brightens (Perotti et al. 1991), as well as possible variations of plasma in the cascade-development region. These factors should give rise to an increase of gamma-ray spectra variations: the former will produce variations at high energies, the latter mainly in the vicinity of 0.5 MeV.

3C 273

The spectrum of quasar 3C 273 is shown in Figure 2. The mean gamma-ray spectrum (solid curve) corresponds to a single power law X-ray spectrum of index $\alpha_x = 1.7$ in the range 1–150 keV and spectrum of an index $\beta_\gamma = 2$ of primary gamma-quanta with energies 100–10^6 MeV. Provided $F_x = 6 \cdot 10^{-4}$ (cm^2 s keV)$^{-1}$ and $d = 860$ Mpc the radius of the cascade-development region was found to be $R = 3 \cdot 10^{16}$ cm.

This quasar is variable in the X-ray band, and the X-ray emission is very decoupled from the emission mechanism in other low-frequency bands (Courvoisier et al. 1990, Dean et al. 1991). Short time variations at hard X-rays of 41 days were observed by SIGMA (Bassani et al. 1992). This is in agreement with our assumption of the existence in the object of a compact region of hard X-ray emission having a dimension $R \leq 10^{17}$ cm.

Twice, in 1976 and 1978, quasar 3C 273 was observed in hard X-ray and gamma-ray regions nearly simultaneously (Swanenburg et al. 1978, Bignami et al. 1981). In the twenty days before the gamma-ray observations in 1976, the X-ray flux may have changed by 40% (Courvoisier et al. 1987). According to our calculations,

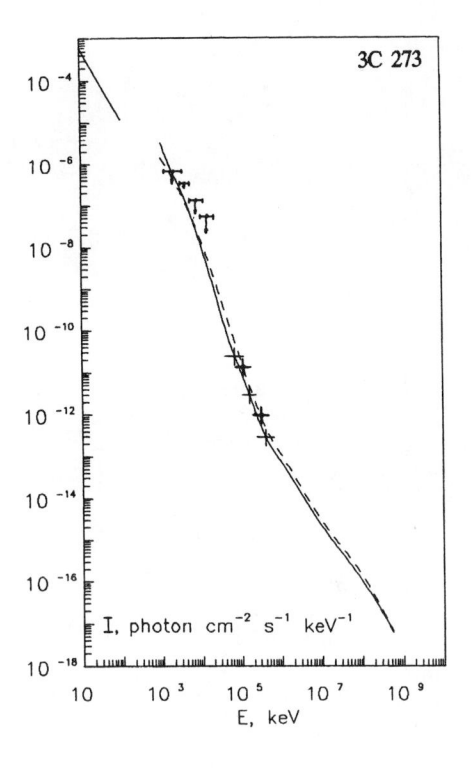

Figure 2. The spectrum of quasar 3C 273, observational data are from White et al. (1980)— ⊤, Swanenenburg et al. (1978) — +, and Bignami et al. (1981) — +.

this should result in the gamma-ray flux changing by a factor of 2 (see dashed curve on Figure 2). No variations of the gamma-ray flux between the two observations in 1976 and 1978 were reported within the 50% uncertainty, but B.N. Swanenburg (in private communication) has indicated that the flux did change—by precisely a factor of 2.

3C 279

The gamma-ray spectrum of quasar 3C 279 detected by GRO during June 1991 exhibited a power law between 50 MeV and 5 GeV with a differential spectral index (2.02 ± 0.07), and was harder than the spectra of many other AGNs (Hartman et al. 1992). Maraschi et al. (1992) have proposed that the gamma-ray spectrum is produced in relativistic jet via a synchrotron self-Compton mechanism, and that it is the high-energy extension of the inverse Compton radiation responsible for lower-energy emission from IR to X-rays.

We should now like to know if it is possible to describe the gamma-ray spectrum of quasar 3C 279 using the cascade model. We begin by supposing that cascade develops on the radiation of accretion disk. Since there is no change in slope of the gamma-ray spectrum up to 5 GeV—which in the cascade model is connected with the existence of the low-energy cutoff of the soft photon spectrum (Ivanenko et al. 1991)—we assume the whole low-energy photon spectrum from radio to X-rays to be the medium of cascade development. Though no information was available on the intensity of the source at other wavelengths during the GRO gamma-observation, it appears likely that intense gamma-emission is associated with an outburst at all wavelengths (Hartman et al. 1992, Maraschi et al. 1992). The

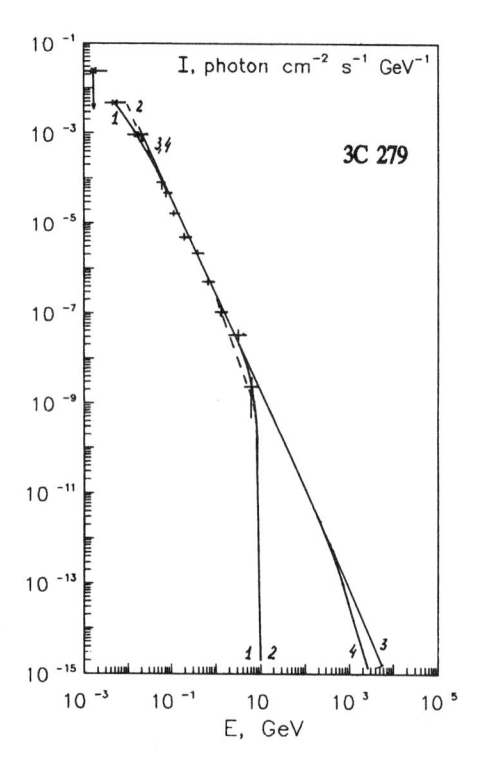

Figure 3. The spectrum of quasar 3C 279: observational data are from Hartman et al. (1992).

multifrequency spectrum of 3C 279 from radio to X-rays in the outburst period (Makino et al. 1989) was taken as continuous one, and represented by a series of single power laws (see Table 1). The resulting absorption cross sections of gamma-quanta and electrons are, respectively, approximately proportional to the energy and nearly constant up to 10^2 TeV.

Table 1. Parameters used to approximate the multifrequency photon spectrum of quasar 3C 279 for the outburst period.

Energy range cutoffs (MeV)	Slope
$4.15 \cdot 10^{-12} - 9.13 \cdot 10^{-10}$	1.0
$9.13 \cdot 10^{-10} - 1.2 \cdot 10^{-7}$	1.8
$1.2 \cdot 10^{-7} - 2.3 \cdot 10^{-6}$	2.0
$2.3 \cdot 10^{-6} - 6.6 \cdot 10^{-6}$	2.28
$6.6 \cdot 10^{-6} - 1.78 \cdot 10^{-3}$	2.03
$1.78 \cdot 10^{-3} - 4.15 \cdot 10^{-2}$	1.58

Different primary high-energy gamma-quanta spectra were taken: 1. high-energy cut off E_{max} = 10 GeV and spectral slope β_γ = 2.2, 2. E_{max} = 10 GeV and β_γ = 2.4, 3. E_{max} = 100 TeV and β_γ = 2.2, 4. E_{max} = 100 TeV and β_γ = 2.4.

The calculated cascade gamma-ray spectra describing the experimental data are shown on Figure 3. As seen from the figure, the gamma quanta spectra may become harder in consequence of the cascade process. It is also evident that different high-energy primary spectra may be involved. This should be taken into account when extrapolating the observed gamma-ray spectrum of 3C 279 into the higher energy region (Stecker *et al.* 1992).

The values obtained for the dimension of the cascade-development region, $10^{19} - 10^{20}$ cm, should be regarded as estimates, since we have taken a homogeneous distribution of photons. In the case under consideration, a cascade process may be associated with the relativistic jet emanating from the nucleus into the ambient radiation field. The different time scales of variations observed in different wavelength bands in quasar 3C 279 (Kanbach *et al.* 1992, Makino *et al.* 1989 and refs. therein) also suggest that the jet associated with this object is the gamma-ray emission region.

It therefore appears that a cascade process may be responsible for forming the observed gamma-ray spectrum of quasar 3C 279.

4. Conclusions

We have considered three different objects of the AGN class in the framework of the cascade model. We have described observed mean gamma-ray spectra and their variations. Assuming spherical symmetry of the ambient soft photon radiation, we have obtained values for gamma-ray production regions which are consistent with the vicinity of a black hole in the case of NGC 4151, with the accretion disk in the case of 3C 273, and with the relativistic jet in the case of 3C 279. It is remarkable that values of the parameter t (see equation (4)) obtained in the cascade model for sources having steep gamma-ray spectra, NGC 4151 and 3C 273, are practically the same. The observed X-ray fluxes F_x are also very close. Due to the presumed different distances to these sources, they are supposed to be of differing dimension and luminosity. According to an alternative view of the extragalactic universe (Arp *et al.* 1990 and this volume), quasars may be associated with nearby galaxies. On this theory, therefore they could be entities resembling the 'microquasars' observed in the center of our Galaxy.

The currently available data do not allow us to determine the structure and size of the emission region precisely. To draw any definite conclusions about the region of gamma-ray emission, we need more abundant data on variations and time scales and simultaneous observations in different wavelength bands. If the regions of emission of soft photons and high-energy gamma-quanta more or less coincide, the time scales of variations would be comparable. In the case of relativistic beaming, the time scales of variations in different wavelength bands can be distinguished.

The cascade model predicts a very characteristic correlation between variations in different wavelength bands, which could be a test of the model. By considering mean gamma-ray spectra and their variations together, one can obtain more detailed information about the sources. We believe that the cascade model merits further development, although any real progress will depend on the availability of more abundant data on variations.

Acknowledgments

The authors thank C. R. Keys for help with the text.

References

Aharonian, F. A., Vardanian, V. V. and Kirillov-Ugrumov, V.G., 1984, *Astrofisika* 20:223.
Aharonian, F. A., Kirillov-Ugrumov, V.G. and Vardanian,V.V., 1985, *Astrophys. Space Sci.* 115:201.
Arp, H. C., Burbidge, G. R., Hoyle, F. *et al.*, 1990, *Nature* 346:807.
Bassani, L. *et al.*, 1992, *Ap. J.* 396:504.
Bednarek, W. and Calvani, M., 1991, *Astron. Astrophys.* 245:41.
Bignami, G. F., Bennet, K., Buccheri, R. *et al.*, 1981, *Astron. Astrophys.* 93:71.
Blandford R. D., Netzer H. and Woltjer L., 1990, in: *Active galactic nuclei*, Courvoisier T. J.-L. and Mayor M. (eds), Springer-Verlag.
Courvoisier, T. J.-L. *et al.*, 1990, *Astron. Astrophys.* 234:73.
Courvoisier, T. J.-L., Turner, M. J. L., Robson, E. I. *et al.*, 1987, *Astron. Astrophys.* 176:197.
Dean, A. J., Bazzano, A., Court, A. J. *et al.*, 1990, *Ap.J.* 349:41.
Fabian, A. C., Blandford, R. D., Guilbert, P. W. *et al.* 1986, *MNRAS* 221:931.
Galper, A. M., Kirillov-Ugrjumov, V. G., Luchkov, B. I., Ozerov, Yu. V., 1973, *Sov. Pisma v JETP* 17:265.
Grindlay, J. E., 1975, *Ap. J.* 199:49.
Hartman, R. C., Bertsch, D. L., Fichtel, C. E. *et al.*, 1992, *Ap. J.* 385:L1.
Herterich, K., 1978, *Nature* 250:311.
Ivanenko, I. P., Lagutin, A. A., Triphonova, S.V. *et al.*, 1991a, in: *Proc. 22 ICRC*, Dublin, 1:121.
Ivanenko, I. P., Lagutin, A. A., Linde, I. A. and Triphonova, S. V., 1991b, *Preprint NPI MSU* 91-24/228, Moscow, 46 p.
Ivanenko, I. P., Lagutin, A. A. and Triphonova, S. V., 1992, *Preprint ASU-92/4*, Barnaul, 52 p.
Kanbach, G. *et al.*, 1992, *IAU Circ.* No 5431.
Lightman, A. P. and Zdziarski, A. A., 1987, *Ap. J.* 319:643.
Makino, F. *et al.*, 1989, *Ap. J.* 347:L9.
Maraschi, L., Ghisellini, G. and Celotti, A., 1992, *Ap. J.* 397:L5.
Oblow, E. M., 1973, *ORNL-TM*-4110.
Oblow, E. M., 1976, *Nuclear Sci. Engin.* 59:187.
Osterbrock, D. E., 1991, *Rep. Prog. Phys.* 54:579.
Perotti, F., Della Ventura, A., Villa, G. *et al.*, 1981, *Ap. J.* 247:L63.
Perotti, F., Maggioli, P., Quadrini, E. *et al.*, 1991, *Ap.J.* 373:75.
Pljasheshnikov, A. V., Lagutin, A.A. and Uchaikin, V.V., 1979, in: *Proceed. 16 ICRC*, Kyoto, 7:7.
Punch, M. *et al.*, 1992, *Nature* 358:477.
Stecker, F. W., De Jager, O. C. and Salamon, M. H., 1992, *Ap. J.* 390:L49.
Svensson, R., 1987, *MNRAS* 227:403.
Swanenburg, B. N., Bennett, K., Bignami, G. F. *et al.*, 1978, *Nature* 275:298.
Turner, T. J., Weaver, K. A., Mushozky, R. F. *et al.*, 1991, *Ap. J.* 381:85.
White, R. S., Dayton, B. and Gibbons, R. 1980, *Nature* 284:608.
Yaqoob, T. and Warwick, R.S., 1989, in: *Proc. 23rd ESLAB Symp. on Two Topics in X-ray Astronomy*, Bologna, 1089.
Yaqoob, T. and Warwick, R.S., 1991, *MNRAS* 248:773.

On the Meaning of Special Relativity
If a Fundamental Frame Exists

F. Selleri

Dipartimento di Fisica, Università di Bari
I.N.F.N., Sezione di Bari

The physical foundations of the theory of special relativity are critically examined from the point of view that a fundamental inertial frame exists in nature. This idea is implicit in the older works of Lorentz and Fitzgerald, and in more recent work by Janossy, Builder, Prokhovnik, Bell and others. Even in the event of modifications to the theory, certain elements will have to be retained: the usual relativistic formulae for energy and momentum, the full equivalence of mass and energy, the idea of c as a limit velocity, Lorentz contraction and time dilation of moving bodies.

1. Motivations

Interest in the foundations of the Special Theory of Relativity (STR) has grown steadily in recent years. A number of very interesting papers on the history of STR (Keswani 1966, Zahar 1973, Tyapkin 1982) have led to a better understanding of the roles of Lorentz and Poincaré and of their conviction—even after 1905—that the existence of an ether needs to be postulated. Lorentz was convinced that the contraction of the length of a moving body is a real phenomenon, and this opinion was shared by Fitzgerald, the other proponent of the contraction effect (Bell 1992). Moreover, it has been convincingly documented that Einstein was opposed to the ether only around 1905, but later reverted to this conception in connection with both General Relativity (GR) and the STR (Kostro 1992). For example, he writes (Einstein 1920):

> There is an important argument in favour of the hypothesis of the ether. To deny the existence of the ether means, in the last analysis, denying all physical properties to empty space.

Likewise, Poincaré, usually considered a conventionalist, wrote (1905):

> We know nothing as to what is the ether, how its molecules are disposed, whether they attract or repel each other; but we know that this medium transmits the optical perturbations and the electrical perturbations at the same time; we know that this transmission should be made in

conformity with the general principles of mechanics and that suffices us for the establishment of the equations of the electromagnetic field.

The previous statement was made at the St. Louis conference, and was printed in the journal *The Monist* on the same page on which the modern formulation of the principle of relativity was given for the first time—one year before Einstein! Other papers by Poincaré document the fact that he remained steadfastly in favour of the ether.

Who opposed the ether, then? Einstein did, until about 1916, and the papers and books he wrote at the time he discovered STR became extremely famous and created the false impression that his theory was based on the "discovery" that the ether does not exist. A naïve understanding of the experimental evidence (Michelson and Morley 1887) strengthened this opinion in the minds of the majority of the scientific community after 1905.

2. Relativistic Energy and Momentum

There is an important but little used physical argument connecting the mathematical expressions for kinetic energy and momentum in any theory based on energy conservation (cf. Lewis 1908). Here we will use it in a simplified one-dimensional notation, writing

$$Fds = dT \qquad (1)$$

where Fds is the work done by the force F in the displacement ds and dT is the variation of the kinetic energy T generated by the application of F. We can think of this as applying to a particle in which the force is always parallel to the velocity. Equation (1) can also be written

$$Fv = \frac{dT}{dt}$$

where dt is the time interval during which the displacement ds and the variation dT take place. According to Newton's law

$$\frac{dp}{dt} v = \frac{dT}{dt}$$

If we now imagine that p and T are functions of time through v, we have

$$\frac{dp}{dv} \dot{v} v = \frac{dT}{dv} \dot{v}$$

Consequently, the following relation must always hold

$$\frac{dT}{dv} = v \frac{dp}{dv} \qquad (2)$$

By integrating (2), we can check, for example, that $p = mv$ implies $T = \tfrac{1}{2}mv^2$, if the kinetic energy is required to vanish for $v = 0$.

In the relativistic case, we can also check that

$$p = \frac{m_o v}{\sqrt{1 - v^2/c^2}} \qquad (3)$$

necessarily leads to

$$T = m_o c^2 \left(\frac{1}{\sqrt{1 - v^2/c^2}} - 1 \right) \qquad (4)$$

provided that T is again required to vanish for $v = 0$. The proof is accomplished by substituting (3) in (2) and integrating. Therefore, the relativistic expressions for momentum and kinetic energy are bound tightly together by a simple, convincing physical argument.

We might then ask ourselves if other reasonable expressions for p and T exist that are consistent with (2). We will next show, however, that (3) and (4) are the necessary forms of momentum and kinetic energy if we assume the following two physical conditions to obtain:

1. Mass-energy equivalence, which can be written

$$E(v) = m(v) c^2 \qquad (5)$$

where $E(v)$ is the total energy, including the energy equivalent of the rest mass

$$E = T + m_o c^2 \qquad (6)$$

and $m(v)$ is the velocity-dependent mass. Mass energy equivalence can be formulated by saying that mass is a totally redundant ingredient of physics, and that it can be completely eliminated. For historical reasons, however, it is prudent to be more conservative, and say only that every physical property of energy is shared by mass multiplied by c^2, and that every function of mass can be adequately performed by energy divided by c^2. We may regard (5) and (6) as consequences of the mass-energy equivalence principle, and anticipate that the inertial mass of a particle will not be given by the rest mass, but by E/c^2, or by $m(v)$, which amounts to the same.

2. Momentum is "quantity of motion". The interpretation of momentum as quantity of motion means that we must multiply a given velocity by the amount of matter having that velocity, or by the total mass:

$$p(v) = m(v) v \qquad (7)$$

Equivalently, we could also write

$$p(v) = \frac{E(v)}{c^2} v$$

because of the mass-energy equivalence. In view of (6), we can write (2) as

$$\frac{dE}{dv} = v \frac{dp}{dv} \qquad (8)$$

where the "initial condition" on E is that it is equal to the rest energy $E_o = m_o c^2$. If we introduce our postulated expressions (5) and (7) into (8) we obtain

$$\frac{dm(v)}{dv} c^2 = \frac{dm(v)}{dv} v^2 + m(v)v \qquad (9)$$

which can easily be integrated to obtain

$$m(v) = \frac{m_o}{\sqrt{1 - v^2/c^2}} \qquad (10)$$

where $m_o = m(0)$. Obviously, inserting (10) into (7) leads to (3), while inserting it into (5) leads to (4) if (6) is also taken into account.

We therefore see that the validity of mass-energy equivalence and of the interpretation of p as quantity of motion lead to the usual expressions for kinetic energy and momentum. This conclusion does not force us to assume that (3) and (4) hold with respect to all conceivable inertial frames. Only if the principle of relativity is assumed can the general validity of (3) and (4) be granted, while in other theories not based on the relativity postulate, the invariance of form of energy and momentum does not generally hold.

3. The Limit Velocity

We have seen that $m(v)$, p and T became infinitely large when $v \to c$. This is a sign that no physical body can overcome the velocity of light, but it is possible to give a physically clearer argument. In fact, the assumption of a complete equivalence between mass and energy is sufficient to ensure that the velocity of light is an upper limit for the velocities of all massive particles. In order to clarify this point, consider a time-dependent force F applied in the direction of the velocity of a particle having rest mass m_o and velocity dependent mass $m = m(v)$. From Newton's law

$$F(t) = \frac{dp}{dt} = \frac{d}{dv}[m(v)v] \frac{dv}{dt} \qquad (11)$$

From (10) and (11) it follows for the acceleration dv/dt that

$$\frac{dv}{dt} = \frac{1}{m_o} F(t) \left(1 - v^2/c^2\right)^{3/2} \qquad (12)$$

whence

$$\frac{dv}{\left(1 - v^2/c^2\right)^{3/2}} = \frac{1}{m_o} F(t) dt \qquad (13)$$

By direct integration of (13) one obtains

$$\frac{v}{\sqrt{1-v^2/c^2}} = I(t) + \gamma \tag{14}$$

where

$$I(t) = \frac{1}{m_o} \int_0^t dt' F(t') \tag{15}$$

and the integration constant γ is given in terms of the initial ($t=0$) velocity v_o as

$$\gamma = \frac{v_o}{\sqrt{1-v_o^2/c^2}}. \tag{16}$$

It follows from (14) that

$$v(t) = c \frac{I(t)+\gamma}{\left[c + (I(t)+\gamma)^2\right]^{1/2}} \tag{17}$$

whence it is clear that $v(t) < c$ at any finite time t, since $I(t)$—the integral of an arbitrarily large but finite force—must remain finite at all times.

It thus appears that there is nothing mystical about the existence of c, which is the limit velocity of physical bodies only because the increase of velocity gives rise to an increase of kinetic energy T, with $T \to \infty$ if $v \to c$ (see equation (4)). Given the complete equivalence of energy and mass, if $T \to \infty$, the inertial properties of the particle become infinitely large, which amounts to saying that the inertial mass grows without limit. Even a very large force would then be unable to accelerate a body beyond c.

4. The Threshold for Inelastic Processes

We will next show that a typically relativistic calculation, such as the calculation of the energy "threshold" for an inelastic particle reaction, can be performed without resorting to the usually employed relativistic conceptions (invariance of the scalar product, c.m. energy, and so on). We will simply rely on the energy and momentum conservation laws. A particle with velocity $\vec{v}_1$ and rest mass m_1 collides with another particle at rest (in the laboratory) having rest mass m_2, producing in the final state $n-2$ particles with rest masses $m_3, m_4,...m_n$ and velocities $\vec{v}_3, \vec{v}_4,...\vec{v}_n$. The conservation laws for energy and the longitudinal component of momentum can be written

$$\frac{m_1 c^2}{\sqrt{1-\beta_1^2}} + m_2 c^2 = \sum_{i=3}^n \frac{m_i c^2}{\sqrt{1-\beta_i^2}} \tag{18}$$

$$\frac{m_1 \beta_1}{\sqrt{1-\beta_1^2}} = \sum_{i=3}^n \frac{m_i \beta_i}{\sqrt{1-\beta_i^2}} \cos\theta_i \tag{19}$$

where $\beta_k = v_k/c$ ($k = 1, 3, \ldots$) and θ_i is the angle between $\vec{v}_i$ and $\vec{v}_1$ ($i = 3, 4, \ldots$). It will be useful to transform (18) and (19) by putting

$$\varepsilon_k \equiv \frac{\beta_k}{\sqrt{1-\beta_k^2}} \tag{20}$$

with $k = 1, 3, \ldots n$. The previous equations can now respectively be written

$$m_1 \sqrt{1+\varepsilon_1^2} + m_2 = \sum_{i=3}^{n} m_i \sqrt{1+\varepsilon_i^2} \tag{21}$$

$$m_1 \varepsilon_1 = \sum_{i=3}^{n} m_i \varepsilon_i \cos \theta_i \tag{22}$$

By squaring the two previous relations and subtracting the second one from the first, we obtain

$$m_1^2 + m_2^2 + 2m_1 m_2 \sqrt{1+\varepsilon_1^2} = \sum_{ij} m_i m_j \left[\sqrt{1+\varepsilon_i^2} \sqrt{1+\varepsilon_j^2} - \varepsilon_i \varepsilon_j \cos \theta_i \cos \theta_j \right] \tag{23}$$

The latter equation fixes ε_1 for given ε_i, θ_i, just as the former equations did. However, it is more physically transparent, as it enables us to easily obtain the minimum value of ε_1 (threshold) for which our reaction can take place. It can readily be shown that every term within parentheses on the right-hand side of (23) cannot be smaller than 1, or that

$$\sqrt{1+\varepsilon_i^2} \sqrt{1+\varepsilon_j^2} \geq 1 + \varepsilon_i \varepsilon_j \cos \theta_i \cos \theta_j$$

In fact, the previous inequality can be transformed to

$$(\varepsilon_i - \varepsilon_j)^2 + 2\varepsilon_i \varepsilon_j (1 - \cos \theta_i \cos \theta_j) + \varepsilon_i^2 \varepsilon_j^2 (1 - \cos^2 \theta_i \cos^2 \theta_j) \geq 0 \tag{24}$$

which is always satisfied, since every ε_i is non-negative. The minimum value of the left-hand side of (24) is obtained for

$$\varepsilon_i = \varepsilon_j$$

and

$$\cos \theta_i \cos \theta_j = +1$$

These same values assign the parenthesis in (23) a minimum value of unity. Notice that the last equality implies $\cos \theta_i = \cos \theta_j = +1$; otherwise, (22) would not be satisfied.

In conclusion, the minimum value of the velocity β_1 for which our reaction can take place is given by the minimum value of ε_1 (in fact, ε_1 is an increasing function of β_1, and vice versa—see equation (20) with $k=1$), which in turn is obtained by giving *all* parentheses in (23) their minimum value of 1. This can only be done by taking:

1. All final velocities $\vec{v}_i$ collinear with $\vec{v}_1$:

$$\cos\theta_3 = \cos\theta_4 = \ldots = \cos\theta_n = +1 \tag{25}$$

2. All final velocities equal to one another:

$$\varepsilon_3 = \varepsilon_4 = \ldots = \varepsilon_n \tag{26}$$

implying

$$\beta_3 = \beta_4 = \ldots = \beta_n \tag{27}$$

Under these conditions, (23) becomes

$$m_1^2 + m_2^2 + 2m_1 m_2 \sqrt{1+\varepsilon_1^2} = M^2 \tag{28}$$

where

$$M = \sum_{i=3}^{n} m_i \tag{29}$$

is the total final mass. Under these conditions, we obtain

$$\sqrt{1+\varepsilon_1^2} = \frac{1}{\sqrt{1-\beta_1^2}} = \frac{M^2 - m_1^2 - m_2^2}{2m_1 m_2} \tag{30}$$

For the "threshold" kinetic energy T_1 of the incoming particle, it follows that

$$T_1 = m_1 c^2 \left[\frac{1}{\sqrt{1-\beta_1^2}} - 1\right] = \frac{M^2 - (m_1+m_2)^2}{2m_2} c^2 \tag{31}$$

We can thus calculate the threshold in T_1 for different inelastic processes, e.g. for the reactions listed in Table 1. The excellent agreement of the experimental observations with the calculated thresholds is proof of the validity of the conservation laws of energy and momentum and of the full equivalence between energy and mass. The above proof of (31), which is a well-known kinematical result in particle physics, has the advantage over the standard proofs that it does not use the relativistic invariance of scalar products and considers only one inertial frame, the laboratory—which for all practical purposes could be assumed to coincide with the fundamental frame (precision rarely exceeds 1 part in 10^3 for such experiments).

Table 1. Thresholds for inelastic reactions calculated from energy and momentum conservation laws

Reaction	Threshold
$PP \to PP\pi^\circ$	280 MeV
$PP \to P\Lambda^\circ K^+$	1585 MeV
$PP \to P\Sigma^+ K^\circ$	1790 MeV
$PP \to P\Xi^- K^+ K^+$	3743 MeV
$\pi^+ P \to \Lambda K^+ \pi^+$	1013 MeV
$\pi^- P \to \Sigma^- K^+$	904 MeV
$K^- P \to \Xi^- K^+$	662 MeV
$K^- P \to \Theta^- K^* K^* \pi^-$	3086 MeV

5. Potentials and Fields of a Moving Charge

It is a well-known fact that the scalar and vector potentials of the electromagnetic fields can be written in the causally meaningful "retarded" form as

$$\varphi(\vec{r},t) = \int d\vec{r}' \frac{\rho(\vec{r}',t-R/c)}{R} \tag{32}$$

$$\vec{A}(\vec{r},t) = \frac{1}{c}\int d\vec{r}' \frac{\vec{j}(\vec{r}',t-R/c)}{R} \tag{33}$$

where

$$R = |\vec{r} - \vec{r}'| \tag{34}$$

We want to calculate the potential generated by a very small (but not pointlike) electric charge in uniform motion with velocity v along the positive x axis (Figure 1). Of course, this is a standard problem, but our proof may have the advantage of being physically clearer. Let Q be the position occupied by q at time t. Two things must be considered:

1. Retarded potentials. The potentials are not generated instantaneously in P by q in Q, but with some delay due to the propagation of information with the speed of light. Therefore one must consider a point Q' occupied by q at an earlier time t' ($t' < t$), such that

$$\Delta t = t - t' = \frac{r'}{c} \tag{35}$$

In this way, the determination of the potential at time t in P occurs simultaneously with the arrival of the particle in Q (namely, at time t). Obviously

$$\overline{Q'Q} = v\Delta t = \frac{vr'}{c} \tag{36}$$

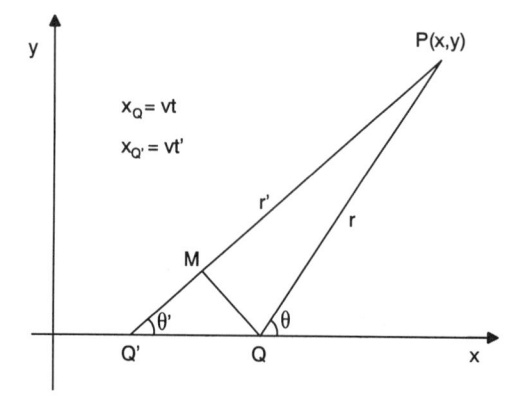

Figure 1. Potential generated by a very small (but not pointlike) electric charge in uniform motion.

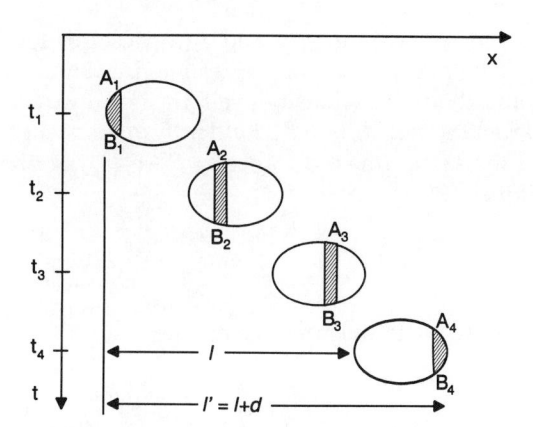

Figure 2. Position of the charge at different times.

2. **Effective charge.** Another factor arises from the finite extension of the charge q (no matter how small it is), but does not depend on its size. We can say that the motion of q and (again) the finite time necessary to propagate information require that we consider not q at time t', but

$$q_{eff}(t') = q\left(1 - \vec{v}\cdot\hat{r}'/c\right)^{-1} \tag{37}$$

where $\hat{r}'$ is a unit vector pointing from Q' to P in Figure 1. Consider the charge q and its different positions at time t_1, t_2, t_3, t_4 (Figure 2).

From Equation (32) we see that all the charge elements existing in space contribute to $\varphi(\vec{r},t)$, provided that they are considered at the different times

$$t' = t - \frac{R}{c}, \tag{38}$$

the difference being due to their different distances from P. In practice, everything happens as if an "exploring" sphere were shrinking at the speed of light, and were contracted to a point in P at time t. The sum of all the charges swept by the exploring sphere (each divided by R, as in (32) above) would then give the required potential $\varphi(\vec{r},t)$.

In our case, if the charged particle is not a geometrical point, but has an extension in space, one must take into account that its parts are at different distances from P, and must, therefore, be considered at different times t'. This conclusion is especially true if the charge is moving: with reference to Figure 2, we see that the potential in the point P (provisionally assumed to be on the x-axis) is generated by the slice A_1B_1 of the electric charge at time t_1, by the slice A_2B_2 at time t_2, etc., by the slice A_4B_4 at time t_4. Obviously the slice AB moves inside the particle with the speed of light, and the charge swept would finally be

$$q = \int_0^d dx \int_{f_1(x)}^{f_2(x)} dy \rho(x,y) \tag{39}$$

if q were at rest, where $f_1(x)$ and $f_2(x)$ describe lower and upper boundaries of the charge. The particle motion represented in Figure 2 stretches the boundaries, as it were, and the charge density itself, in such a way that the total length swept inside the particle is not d, but l'. Evidently, l' is swept at the speed of light c, while, during the same time a distance $l = l' - d$ is described by the particle with speed v. Therefore

$$l' = l \frac{c}{v} \tag{40}$$

We can now write the effective charge

$$q_{\text{eff}}(t') = \int_0^{l'} dx \int_{f_1(x/\mu)}^{f_2(x/\mu)} dy \rho\left(\frac{x}{\mu}, y\right) \tag{41}$$

where $\mu = l'/d$ is the stretching factor. By writing $x' = x/\mu$, it follows from (41) that

$$q_{\text{eff}}(t') = \frac{l'}{d} \int_0^d dx' \int_{f_1(x')}^{f_2(x')} dy \rho(x', y) = \frac{l'}{d} q \tag{42}$$

Remembering (40) and the relation $d = l' - l$, which is evident from Figure 2, we finally arrive at

$$q_{\text{eff}}(t') = q \frac{l'}{l' - l} = q \frac{1}{1 - v/c} \tag{43}$$

If the point P is not on the x-axis, we must consider only the projection of $\vec{v}$ over $\vec{r}'$, and (43) becomes (37). We may conclude that the scalar potential is given by

$$\varphi(\vec{r}, t) = \frac{q_{\text{eff}}(t')}{r'} = q\left(r' - \frac{1}{c} \vec{v} \cdot \vec{r}'\right)^{-1} \tag{44}$$

This expression can be cast in a slightly different form by observing, from Figure 1, that

$$r' - \frac{1}{c} \vec{v} \cdot \vec{r}' = r' - \frac{r'v}{c} \cos \theta'$$

$$= \overline{Q'P} - \overline{Q'M} = \overline{MP}$$

$$= \left(r^2 - \frac{r'^2 v^2}{c^2} \sin^2 \theta'\right)^{1/2}$$

$$= \left(r^2 - \frac{r^2 v^2}{c^2} \sin^2 \theta\right)^{1/2}$$

$$= r\sqrt{1 - \beta^2 \sin^2 \theta} \tag{45}$$

where $\beta = v/c$. It then follows for φ and for the similarly calculated vector potential $\vec{A}$ that

$$\varphi(\vec{r},t) = \frac{q}{r}\left(1 - \beta^2 \sin^2 \theta\right)^{-\frac{1}{2}}$$
$$\vec{A}(\vec{r},t) = \frac{q\vec{v}}{rc}\left(1 - \beta^2 \sin^2 \theta\right)^{-\frac{1}{2}} \qquad (46)$$

The electric and magnetic fields can be calculated as follows

$$\vec{E} = -\vec{\nabla}\varphi - \frac{1}{c}\frac{\partial \vec{A}}{\partial t}$$
$$\vec{H} = \vec{\nabla} \times \vec{A} \qquad (47)$$

and after a straightforward calculation, we obtain

$$\vec{E} = \frac{q(1 - \beta^2)}{r^2\left[1 - \beta^2 \sin^2 \theta\right]^{\frac{3}{2}}}\hat{r} \qquad (48)$$

$$H_x = 0; \quad H_y = -\beta E_z; \quad H_z = \beta E_y \qquad (49)$$

where $\hat{r} = \vec{r}/r$ is a unit vector directed from Q to P.

We can see that the retarded field $\vec{E}$ depends on θ, as well as r, but is nevertheless central, since it is proportional to $\hat{r}$. This feature guarantees that angular momentum conservation will hold for the motion of a charge in $\vec{E}$.

6. Length Contraction of Moving Bodies

In the words of John Bell (1992):

...the main point to be stressed is that electrical forces change with motion, and if they are important in matter it is simply unreasonable to think that matter will keep the same shape when in motion.

In fact, it was shown by Lorentz, and later confirmed by other authors, that electric charges in motion in the field of a moving charge give rise to bound states that are contracted in the direction of motion by the Lorentz factor $(1 - \beta^2)^{\frac{1}{2}}$. Therefore, the explanation of the negative result of the Michelson-Morley experiment is immediately available even in a theory in which Maxwell's equations hold in a preferred frame.

The Lorentz-Fitzgerald contraction of moving bodies must be a physically real phenomenon if it is due to the mechanism outlined in the previous section. There is, however, another argument, due to John Bell (1987), that points in the same direction. Consider two identical spaceships A and B, moving in space with the same velocity v_o, one behind the other on the same straight line. Their velocity v_o is measured with respect to some inertial frame I, in which time is defined and many synchronized clocks are placed at rest near the path of A and B. When the clocks near the two spaceships show time $t = 0$, their engines are ignited by the pilots. The two engines are assumed identical and produce exactly the same force $F(t)$ acting on A and B, starting at the same I-time $t = 0$. We can thus write

$$\begin{cases} x_A(t) = x_A(0) + \int_0^t dt' v(t') \\ x_B(t) = x_B(0) + \int_0^t dt' v(t') \end{cases} \quad (50)$$

where $v(t')$ is given by (17) with the same γ for A and B because of (16) and the existence of the common initial velocity v_o. It follows from (50) that

$$x_A(t) - x_B(t) = x_A(0) - x_B(0)$$

meaning that the distance between the two spaceships, as seen from I, is a constant. Suppose now that a fragile thread is tied initially between A and B. If it is just long enough to span the distance AB initially, then as the rockets speed up, it will become too short, because of its Lorentz-Fitzgerald contraction, and must finally break. The breaking takes place when, at a sufficiently high velocity, the artificial obstacle to contraction gives rise to a large internal stress. What is remarkable is that such a conclusion is exactly the same as what one would reason in STR, or, in the spirit of Lorentz-Fitzgerald, within classical physics with the contraction mechanism.

Consider next a different problem, that of a rod AB moving with velocity $\vec{v}$ with respect to the fundamental system I. We assume it to be inclined at an angle θ with respect to $\vec{v}$ (Figure 3), where θ is also measured in I.

Suppose that the Lorentz-Fitzgerald contraction has taken place for the component $l \cos \theta$ of the rod along v, while $l \sin \theta$ has remained unmodified. The length l_o of the rod when stationary in I can then be calculated from the equation

$$l_o^2 = \frac{l^2 \cos^2 \theta}{1 - \beta^2} + l^2 \sin^2 \theta \quad (51)$$

where $\beta = v/c$ and the factor $(1 - \beta^2)^{-1}$ corrects for the contraction of $l \cos \theta$. Obviously, we are also assuming that the particular acceleration used by the rod to reach the velocity v has no effect on the length l: only the final velocity matters. By inverting (51) we obtain

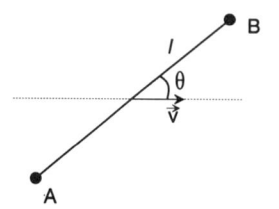

Figure 3. Lorentz-Fitzgerald contraction of a moving rod.

$$l = \frac{l_o(1-\beta^2)^{1/2}}{(1-\beta^2\sin^2\theta)^{1/2}} \tag{52}$$

which gives the contraction effect for a rod of arbitrary inclination. Just like the elementary contraction effect (case $\theta = 0$), the value (52) of l must be imagined as a real contraction due to translation with respect to I. Note, however, that l_o is the length when the rod is at rest in I, while θ is the angle of the moving rod with its velocity with respect to I. The quantities in the right-hand side of equation (52) are, then, in a sense, not homogeneous.

7. The Velocity of Light in a Moving Frame

If the velocity of light is c in all directions of the fundamental system I, the Galilei transformation gives

$$\vec{c}' = \vec{c} + \vec{v} \tag{53}$$

for the velocity of light $\vec{c}'$ with respect to a system I' moving with velocity $\vec{v}$ with respect to I. By inverting and squaring we obtain

$$c^2 = c'^2 + v^2 - 2c'v\cos\theta \tag{54}$$

where θ is the angle between $\vec{c}'$ and $-\vec{v}$ (see Figure 4).

By solving (54) as an equation in c', we obtain

$$c' = v\cos\theta + \sqrt{c^2 - v^2\sin^2\theta} \tag{55}$$

the positive sign being a consequence of the condition $c' = c$ if $v = 0$. Although in general, $c' \neq c$, the consequences of (55) are often startlingly similar to the results of STR, as Builder (1958) and Prokhovnik (1985) have pointed out (cf. also Janossy 1964).

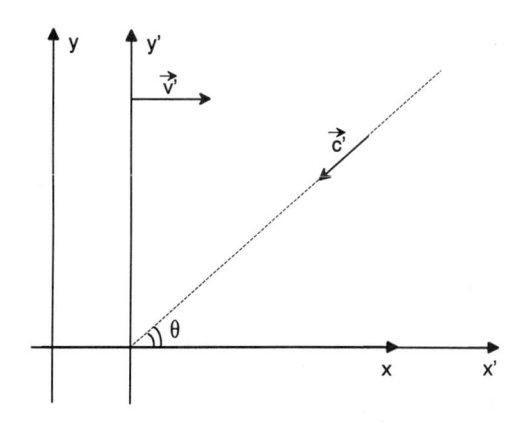

Figure 4. Velocity of light in a moving frame.

8. The Time Dilation Effect

Consider now one of the most elementary clocks that can be conceived, consisting of the rigid moving rod of Figure 3, having mirrors inserted at the two ends A and B to reflect a beam of light backwards and forwards along the length of the rod. The unit of time T_o will be defined as the time required to describe the trajectory ABA. We would then have

$$T_o = \frac{2L_o}{c} \tag{56}$$

if the rod were at rest in I with length L_o, since the light velocity would, in this case, be simply c.

In the general case of a moving rod, the unit of time is

$$T = \frac{L}{c'_{AB}} + \frac{L}{c'_{BA}} = \frac{2L\sqrt{c^2 - v^2 \sin^2 \theta}}{c^2 - v^2} \tag{57}$$

since c'_{AB} is given by (55), while in c'_{BA} we must regard θ as being increased by π, which amounts to a change of sign of the first term in (55).

If we use the fundamental result (52) for L we get

$$T = \frac{T_o}{\sqrt{1 - v^2/c^2}} \tag{58}$$

This is a truly striking consequence of the anisotropic formulae for length and velocity of light in a moving system: in spite of the fact that L, c'_{AB} and c'_{BA} in (57) are all θ-dependent, it so happens that their combination is completely independent of θ. It is only because of this that we can think of T as a unit of time.

Due to the factor $(1 - v^2/c^2)^{-\frac{1}{2}}$ in (58), we can say that the unit of time in the moving system is longer than the corresponding unit of time T_o in I. Given a certain time interval measured as Δt in units of T_o, it will instead be measured as

$$\Delta t' = \Delta t \sqrt{1 - \frac{v^2}{c^2}} \tag{59}$$

if units of T'_o are used. This is like saying that time is running slower in the moving system: the traveling twin will be *younger* than the stationary one when he comes to rest in the same inertial system.

From a theoretical point of view, it can be stressed that the same result obtained above was deduced by Bell (1987) by studying an electron bound to a proton in motion with the use of the retarded fields in equations (32) and (33). His conclusion was that the period of rotation of the electron in a hydrogen atom in motion exceeds the same period for an atom at rest by a factor $(1 - v^2/c^2)^{-\frac{1}{2}}$. Therefore, the choice of the unit of time given by the hydrogen atom also leads to the result (59) obtained above with the light bouncing back and forth between the extremities of a moving rod.

Notice, however, that our conclusion would have been different if we had assumed we possessed a previously defined unit of time, because (58) would then have represented a relationship between time intervals (rather than between time units): the interval measured from the moving system would have been longer, since (58) and (59) have an opposite structure in terms of the factor $(1-v^2/c^2)^{-1/2}$. The correctness of one of the two choices can only be suggested by nature, and it is well known that high-velocity unstable particles live longer than similar particles at rest. The time dilation in moving systems is, in fact, one of the best established facts of physics, given the very considerable experience of particle physicists with beams of unstable objects (pions, kaons, hyperons, *etc.*) in high-energy accelerators. The situation is analogous in cosmic ray physics: without time dilation, the muons that decay with a lifetime of 2×10^{-6} sec and are produced about 20 km above sea level would all disappear within 2 km of the production site, while it is experimentally known that a considerable fraction of them reach sea level.

But time dilation has also been established at the macroscopic level. Hafele and Keating (1972) placed several cesium atomic clocks aboard ordinary commercial around-the-world jet flights. One flight circumnavigated the earth traveling in the eastward direction; the other in the westward direction. The clocks were compared with similar clocks that remained on the ground. Relative to the ground clocks, the clocks that traveled eastward lost 59±10 nanoseconds, while those traveling westward gained 273±7 nanoseconds. These results are in excellent agreement with the predictions of standard relativity theory, predictions depending both on the flight velocity and on the earth's gravitational field.

9. The Velocity of Light

We again consider formula (52) for the contraction of a rod forming an angle θ with respect to its velocity.

Obviously, this formula applies to all conceivable rods, and in particular to one of rest length L_o taken as the unit of length (*e.g.* 1 meter).

$$L = \frac{L_o(1-\beta^2)^{1/2}}{(1-\beta^2 \sin^2 \theta)^{1/2}} \tag{60}$$

Therefore, the contracted rod measured with the contracted unit of length having the same inclination will necessarily give the same result as obtained by measuring l_o with L_o:

$$\frac{l}{L} = \frac{l_o}{L_o} \tag{61}$$

In other words, the moving observer will be unable to observe his motion by means of length contraction: all the objects at rest in his system will appear to him as having the same length they had at rest in the fundamental system. The situation is similar if one tries to measure time with the time-dilated unit T given by (58). This unit is similarly unable to allow a moving observer to detect the time dilation of the physical process taking place in his inertial system.

Therefore, if an observer O is initially at rest in the fundamental frame I and then accelerates until he acquires velocity v with respect to I, he will continue to find the same results for lengths and time intervals before, during and after the acceleration. For him, the rod AB always has the length l_o, whatever its inclination with respect to the x-axis. A light pulse moving forwards or backwards on AB, according to the observer, always takes the time $2l_o/c$. But this amounts to saying that he continues to observe c as the velocity of light, and in all directions!

References

Bell, J.S., 1992, George Francis Fitzgerald, *Physics World* September 1992.
Bell, J.S., 1987, How to teach special relativity, in: *Speakable and Unspeakable in Quantum Mechanics*, Cambridge University Press.
Builder, J., 1958, *Australian J. Phys.* 11:279.
Einstein, A., 1920, *Aether und Relativitätstheorie*, Springer Verlag.
Hafele, J., and Keating, R., 1972, *Science* 177:166.
Janossy, L., *Acta Physica Hungarica* 17:421.
Keswani, G.H., 1966, *Brit. J. Phil. Sci.* 17:235.
Kostro, L., 1992, Relativistic ether conception, in: *Studies in the History of General Relativity*, J. Eisenstaedt and A.J. Kox, eds., Birkhäuser.
Larmor, J., 1900, *Aether and Matter*, Cambridge Univ. Press.
Lewis, G.N., 1908, *Philosophical Magazine* 16:705.
Lorentz, H.A., 1892, The relative motion of the earth and the ether, reprinted in: *Collected Papers* 4:219.
Michelson, A.A. and Morley, E.W., 1887, *Am. J. of Science* 34:333.
Poincaré, H., 1905, *The Monist* 15:1.
Prokhovnik, S.J., 1985, *Light in Einstein's Universe*, Reidel.
Selleri, F., 1990, *Z. Naturforsch.* 46a:419.
Tyapkin, A.A. and Shybanov, A.C., 1982, *Poincaré*, Molodaia Gvardia.
Winterberg, F., 1987, *Z. Naturforsch.* 42a:1428.
Zahar, E., 1973, *Brit. J. Phil. Sci.* 24:95 and 233.

Galilean Gravitation on a Manifold

D. F. Roscoe

Department of Applied Mathematics
University of Sheffield
Sheffield S10 2TN, UK.
Fax: 742-739826, Email: D.Roscoe@uk.ac.shef.pa

This paper argues that except for a chance event—the early death of Heinrich Hertz at the turn of the century—the development of 20th century physics might have been quite different from its actual course. In particular, it is suggested that curved spacetime gravitational physics would never have been formulated, being replaced, instead, by formalisms describing gravitational action within which, ultimately, quantization would have been a relatively routine process.

1. Introduction

Perhaps the pivotal problem of late 19th Century physics was the fact that Maxwell's equations were not Galilean invariant. As is well known, the dilemma was considered finally resolved when these equations were found to be invariant with respect to motional transformations in what is now called the Lorentz group, and the shape of modern physics has been largely determined by this event. However, prior to this, Hertz (1962) had discovered that Maxwell's equations could be made Galilean invariant by the simple formal step of replacing partial time derivatives, wherever they occurred in Maxwell's theory, by total time derivatives. This new set of equations, referred to here as Hertz's equations, introduced an uninterpreted v_c say, into electromagnetic theory; Maxwell's equations are recovered when $v_c = 0$. For Hertz, an obvious interpretation—given the mood of the time—of this *convection velocity* was as a description of the motion of the laboratory frame, within which Maxwell had derived his equations, with respect to the hypothesised luminiferous aether. This interpretation had the positive advantage of predicting an observable effect: that the motion of a dielectric in the laboratory would create a magnetic field detectable by an instrument at rest in the laboratory. This prediction was experimentally falsified in 1903 by Eichenwald (1903), and it was this event which effectively killed off Hertzian electromagnetism, opening the way to the modern Lorentzian interpretation.

By the time of Eichenwald's experiment, Hertz was already dead, and so had no chance to review the status of his Galilean invariant electromagnetism in the light of Eichenwald's null result. However, it can be speculated, with profit, what might have happened had he lived. Since $\mathbf{v}_c$ does not describe an 'aether velocity', an obvious next question would be: do Maxwell's equations adequately parametrize all the possible relationships that exist between the laboratory elements which go to make up the experiments which the theory was constructed to describe? In electromagnetism experiments there are basically two elements a distribution of charges, ρ, which create the Maxwellian fields, and a system of detectors, which detect these fields. Arbitrary motions of the charges are parametrized in the $\mathbf{j} = \rho\mathbf{v}$ current term, but, as Phipps (1986) points out, in the classical experiments, detectors are always assumed to be at rest in the laboratory frame. Consequently, Phipps suggests that an alternative interpretation of the Hertz convection velocity, $\mathbf{v}_c$, is that it parametrizes arbitrary motions of the laboratory detectors. From a pragmatic point of view this is reasonable since, if the process of accelerating charges has observable consequences, why should not the process of accelerating the detectors also have observable consequences? Of course, the acid test of such an interpretation must always be found in the laboratory, and it might very well be that the Hertz theory would fail under such scrutiny. However, let us speculate that, in fact, the Hertz theory with the foregoing interpretation for $\mathbf{v}_c$ was confirmed—so far as the present author is aware, it has never been put to the test. In this case, a very interesting situation would have emerged: basically, the Galilean group would have been likely to retain its pre-eminence, although the Lorentz group would have still been lurking, hidden away, as the motion group for the Maxwellian $\mathbf{v}_c = 0$ special case of the Hertz theory, maybe to emerge at some later date—albeit interpreted rather differently (see postscript).

The retention of the Galilean group, as the primary motion group of physics, into the first part of this century, would almost certainly have had profound effects on the subsequent evolution of gravitation theory. The problem posed by the excess advance of Mercury's perihelion to Newtonian gravitation theory would still have stimulated a fundamental review of gravitation theory. However, in the absence of the new Lorentzian physics, and therefore of the concepts of spacetime manifolds awaiting curvature, this review would, in the first instance, have taken place within the context of Galilean physics, with a natural first stop being an analysis of the assumptions built into the Newtonian gravitational framework; subsequent progress would have been determined largely by the results of such an analysis. This paper sets out to reconstruct gravitation theory according to this general pattern.

2. The Overview

In this paper, the phrase 'inertial interaction' is used to categorize any interaction in which the basic determining factor is the relative inertial masses of the particles involved. With this understanding, the following section, Section 3, analyses Newtonian gravitational action, showing it to be interpretable as a special case of inertial interaction, and an approximation predicated upon the assumption $M \gg m$, where M is the inertial mass of the gravitational source, and m is the inertial mass of the gravitating particle. Section 4 completes the analysis of the Newtonian theory by analysing the concept of 'potential energy', showing it to be simply a convenient metaphor used to represent the dynamical behaviour of the gravitating source. From the constructive point of view, the fact that these analyses provide the Newtonian gravitational approximation with a purely *inertial* context suggests the existence of an exact Galilean invariant formalism, describing gravitational action, to be obtained from an appropriate consideration of Newtonian at-a-distance inertial interaction.

Galilean gravitational interactions can be considered as belonging to that generic class of interactions in which linear momentum is conserved and through which particle trajectories are independent of specific mass values; this class is termed, for convenience, as the class of *generalized gravitational interactions*. Subsequent to the Newtonian analysis of sections 3 and 4, it is shown how this generic class can be modelled purely in terms of the constraints imposed by the physics of momentum conservation, together with a naturally arising minimum principle. Specifically, it is possible to show how any momentum conserving interaction, involving three or more particles, can be given an abstract representation in terms of an unspecified class of Galilean manifolds (which have nothing to do with *space-time manifolds*). It is then noted that the freedom exists to constrain the interactions concerned so that the corresponding trajectories are coincident with geodesics in an abstract space in which these manifolds act as coordinate surfaces. This latter process makes trajectories through such interactions independent of specific inertial mass values, and so can be interpreted as a modelling step which identifies these interactions with the generic class of generalized gravitational interactions. However, because the Galilean manifolds over which these interactions take place are unspecified, this generic class cannot be *a priori* identified with actual gravitational interactions, but can only be said to include them as special cases; even so, this is sufficient to identify actual gravitational process as a special case of inertial process, and, therefore, make the Equivalence Principle formally redundant. The final step of identifying that subclass of Galilean manifolds over which actual gravitational processes occur requires the additional knowledge arising from Newtonian experience which suggests the concept of 'potential' to be central to gravitational physics; this additional knowledge can be used as a constraint on the class of admissible Galilean manifolds so that they become identified with the level surfaces of an unspecified potential function defined on a velocity space.

The paper continues, in Section 11, by suggesting one particular approach for determining appropriate potential functions which is worked through to obtain approximate equations of motion. These equations of motion are distinguished in being *independent* of externally determined parameters, such as the gravitational constant of classical theory, and gravitational dynamics are then determined by the simple specification of initial dynamic conditions. The theory is therefore *complete* in the sense of requiring no exterior determinations of arbitrary parameters. The paper finishes by applying the (parameter free) approximate equations of motion to predicting periods for planetary orbits in the solar system, and the resulting numerical integrations are presented in Section 13.

However, perhaps the most significant property of the analysis which leads to the exact Galilean formalism, presented here, is it is not 'Galilean specific'—in other words, the general approach can be applied directly to analyse momentum conservation within the Lorentzian context. In this latter context, it can be expected to lead to a formalism describing gravitational process in which the instantaneous action-at-a-distance of the Galilean formalism is replaced by a retarded local action in which momentum transfer takes place between retarded fields and the particles which act as sources for these fields.

3. Newtonian Gravitation and the Third Law

One of the enduring mysteries of classical and modern physics arises from the fact that the value obtained for the gravitational-mass ratio of two particles compared in a weighing experiment is identical to the value obtained for the inertial-mass ratio of the same two particles compared in a collision experiment (to within experimental error). For this reason it is common to speak of the equivalence between inertial and gravitational mass, and there is a tendency to use the concepts interchangeably. However, an appropriate consideration of the general nature of Newtonian instantaneous action-at-a-distance shows that the gravitational-mass ratio of two particles determined in a weighing experiment can be interpreted *directly* as the inertial-mass ratio of the same two particles determined in a particular kind of 'at-a-distance collision experiment'.

A common formulation of Newton's Third Law states 'for every action there is an equal and opposite reaction' and so, for two particles in collision (observed from an inertial frame), there necessarily follows

$$m\Delta v = -M\Delta V,$$

where m, M are the respective inertial masses and Δv, ΔV are the respective velocity changes through the collision. However, this formal rendering of the Third Law for a two-particle system is specific to the case of *collision*, in which interaction occurs through direct contact. Now, the various phenomena of gravitation and electromagnetism impose the notion of a more general case in which the interaction between two particles can be characterized as being continual, and 'at-a-distance'. In this case, the foregoing formal statement which summarizes the case of 'collision' is inadequate, and must be generalized to become

$$m\ddot{x} \equiv +G(t); \quad M\ddot{X} = -G(t), \tag{1}$$

where $G(t)$ can be interpreted as the programme of Newtonian forces exerted on each particle in the system. However, (1) by itself is not sufficient to express the full meaning of the Third Law for the at-a-distance case since it merely asserts the equality of the opposing forces, but neglects to say these forces act along the straight line which connects these particles—a condition which is generally taken to be implied in any statement of the Third Law. This condition of colinearity can be expressed as

$$\mathbf{x} - \mathbf{X} = \lambda G(t) \tag{2}$$

where λ is an unknown Galilean scalar. If it is noted from (1) that $\ddot{x} = G/m$ and $\ddot{X} = -G/m$, then there follows

$$\ddot{\mathbf{x}} - \ddot{\mathbf{X}} = \left(\frac{m+M}{mM}\right) G(t),$$

so that writing $\mathbf{Y} \equiv (\mathbf{x} - \mathbf{X})$, then $G(t)$ can be eliminated between this latter statement and (2) to obtain

$$\mathbf{Y} = \lambda \left(\frac{mM}{m+M}\right) \ddot{\mathbf{Y}} \tag{3}$$

Now consider the general nature of a procedure which is required to compare two test masses, m_1 and m_2, *for equality only* via measurements made on their respective interactions with the mass M, which occur according to (3). Suppose these latter measurements are made simultaneously and in the same location—implying Y and λ are identical for both cases—and suppose they simply consist of instantaneous measurements of the two relative acceleration vectors, denoted as $\ddot{Y}_1$ and $\ddot{Y}_2$ respectively. Now taking the scalar product of (3) with itself there follows, for each set of measurements, the scalar equations

$$(Y,Y) = \lambda^2 \left(\frac{m_1 M}{m_1 + M} \right)^2 (\ddot{Y}_1, \ddot{Y}_1)$$

$$(Y,Y) = \lambda^2 \left(\frac{m_2 M}{m_2 + M} \right)^2 (\ddot{Y}_2, \ddot{Y}_2),$$

where the notation (a, b) denotes the scalar product of a with b, and it follows immediately that

$$\left(\frac{m_1}{m_2} \right) \left(\frac{m_2 + M}{m_1 + M} \right) = \sqrt{\frac{(\ddot{Y}_2, \ddot{Y}_2)}{(\ddot{Y}_1, \ddot{Y}_1)}}.$$

It is clear from this latter expression that m_1 and m_2 can only be equal when the magnitudes of the two relative acceleration vectors are also equal. In this general way, it can be seen how Newtonian action-at-a-distance can be used to calibrate the relative *inertial* masses of particles, and how—significantly—the process is independent of the specific form of λ which defines the quantitative nature of the action; that is, the process could be electromagnetic, gravitational, or anything else one cares to imagine.

Newton's Law of Universal Gravitation can now be seen as simply a special case which starts with the assumption M (say) is so large compared to m that, by (1), $\ddot{x} \gg \ddot{X} \approx 0$ with the consequence x can be considered fixed in some subclass of inertial frames; consequently, relative to those particular frames in this subclass for which $\ddot{X} \approx 0$, (3) becomes $x \approx \lambda m\ddot{x}$. Consistency with Kepler's Laws is then obtained if $\lambda \equiv -\lambda_0 |x|^3$, where λ_0 is an appropriately determined positive constant. With this perspective of Newtonian gravitation, it is possible to see how the process described above to test for equality between inertial masses in a Newtonian at-a-distance interaction defines the essential features of the classical pan-scale weighing experiment. In other words, relative masses determined in weighing experiments can be interpreted simply as measures of relative *inertial* masses determined in at-a-distance 'collision' experiments, and Newtonian gravitation can be interpreted as simply a special case of an 'at-a-distance inertial interaction'.

4. The Nature of Newtonian Potential

The foregoing discussion on the general nature of Newtonian gravitation leads immediately to questions concerning the nature of Newtonian 'potential energy' and, in the following, it is shown how this concept can be viewed as simply a modelling device which allows the dynamical behaviour of the gravitating source to be ignored and that, in fact, it is essentially the negative kinetic energy of the source.

Consider a binary inertial interaction, described from a frame which is at rest with respect to the system's centre of gravity, and suppose the particles concerned have masses and velocities $(m, \mathbf{v})$, $(M, \mathbf{V})$. Then

$$m\mathbf{v} + M\mathbf{V} = 0,$$

from which

$$\frac{1}{2}m(\mathbf{v}, \mathbf{v}) = \left(\frac{M}{m}\right)\frac{1}{2}M(\mathbf{V}, \mathbf{V}),$$

so that, using $(KE)_m$ and $(KE)_M$ to denote the respective kinetic energies, there finally follows

$$(KE)_m - \left(\frac{M}{m}\right)(KE)_M = 0. \qquad (4)$$

The important point to remember is, since this relationship follows directly from the Third Law, it *must* be true for any Newtonian binary interaction including, specifically, a classical Newtonian gravitating pair for which $M \gg m$. In such a case, m is the mass of the gravitating particle, and M is the mass of the 'gravitational source'. The kinetic-energy/potential-energy relationship for the gravitating particle in such a system, given in the form $(KE)_m + (PE)_m = 0$, is, like (4), only true in that class of frames which is at rest with respect to the system centre of gravity; consequently, it can be compared directly with (4) with the result

$$(PE)_m = -\left(\frac{M}{m}\right)(KE)_M.$$

That is, the potential energy of a gravitating-mass is proportional to the *negative* kinetic energy of the source-mass, and this conclusion is based purely on the physics of momentum conservation. In other words, the Newtonian concept of 'potential energy' is simply a modelling device which, in cases for which $M \gg m$, allows the dynamical behaviour of the gravitating source to be ignored.

5. Summary of Newtonian Analysis

It was shown, in Section 3, how Newtonian gravitation theory is an approximation based on the assumption that the gravitational source is at rest in some inertial frame $(m/M \approx 0)$, and that a weighing experiment can be interpreted directly as an 'at-a-distance' determination of inertial mass whilst, in Section 4, it was shown how the Newtonian concept of 'potential energy' of a particle in a binary interaction can be interpreted directly as a measure of the negative kinetic energy of the second particle involved. Reasonable conclusions following from these simple analyses are

- there exists an exact Galilean gravitation theory;
- this exact theory will follow from an appropriate consideration of 'at-a-distance' inertial interaction;
- any concept of 'potential' in this exact theory will take the form of some kind of kinetic-energy function satisfying some kind of potential equation.

The remaining sections are devoted to obtaining this theory, and to demonstrating its validity.

6. The Galilean Model Universe

Suppose a model Universe consists of $N < \infty$ material particles, $P_1 \ldots P_N$, having inertial masses $m_1 \ldots m_N$, and suppose these N particles have respective velocities $\mathbf{v}^1 \ldots \mathbf{v}^N$ at time t, and $\mathbf{v}^1 + d\mathbf{v}^1 \ldots \mathbf{v}^N + d\mathbf{v}^N$ at time $t + dt$ measured from within some inertial frame. Then, according to Newton's Third Law, there is the relationship

$$m_1 d\mathbf{v}^1 + m_2 d\mathbf{v}^2 + \ldots + m_N d\mathbf{v}^N = 0$$

between the velocity-change vectors of the whole class of Universal particles. Now suppose the whole model Universe of N particles is partitioned, arbitrarily, into exactly three non-trivial agglomerations of particles, and that these agglomerations have respective masses M_1, M_2 and M_3, respective centre-of-mass velocities $\mathbf{V}^1$, $\mathbf{V}^2$ and $\mathbf{V}^3$ at time t, and respective centre-of-mass velocities $\mathbf{V}^1 + d\mathbf{V}^1$, $\mathbf{V}^2 + d\mathbf{V}^2$ and $\mathbf{V}^3 + d\mathbf{V}^3$ at time $t + dt$. The foregoing statement of global momentum conservation in the model Universe can then be expressed as

$$M_1 d\mathbf{V}^1 + M_2 d\mathbf{V}^2 + M_3 d\mathbf{V}^3 = 0. \tag{5}$$

The particular representation, (5), of global momentum conservation in the model Universe is the primary form upon which the whole of the following analysis is predicated.

7. The Galilean Reference Scalar

In the following, it is shown how (5) provides the basis for a geometric representation of gravitational action within the Galilean model Universe leading, finally, to the identification of gravitation as a special kind of inertial process, and making the Equivalence Principle redundant to physics.

According to (4), the trajectories, represented by the velocity-change vectors $d\mathbf{V}^1$, $d\mathbf{V}^2$, $d\mathbf{V}^3$, of the three agglomerations involved in the global momentum conserving interaction, must lie in some Galilean invariant 2-dimensional subspace of the velocity space.

Analogy 1: consider three non-colinear lines which lie in the same plane in ordinary Euclidean space; the direction of any one of these lines can be expressed purely in terms of the directions of the other two; these lines are then said to be linearly dependent, and to lie in a two-dimensional subspace—the plane which contains them—of the general three-dimensional Euclidean space.

Any such Galilean subspace can, in turn, be considered to define a tangent plane to a level surface of some Galilean scalar function, U, for example, defined on the velocity space.

Analogy 2: the Euclidean plane of analogy 1 can be thought of as tangential to the surface of an ordinary sphere which, in turn, can be referred to as a level surface of the Euclidean distance function, $x^2 + y^2 + z^2$.

In this way, it is clearly seen that the level surfaces of the scalar function U are simply abstract representations of the agglomeration trajectories into, and out of, the global momentum conserving interaction; they have no meaning beyond this. However, in the following it will be shown that these level surfaces provide a very powerful means for referencing the motion of an arbitrary particle which belongs to any of the three agglomerations; for this reason, the scalar U will be referred to as the 'reference scalar' and the level surfaces themselves will be referred to as the 'interaction manifolds'.

8. Geometry and the Reference Scalar

Consider now the trajectory, $\mathbf{v}$, of a single particle chosen arbitrarily from one of the three agglomerations; in general, such a trajectory will not be confined to any particular level surface of U, but will cut across such surfaces with the consequence that it can be parameterized by changes in U.

To see how this can be done, first note that, since $n_a \equiv \nabla_a U$, $a = 1, 2, 3$, defines normal vectors to the level surfaces of the reference scalar, U, then

$$dn_a = \nabla_j(\nabla_a U) dv^j \equiv g_{ja} dv^j, \quad \text{where} \quad g_{ab} \equiv \nabla_a \nabla_b U, \qquad (6)$$

gives the covariant change in n_a along an arbitrary infinitesimal arc $d\mathbf{v}$. Effectively, therefore, if $d\mathbf{v}$ is confined to the particle trajectory, $\mathbf{v}$, then dn_a provides a vector measure of reference-scalar change along an infinitesimal section of trajectory and, in particular, the scalar product

$$dn_i dv^i \equiv g_{ij} dv^i dv^j \equiv ds^2$$

will give a scalar measure of the invariant resolution of this change along $d\mathbf{v}$.

Now suppose p and q are two arbitrarily chosen point velocities, and suppose the latter expression is integrated to give the scalar invariant

$$I(p,q) = \int_p^q \sqrt{dn_i dv^i} \equiv \int_p^q \sqrt{g_{ij} dv^i dv^j} \equiv \int_p^q \sqrt{g_{ij} \dot{v}^i \dot{v}^j}\, d\tau, \qquad (7)$$

where τ is some parameter, and further suppose that $I(p,q)$ is minimized with respect to choice of the trajectory connecting p and q. This minimizing trajectory is then geodesic in the velocity space formed by the union of all level surfaces of the reference scalar, U, and within which g_{ab} provides the metric tensor; it is to be noted that these trajectories are specified by the Euler-Lagrange equations defined using the Lagrangian density $\mathcal{L} \equiv \sqrt{g_{ij} \dot{v}^i \dot{v}^j}$, of (7).

Analogy 3: a level surface of the Euclidean distance function, $x^2 + y^2 + z^2$, is the set of all points which are a fixed distance away from the origin; the union of all such level surfaces is the set of all points in an ordinary Euclidean space, and a geodesic in this space is simply the shortest distance between any two points.

Since the existence of this abstract velocity space—and hence of geodesics within it—is independent of specific inertial-mass values, then this latter step makes particle trajectories independent of particle inertial masses, and is the modelling step defining that general subclass of trajectories which conform to the characteristic phenomonology of gravitation.

The abstract velocity space with metric tensor g_{ab} is, of course, not yet sufficiently specified since the formal expression of this tensor, given in (6), has the explicit form

$$g_{ab} = \frac{\partial^2 U}{\partial v^a \partial v^b} - \Gamma^k_{ab} \frac{\partial U}{\partial v^k}, \tag{8}$$

where Γ^k_{ab} are the objects (connection coefficients) which describe the velocity-space geometry, and this has yet to be specified. However, before these connection coefficients are finally specified, it is easier to first consider the nature of the most simple possible circumstance described by (8).

9. The No-Interaction Solution

The most simple case which the presented formalism might be expected to describe is that which contains the class of all possible non-interacting inertial particle trajectories. In this particular case, the interaction manifold metric tensor, g_{ab}, and the corresponding connection coefficients, Γ^a_{bc}, would be expected to assume their most simple possible forms, and therefore to be given by

$$g_{ab} = \delta_{ab}, \text{ and } \Gamma^a_{bc} \equiv 0, \tag{9}$$

where δ_{ab} is the 3×3 unit matrix. It is easily seen that this prescription represents a consistent resolution of (8).

In order to see that this formal resolution of the given formalism actually does contain the class of non-interacting inertial trajectories, first note that (9) in (7) gives the Lagrangian density $\mathcal{L} \equiv \sqrt{\dot{v}^i \dot{v}^j \delta_{ij}}$. The Euler-Lagrange equations can now be written down as $\dot{v} = k\mathcal{L}$, for arbitrary vector constant k, and it is easily shown that this has the unique solution $\dot{v} = 0$, which represents the class of all possible inertial trajectories.

In order to determine the form of the reference scalar, U, corresponding to this case, simply note that (8) with (9) can be written

$$\delta_{ab} = \frac{\partial^2 U}{\partial v^a \partial v^b},$$

from which the most general solution is found to be given by

$$U = \frac{1}{2} v^i v^j \delta_{ij} + \alpha^i v^j \delta_{ij} + \beta,$$

where α^a is a constant vector and β is a constant scalar. Writing $\alpha^a = -v_0^a$, such a solution can always be written as

$$U = \frac{1}{2}(v^i - v_0^i)(v^j - v_0^j)\delta_{ij} + \gamma \qquad (10)$$

where γ is some constant scalar. Since v_0^a is constant, it can be interpreted as the velocity of some arbitrarily chosen inertially moving reference point—for example, the system centre of mass (c of m) is a natural choice. In this case, $\mathbf{w} \equiv \mathbf{v} - \mathbf{v}_0$ is the velocity of the particle relative to this reference point and, since this latter vector is a vector with respect to the Galilean group (in fact, it is a vector with respect to transformations between arbitrarily accelerating frames), then U is a Galilean invariant scalar function of velocities. With the foregoing interpretation of $\mathbf{v}_0$, the first term of (10) has an obvious interpretation as the kinetic-energy/unit mass defined relative to the arbitrarily chosen reference point; it is subsequently referred to as the 'relative kinetic-energy density', and is denoted by Φ.

10. The Connection Coefficients

The identification of the velocity-space metric, g_{ab}, with the 3×3 unit matrix, δ_{ab}, for the case of non-accelerated motions makes the task of identifying the connection coefficients a relatively routine procedure.

The velocities in relation (8) are particle velocities defined relative to an arbitrarily defined inertial frame. Since inertial motion within the context of a Galilean model Universe can only be defined as motion which is unaccelerated with respect to the global centre-of-mass, then a natural frame of reference is any which is stationary with respect to this global centre. It follows that, without loss of generality, the velocity in (8) can be identified with the velocity $\mathbf{w}$ of the previous section, and therefore be considered as being defined relative to this absolutely defined centre. Since $\mathbf{v}$ can now be seen as a *relative* velocity defined between two physical points, and since the reference scalar, U, is defined in terms of $\mathbf{v}$, then (8) is invariant with respect to transformations between arbitrarily *accelerating* frames, and not just between inertial frames. This means that a coordinate transformation can *always* be made such that, locally, particles appear to be unaccelarated, and therefore—by the result of the previous section—to have local dynamics prescribed by the relation $g_{ab} \approx \delta_{ab}$. It follows, as a standard result of differential geometry, that the Ricci condition, $\nabla_c g_{ab} = 0$, is necessarily true everywhere; from this it follows, again as a standard result, that the connection coefficients must be the Christoffel symbols, and given by

$$\Gamma_{ab}^k \equiv \frac{1}{2} g^{kj} \left(\frac{\partial g_{bj}}{\partial v^a} + \frac{\partial g_{ja}}{\partial v^b} - \frac{\partial g_{ab}}{\partial v^j} \right).$$

The general formalism is now completely defined to within the specification of the reference scalar, U.

11. The Potential Structure of the Reference Scalar

So far, the presented formalism has been obtained as a pure description of that generic class of at-a-distance inertial interactions through which particle trajectories are independent of specific particle masses and so, *a priori*, the formalism includes actual gravitational interactions as a special case; it cannot yet be identified with them since, whilst (8) can be considered as a constraint on the structure of the metric tensor, g_{ab}, which determines particle trajectories through gravitational interactions, it does not determine this tensor since there is, as yet, no rule for the determination of the reference scalar, U. However, as was noted in Section 2, the success of Newtonian gravitational physics can be interpreted as providing very strong circumstantial evidence for the importance of the potential concept to gravitational physics in general; Section 4 showed the Newtonian potential in a binary interaction to be closely related to a kinetic energy in such an interaction whilst, in Section 8, it was shown how the reference scalar U (the level surfaces of which define the interaction manifolds of the given formalism) could—in very simple circumstances—be interpreted as a 'relative kinetic energy density'. Altogether, these circumstances suggest that, for gravitational interactions, U must be constrained to have some form of potential content or, equivalently, constrained to satisfy some form of Poisson-type equation.

Whilst there appears to be no unique way of determining which Poisson-type equation should be imposed on the system, an obvious starting point is to consider the possibility of a 'weak field' case, defined as a small perturbation of the no-interaction case for which, from (10),

$$U = \frac{1}{2}[(\mathbf{v}-\mathbf{v}_0),(\mathbf{v}-\mathbf{v}_0)] + \gamma \equiv \frac{1}{2}(\mathbf{w},\mathbf{w}) + \gamma \equiv \Phi + \gamma$$

$$\delta_{ab} = \frac{\partial^2 U}{\partial w^a \partial w^b}.$$

Taking the inner product of this equation with δ^{ab} shows that $U = \Phi + \gamma$ also satisfies

$$\delta^{ij}\frac{\partial^2 U}{\partial w^i \partial w^j} \equiv \nabla^2 U = 3,$$

which is a Poisson-type equation, as required. However, this latter equation has the more general solution

$$U = \Phi + \gamma + \frac{\beta}{\sqrt{\Phi}} \qquad (11)$$

which has the required potential component and which, for large values of Φ, can be considered a small perturbation of the no-interaction solution. If the level surfaces of this Poisson U are assumed to define interaction manifolds, then the corresponding metric tensors on these manifolds can be found by putting this U in (8), and solving the resulting differential equations for the g_{ab}. These equations are non-linear, and probably only resolvable in a numerical sense; however, a *closed* approximate solution can be be easily obtained if $g_{ab} \approx \delta_{ab}$ is assumed, and used in connection coefficients of (8). In this case, $\Gamma^a_{bc} \approx 0$, and (8) leads to

$$g_{ab} \approx \frac{\partial^2 U}{\partial w^a \partial w^b},$$

where U is defined at (10), so finally

$$g_{ab} \approx U' \delta_{ab} + U'' w^a w^b \qquad (12)$$

where $U' \equiv dU/d\Phi$ etc. Notice that, since g_{ab} is defined purely in terms of derivatives of U, γ in (10) is arbitrary, and it is only necessary to determine β.

12. Equations of Motion

The geodesic trajectories in reference scalar space corresponding to g_{ab} defined at (12) are calculated as those trajectories which minimize the integral (7), and these arise as the solutions of the Euler-Lagrange equations defined by the Lagrangian density

$$\mathcal{L} \equiv \sqrt{\dot{w}^i \dot{w}^j g_{ij}}.$$

However, this density function is homogeneous, of degree one, in the acceleration vector, $\dot{w}$, and so leads to equations of motion which are invariant with respect to arbitrary transformations of the independent parameter, τ ($\dot{w}^i \equiv dw^i/d\tau$). For example, the particular choice of parameter defined by

$$d\tau = \sqrt{dw^i dw^j g_{ij}},$$

implies $\mathcal{L}$ = constant, and leads to the relatively simple set of equations for geodesic trajectories in reference scalar space given by

$$2U'\ddot{\mathbf{w}} + 2\frac{dU'}{d\tau}\dot{\mathbf{w}} + \left(\frac{d}{d\tau}[U''(\mathbf{w},\dot{\mathbf{w}})] + U''(\mathbf{w},\ddot{\mathbf{w}})\right)\mathbf{w} = 0 \qquad (13)$$

where U and Φ are defined at (10), and the notation $U' \equiv dU/d\Phi$ has been used. However, we are generally only concerned with that choice which identifies the parameter, τ, with Galilean time, t, and there is one particular circumstance in which the parameter τ coincides with physical time, t: using (12), and the definition of τ, above, it is easily seen that

$$\left(\frac{d\tau}{dt}\right)^2 \approx U'(\dot{\mathbf{w}}, \dot{\mathbf{w}}) + U''(\mathbf{w}, \dot{\mathbf{w}})^2.$$

In the particular case of circular motion, $(\mathbf{w}, \dot{\mathbf{w}}) = 0$, and $(\dot{\mathbf{w}}, \dot{\mathbf{w}}) =$ *constant* and $U' =$ constant since U' is a function of energy only. It follows that $d\tau/dt =$ constant and so τ and t are effectively identical. Consequently, the theory can be tested for the dynamics of *circular* gravitational orbits using the approximate equation of motion, (13).

The first thing to notice about this equation is that it contains a single undetermined constant, β, in the functions U' and U'' arising from (11). In Appendix A, it is shown that the system (13) has a particular class of solutions for which the relative kinetic-energy density, Φ, is constant, which corresponds to the cases of unaccelerated and circular motions. It is further shown, in Appendix A, how the system (13) can only support solutions which are small perturbations of these particular solutions if $\beta = -4(\Phi_0)^{3/2}$, where Φ_0 is the relative kinetic-energy density at the initial time. This effectively means that the system (13) is independent of all externally measured parameters, and is, therefore, an internally complete theory.

13. Results

The basic details of the procedure used to solve equations of motion, (13), are given in Appendix B and C whilst, in this section, the particular results which arise when they are used to model planetary motions in the solar system are presented. First, however, it is shown how, for the special case of circular orbits, the approximate equations of motion reproduce *exact* Newtonian dynamics.

Particles in circular orbits are characterized by constant kinetic-energies $((\mathbf{w},\mathbf{w}) = constant)$ and magnitudes of acceleration $((\dot{\mathbf{w}},\dot{\mathbf{w}}) = constant)$ and it is shown in Appendix D how, for such orbits, the equations of motion (13) reduce to

$$\ddot{\mathbf{w}} + \left[\frac{(\dot{\mathbf{w}},\dot{\mathbf{w}})}{(\mathbf{w},\mathbf{w})}\right]\mathbf{w} = 0 \tag{14}$$

Since the coefficient of $\mathbf{w}$ in this equation is constant, then it describes a simple harmonic motion with period, T, given by

$$T^2 = 4\pi^2\left[\frac{(\mathbf{w},\mathbf{w})}{(\dot{\mathbf{w}},\dot{\mathbf{w}})}\right] \tag{15}$$

In order to compare this prescription for orbital periods with the Newtonian prescription, note first that the Newtonian formulation gives circular orbits according to

$$\frac{1}{r} = \frac{\gamma M_s}{h^2}, \quad r\dot{\theta} = \frac{\gamma M_s}{h},$$

where γ is the gravitational constant, M_s is the solar mass, and h is angular momentum/unit mass of orbiting particle. Since, in (15), $(\mathbf{w},\mathbf{w})$ is the magnitude of the square velocity, and $(\dot{\mathbf{w}},\dot{\mathbf{w}})$ is the magnitude of the square acceleration, then there follows immediately, for circular orbits, the relations

$$(\mathbf{w},\mathbf{w}) = \left(r\dot{\theta}\right)^2 = \left(\frac{\gamma M_s}{h}\right)^2,$$

$$(\dot{\mathbf{w}},\dot{\mathbf{w}}) = \left(r\dot{\theta}^2\right)^2 = \frac{(\gamma M_s)^6}{h^8}.$$

Substitution into (15) gives

$$T^2 = 4\pi^2 \left[\frac{h^6}{(\gamma M_s)^4} \right],$$

which is the Newtonian period arising from Kepler's Third Law. It follows that if (13) is integrated using $\mathbf{w}_0$ and $\dot{\mathbf{w}}_0$ taken at a point on a Newtonian circular orbit, then the resulting orbit will have the Newtonian period, given above. This, of course, is exactly what happens and, as a check, the full system (13) was integrated to compute the orbital periods of three hypothetical planets having circular orbits, but different angular momenta/unit mass, h. These had Newtonian periods of 4.124 days, 3.872 years and 304.8 years respectively. The integrated periods were computed as 4.124 days, 3.872 years and 304.8 years respectively, as expected.

Finally, the equations (13) were integrated by specifying initial conditions chosen from the orbits of real planets (see Appendix C); the approximate nature of g_{ab} defined at (12) has the consequence that the resulting orbits are generally quite unlike the real orbits and—except for the case of exactly circular orbits—cannot be represented by simple repeated closed curves. However, by defining the 'quasi-period' of an orbit as the elapsed time from the initial point until the first moment of closest approach to this initial point, it is possible to compute a 'quasi-period' for each of the planets, and the results of these integrations are given in the table below. The first column records the name of the planet concerned, the second column records its observed period, the third column records its computed quasi-period and the fourth column records the computed period which would arise if the planet concerned had unchanged angular momentum, but a circular orbit.

Table 1

Planet	Observed Period	Computed Period	Computed Period Circular Orbits
	years	years	years
Mercury	0.241	0.239	0.230
Venus	0.615	0.615	0.615
Earth	1.000	1.000	1.000
Mars	1.881	1.878	1.865
Jupiter	11.86	11.86	11.83
Saturn	29.46	29.45	29.37
Uranus	84.02	84.00	83.83
Neptune	164.8	164.8	164.8
Pluto	247.7	244.4	231.4

14. Conclusions

A Galilean invariant formalism describing a generic gravitational action has been derived which is predicated upon the concept of Galilean invariant inertial mass together with the results of the Eotvös experiment, which imply that the trajectories of material particles in a gravitational field appear to be independent of their specific mass values; this formalism is exact in the sense that it provides the formal structure of an exact (but unknown) Galilean theory of gravitation, to which the Newtonian theory is only an approximation. This unknown theory is determined to within a Poisson-type equation, and it is the determination of this which provides the hypothetical content of the presented work. An appropriate equation has been suggested and used to obtain approximate equations of motion which are complete in the sense of being independent of all externally measured parameters. These approximations were applied to the estimation of orbital periods within the solar system, and the results of these calculations are so accurate it is almost inconceivable that the formalism presented is not giving a fundamental insight into gravitational phenomena.

From the point of view of modern physics, the most significant property of the presented formalism is that the methods by which it is derived are independent of any specific motion-group. For example, when applied to the concept of Lorentz-invariant inertial mass (that is, the rest-mass of Special Relativity) it can be expected that a formalism describing gravitational process will arise in which the instantaneous actions-at-a-distance of the Galilean formalism will be replaced by a retarded local action in which momentum transfer takes place between retarded fields and the particles which act as sources for these fields. Such a formalism will be exact in the same sense that its Galilean equivalent is exact and, by virtue of its general structure, will provide a natural correspondence limit for a conventional field quantization of gravitational action.

Postscript

Independent of the empirical validity of the Hertz theory, the fact that the full Hertz theory is Galilean invariant whilst its Maxwellian special case in Lorentz invariant, suggests, somehow, the existence of a connection between the Lorentz group and the Galilean group, with the Lorentz group being the subordinate group. Certainly, such a concept is entirely inconsistent with the standard interpretations given to the Lorentz group, but a recent paper by Trempe (Trempe 1992) shows how the formal structure of the Lorentz group (that is, without its kinematic content) can be obtained purely by considering the geometry of an ordinary ellipse. This paper then goes on to provide kinematic structure by considering an ellipse expanding homothetically in Galilean space and time. In this way, Trempe links the Lorentz group in a subordinate fashion to the Galilean group, noting, of course, the necessity for a re-interpretation of the Lorentz group, which he duly provides.

Appendix A. A Determination of the Constant β in the Function U

The function U, the level surfaces of which define the interaction manifold geometry, is defined according to

$$U \approx \Phi + \gamma + \frac{\beta}{\sqrt{\Phi}} \qquad (16)$$

where γ and β are undetermined constants. Since only *derivatives* of U are required for the equations of motion, then it is only necessary to have knowledge of β. This latter constant can be determined by the requirement that the given formalism supports solutions which can be expressed as small perturbations of the special case solutions for which the relative kinetic-energy density is always constant. Since these latter solutions correspond to cases of unaccelerated motion, and to cases of circular motion, and since experience shows that small perturbations from these particular states do exist in gravitationally interacting systems, then the requirement is necessary if the formalism is to successfully describe the world.

The simplest way to proceed is to form the energy equation from the equations of motion, (13). Firstly, however, note that since $U' \equiv dU/d\Phi$, then the relation

$$U''(\mathbf{w},\dot{\mathbf{w}}) = U'' \frac{d\Phi}{dt} = \frac{dU'}{dt}$$

holds. With this, then (13) can be expressed in a slightly more convenient form as

$$2U'\ddot{\mathbf{w}} + 2\frac{dU'}{dt}\dot{\mathbf{w}} + \left[\frac{d^2U'}{dt^2} + U''(\mathbf{w},\ddot{\mathbf{w}})\right]\mathbf{w} = 0 \qquad (17)$$

To obtain the energy equation from this latter equation then:

- Take the scalar product of (17) with $\dot{\mathbf{w}}$, and integrate with respect to t to obtain an equation involving Φ and $(\dot{\mathbf{w}},\dot{\mathbf{w}})$ only;
- Take the scalar product of (17) with $\mathbf{w}$ to obtain a further equation involving Φ and $(\dot{\mathbf{w}},\dot{\mathbf{w}})$ only;
- Eliminate Φ between these two equations, and use the definition of U at (16), to obtain

$$\frac{1}{2}\left(2 + \frac{\beta}{\Phi^{3/2}} - \frac{\beta^2}{\Phi^3}\right)\frac{d^2\Phi}{dt^2} - \frac{3}{2\Phi}\left(\frac{1}{4}\frac{\beta}{\Phi^{3/2}} - \frac{1}{2}\frac{\beta^2}{\Phi^3}\right)\left(\frac{d\Phi}{dt}\right)^2 = K\left(1 + \frac{1}{4}\frac{\beta}{\Phi^{3/2}}\right),$$

which is the equation describing the evolution of the relative kinetic energy density along a particle's trajectory. The constant K is a constant of integration.

It is clear that a special solution of this latter equation is given by $\Phi^{3/2} = -\beta/4$, so $\beta < 0$ for real solutions, and that these solutions correspond to the cases of unaccelerated and circular motions.

If solutions are searched for which exist as small periodic perturbations of these latter special case solutions, then it is found

$$\Phi(t) \approx \Phi_0 + \dot{\Phi}_0 \sqrt{\frac{6\Phi_0}{K}} \sin\sqrt{\frac{K}{6\Phi_0}} t$$

where Φ_0 is the initial value of Φ and is necessarily given by

$$\Phi_0 = \left(\frac{-\beta}{4}\right)^{2/3}.$$

In this way, it can be seen that $\beta = -4(\Phi_0)^{3/2}$ so that, finally,

$$U \approx \Phi + A - 4\Phi_0\sqrt{\frac{\Phi_0}{\Phi}} \qquad (18)$$

Since β is the only parameter in the equations of motion, it follows that these equations are independent of externally measured parameters, and are therefore self-contained, and complete.

Appendix B. Solving the Equations of Motion

In this appendix, the general procedure used to generate trajectories from the equations of motion is briefly described.

Following (17), the approximate equations of motion arising from the given formalism for any particle involved in an interaction is given by

$$2U'\ddot{\mathbf{w}} + 2\frac{dU'}{dt}\dot{\mathbf{w}} + \left[\frac{d^2 U'}{dt^2} + U''(\mathbf{w},\ddot{\mathbf{w}})\right]\mathbf{w} = 0.$$

For the purpose of numerical integration, it is convenient to express this latter form of the equations of motion in the form

$$A\ddot{\mathbf{w}} + B\dot{\mathbf{w}} + C\mathbf{w} = 0 \tag{19}$$

where A, B and C are functions of $\mathbf{w}$ and $\dot{\mathbf{w}}$ only. This is done in the following way: firstly, the relations

$$\frac{dU'}{dt} \equiv U''\Phi, \quad \frac{d^2 U'}{dt^2} \equiv U''\dot{\Phi} + U'''\Phi^2 \tag{20}$$

are used to express the equations of motion as

$$2U'\ddot{\mathbf{w}} + 2\frac{dU'}{dt}\dot{\mathbf{w}} + \left[U''(\dot{\mathbf{w}},\dot{\mathbf{w}}) + U'''\Phi^2 + 2U''(\mathbf{w},\ddot{\mathbf{w}})\right]\mathbf{w} = 0 \tag{21}$$

Now form the scalar product of this latter equation with $\mathbf{w}$, and rearrange the resulting equation in terms of $(\mathbf{w},\ddot{\mathbf{w}})$ to get, finally,

$$(\mathbf{w},\ddot{\mathbf{w}}) = -\left[\frac{\dot{\Phi}^2 + \Phi(\dot{\mathbf{w}},\dot{\mathbf{w}})}{U' + 2U''\Phi}\right]U'' \tag{22}$$

The equations of motion for each particle in the binary interaction are then solved in the form of (21) with (22), where U is defined at (18).

Appendix C. A Technical Difficulty

The results, given in Section 13, for the orbital periods of solar system bodies were obtained simply by choosing, as initial conditions, the known velocity and acceleration at a particular point on the orbit derived from standard Newtonian theory. In principle, given the general validity of the given formalism, together with the availability of an exact representation of its equations of motion, then, that choosing initial conditions in this way should be sufficient for these equations of motion to recover the whole Newtonian orbit. However, in the present case, the equations of motion obtained here are derived as equations defined on an *approximate* interaction manifold, and are therefore, themselves only approximate. It follows that two distinct sets of initial conditions which belong to the *same* Newtonian orbit will, in practice, lead to *distinct* trajectories here. Since there is no *a priori* reason to suppose that such distinct trajectories will have the same periods then, from the point of view of calculating approximate predictions for these periods, there is a problem concerning the choice of initial conditions. To resolve this problem in a consistent way for all orbits, it is argued that since the relative kinetic-energy density is the basic object of the given formalism, then the initial conditions for integration of the corresponding equations of motion should be taken from that point on the Newtonian orbit where the orbital kinetic-energy attained its mean value for the orbit. That is, since the relative kinetic energy density on the Newtonian orbit given by

$$\frac{1}{r} = \frac{\mu}{h^2}(1+\epsilon\sin\theta), \quad r^2\dot{\theta} = h,$$

is given by

$$KE = \frac{1}{2}\left[\left(\frac{\mu\epsilon}{h}\right)^2 + \left(\frac{\mu}{h}\right)^2\right] + \epsilon\left(\frac{\mu}{h}\right)^2\sin\theta,$$

where ϵ, μ and h have their usual Newtonian meanings, and θ is the angular displacement on the orbit, and since this KE attains its mean value at $\theta = 0$, then the velocities and accelerations supplied to the equations of motion as initial conditions should be calculated at the same point. Consequently, the integrations of Section 12 all had their initial conditions computed from

$$\dot{r} = -\frac{\mu\epsilon}{h}, \quad r\dot{\theta} = \frac{\mu}{h}.$$

Appendix D. The Special Case of $\Phi = $ *constant* Motion

The equations of motion for a single trajectory are given, from (21) and (22), by

$$2U'\ddot{\mathbf{w}} + 2\frac{dU'}{dt}\dot{\mathbf{w}} + \left[U''(\dot{\mathbf{w}},\dot{\mathbf{w}}) + U'''\dot{\Phi}^2 + 2U''(\mathbf{w},\ddot{\mathbf{w}})\right]\mathbf{w} = 0.$$

$$(\mathbf{w},\ddot{\mathbf{w}}) = -\left[\frac{\dot{\Phi}^2 + \Phi(\dot{\mathbf{w}},\dot{\mathbf{w}})}{U' + 2U''\Phi}\right]U''$$

If $\Phi = \Phi_0$ where Φ_0 is constant then, immediately,

$$2U'\ddot{\mathbf{w}} + \left[U''(\dot{\mathbf{w}},\dot{\mathbf{w}}) - 2U''\left(\frac{\Phi(\dot{\mathbf{w}},\dot{\mathbf{w}})}{U' + 2U''\Phi}\right)U''\right]\mathbf{w} = 0.$$

Now use the expressions for U' and U'' derived from (18), evaluating these at $\Phi = \Phi_0$, to obtain

$$\ddot{\mathbf{w}} + \left[\frac{(\dot{\mathbf{w}},\dot{\mathbf{w}})}{2\Phi_0}\right]\mathbf{w} = 0.$$

Since this equation is only true for the case $2\Phi \equiv (\mathbf{w},\mathbf{w}) = $ *constant*, then $\mathbf{w}$ and $\dot{\mathbf{w}}$ must be specified to satisfy $(\mathbf{w},\dot{\mathbf{w}}) = 0$ at the initial point, and this implies the above equation can be written, finally, as

$$\ddot{\mathbf{w}} + \left[\frac{(\dot{\mathbf{w}},\dot{\mathbf{w}})}{(\mathbf{w},\mathbf{w})}\right]\mathbf{w} = 0,$$

which is the form given in the main text at (14). To see this, it is only necessary to form the scalar product of this latter equation with $\mathbf{w}$, and integrate, to find $(\mathbf{w},\dot{\mathbf{w}}) = $ *constant*. However, since this latter constant is specified to be zero at the initial point then, immediately, $(\mathbf{w},\mathbf{w}) \equiv 2\Phi = $ *constant*, which was the basic assumption.

References

Hertz, H. R., 1962, *Electric Waves*, Dover.
Eichenwald, A., 1903, *Annalen der Physik* 11:1 and 421.
Phipps, T. E., 1986, *Heritical Verities: Mathematical Themes in Physical Description*, Classic Nonfiction Library.
Rund, H., 1973, *The Hamilton-Jacobi Theory in the Calculus of Variations*, Krieger.
Trempe, J., 1992, Light Kinematics in Galilean Space-Time, *Physics Essays* 5:1.

Astrophysical and Cosmological Consequences of Velocity-Dependent Inertial Induction

Amitabha Ghosh

Department of Mechanical Engineering
Indian Institute of Technology Kanpur
India 208016

We discuss various astrophysical and cosmological consequences of the introduction of a velocity dependent term in a phenomenological model of dynamic gravitational interaction. In particular, we show that the observed cosmological redshift is consistent with a stationary universe, although other aspects of a quasistatic cosmological model have not been touched upon.

Without incorporating any free adjustable parameters in the model, the value for the observed Hubble constant has been obtained. The model further shows that the gravitational and inertial masses are exactly equal, and G decreases with distance exponentially, the coefficient of distance in the exponent being equal to $(-H_o/c)$. Local interactions are shown to produce a number of effects which are detectable and can explain a number of ill explained and unexplained phenomena. Furthermore, the proposed mechanism can act as a servomechanism to distribute matter in spiral galaxies in a unique way which results in flat rotation curves. Suggestions for further work are put forward, including an investigation of the filamentary nature of the large scale structure.

Introduction

At a meeting of the Institution of Mechanical Engineers of the United Kingdom on June 18, 1943, a very interesting observation was made by Dr. E. Orowan, F.R.S. of University of Cambridge during a discussion on a paper entitled "The Significance of Tensile and other Mechanical Test Properties of Metals" by Dr. Hugh O'Neill. To quote him:

The extension of a piece of metal was, in a sense, more complicated than the working of a pocket watch, and to hope to derive information about its mechanism from two or three data derived from measurements during the tensile test was perhaps as optimistic as would be an attempt to learn about the working of a pocket watch by determining its compressive strength.

The basic idea behind the above is that, even though the overall characteristics of a physical phenomenon can be approximately described by a simple phenomenological model, all physical phenomena are extremely complicated in their detailed structure. As a result, the descriptive laws which are supposed to represent the behaviour of the phenomena are also very complicated in their details. In other words, there is always more detail to be found in the fine structure of the phenomenological rules. Thus, while the gravitational interaction between two particles is well described by Newton's simple inverse square law, there is no reason to believe that this law provides a complete description of the phenomenon.

In a recent article in *Nature*, Lindley (1992) has remarked that Milgrom's attempt to explain the galactic rotation curves by adding extra terms to Newton's inverse square law without invoking missing mass amounts to messing around with gravity, since it is *ad hoc* and destroys the simplicity of the rules. However, he remains silent on the many *ad hoc* free parameters in the standard model of cosmology and the introduction of inflation, which definitely complicate a simple expanding model of the universe. The real issue is perhaps reflected in the following comments (Lizhi and Yaoquan 1987).

The awe-inspiring fame that accompanies every great success often so affects the successors that they may refuse to consider, or dislike to consider what among the consequences is really the proved truth and what remains a conjecture or a hypothesis.

Though the above remark was made in connection with the transition from Newtonian to Einsteinian era, it holds true for the expansion hypothesis also. Since no mechanism for the observed cosmological redshift other than the Doppler effect has been found within the framework of Newtonian and relativistic mechanics, most scientists are unwilling to abandon the expansion hypothesis. Yet there has not been a single proof in favour of cosmological expansion and the non-Euclidean nature of the space. Both are presumed to have their origin in General Theory of Relativity. On the other hand, most of the proposed "tired light" mechanisms cannot be verified through alternative tests and also bear the mark of *ad hoc*ism.

The main objectives of this paper are as follows:

1. The gravitational interaction between two particles will be expanded to include the dynamical terms depending on relative velocity and acceleration. This model is also phenomenological, but free parameters will be avoided.
2. It will be shown that such an interaction of a particle with the rest of the universe (which is taken to satisfy the perfect cosmological principle *i.e.* infinite, homogeneous, non-evolving and quasistatic) results in a small velocity-dependent drag force over and above the acceleration-dependent force m**a**. This also yields an exact equivalence between gravitational and inertial masses. Furthermore, the small velocity dependent drag force will be shown to be responsible for the cosmological redshift.
3. It will be further shown that G is not a constant, but decreases exponentially with distance. This eliminates both the problem of an infinite potential at a point and the paradox in Newtonian gravitational law whereby a particle in a universe is acted upon by an arbitrary force.
4. Unlike most other cases, this model will be applied to a host of other problems involving both material bodies and photons. All the predicted effects are shown to be present, while the model can also explain a number of unexplained or poorly-explained phenomena.
5. During the last nine years, the terminology involved in the model of inertial

induction has not been rationalized. Here we will attempt to arrive at a rationalized nomenclature.

Cosmic Drag and Exact Equivalence of Gravitational and Inertial Mass

The concept of a dynamic gravitational interaction, in which the interacting force between two bodies depends not only on separation, but also on relative motion, was first suggested by George Berkeley thirty years after the publication of the *Principia*. However, the tremendous success of Newton's law of gravitation did not allow any progress in this direction until 1872, when the point was raised again by Ernst Mach. He stated that the inertial property of any given object depends upon the presence of other material bodies in the universe. In other words, the property of inertia is nothing but the manifestation of a dynamic gravitational interaction. This idea influenced many contemporary scientists, including Einstein, and it eventually became known as *Mach's Principle*. The question whether inertia is an intrinsic property of matter or represents an interaction with the matter in the rest of the universe is still not resolved.

The main difficulty in coming to a definite conclusion about *Mach's Principle* was due to the absence of any quantitative model of the theory. Sciama (1953, 1961, 1969) was the first to propose a quantitative model of dynamic gravitation incorporating *Mach's Principle*. In his model, the force between two bodies has one static part, which is the usual Newtonian gravitation, and another part dependent on the acceleration between the bodies which was termed inertial induction. Sciama proposed that the interactive force between two bodies of mass m_1 and m_2 at a distance r will be given by

$$\frac{Gm_1m_2}{r^2} + \frac{Gm_1m_2}{c^2r}a$$

where a is the acceleration between the bodies, G is the constant of gravitation and c is the velocity of light. Assuming the universe to be in a quasistatic state, a body of mass m moving with an acceleration a with respect to the mean rest-frame of the universe, will experience a resistance

$$F = ma \sum_{universe} \frac{GM}{c^2r} \qquad (1)$$

where M represents the mass of a body in the universe at a distance r from the mass m. Sciama and others (Cook 1976, Selak 1984) have shown that

$$\sum_{universe} \frac{GM}{c^2r}$$

becomes of the order of unity when the presently estimated values of the average matter density in the universe and the radius of the observable universe are used. Thus, it has been demonstrated that inertia is nothing but a manifestation of the acceleration dependent gravitational interaction of a body with the matter in the rest of the universe. However, the model fails to show why the coefficient of ma in (1) is *exactly equal* to unity!

Ghosh (1984, 1986a, 1990) has proposed a more comprehensive model of dynamic gravitational interaction according to which the interacting force between two objects depends on separation and relative acceleration, as well as relative velocity. The structure of the phenomenological model of the dynamic gravitational interaction between two objects can be represented as follows:

$$\mathbf{F} = \mathbf{F}_s + \mathbf{F}_v + \mathbf{F}_a + \ldots \tag{2}$$

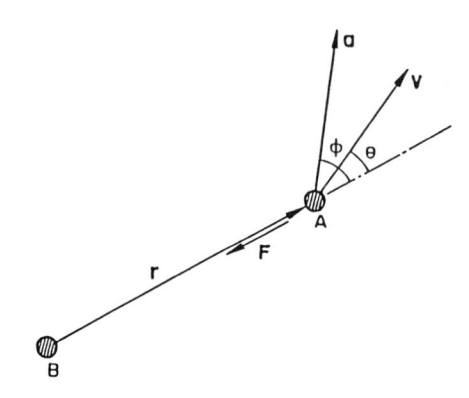

Figure 1. Dynamic gravitational interaction between two particles.

$\mathbf{F}$ is the force on A due to B (Figure 1), $\mathbf{F}_s$ is the force depending on the relative separation of the two bodies, $\mathbf{F}_v$ is the force depending on the relative velocity and $\mathbf{F}_a$ is the force depending on the relative acceleration. There may be other terms depending on the higher order derivatives, but for the present we will ignore them. Each component may have a complicated structure. For example the first term in $\mathbf{F}_s$ can represent the usual Newtonian law, but may also contain other position-dependent terms, as proposed by other researchers (Milgrom 1983, Kuhn and Kruglyak 1987 and references therein). This is also the case with $\mathbf{F}_v$ and $\mathbf{F}_a$. We can write

$$\mathbf{F}_s = -\frac{Gm_A m_B}{r^2}\mathbf{u}_r + \mathbf{F}'_s(r) \tag{3a}$$

$$\mathbf{F}_v = -\frac{Gm_A m_B}{c^2 r^2} v^2 \mathbf{u}_r f(\theta) + \mathbf{F}'_v(\mathbf{r}, \mathbf{v}) \tag{3b}$$

$$\mathbf{F}_a = -\frac{Gm_A m_B}{c^2 r} a \mathbf{u}_r f(\phi) + \mathbf{F}'_a(\mathbf{r}, \mathbf{v}, \mathbf{a}) \tag{3c}$$

where $\mathbf{r}(= r\mathbf{u}_r)$, $\mathbf{v}(= v\mathbf{u}_v)$ and $\mathbf{a}(= a\mathbf{u}_a)$ are the position, velocity and acceleration of body a with respect to B ($\mathbf{u}_r$, $\mathbf{u}_v$ and $\mathbf{u}_a$ are the unit vectors); $f(\theta)$ and $f(\phi)$ (with $\cos\theta = \mathbf{u}_r \cdot \mathbf{u}_v$ and $\cos\phi = \mathbf{u}_r \cdot \mathbf{u}_a$) represent the inclination effects; m_A and m_B are the gravitational masses of bodies A and B (in fact m_A and m_B are the relativistic masses of A and B). However, as in the first term of $\mathbf{F}_s$ (which is well established), the relativistic effect is negligible except in special situations. $f(\theta)$ is a symmetric function of θ and satisfies the following conditions:

$$f(\theta) = 1 \text{ for } \theta = 0$$
$$\text{and } f(\theta) = -1 \text{ for } \theta = \pi$$

The primed terms F'_s, F'_v and F'_a represent the higher order terms. If we neglect the higher order terms in (2) and those in (3a), (3b) and (3c), the simplest form of dynamic gravitation can be written as follows:

$$\mathbf{F} = -\frac{Gm_A m_B}{r^2}\mathbf{u}_r - \frac{Gm_A m_B}{c^2 r^2}v^2\mathbf{u}_r f(\theta) - \frac{Gm_A m_B}{c^2 r}a\mathbf{u}_r f(\phi) \tag{4}$$

It should be further stressed that G is not a constant, but decreases with distance. In conventional mechanics, the influence of gravitation is assumed to decrease as $1/r^2$ because the surface area of a sphere increases as r^2. But the decrease in force is due both to the depletion of flux as $1/r^2$ and to a decline in the strength (or energy) of the agent which is responsible for the gravitational interaction. Laplace and others (Laplace 1880, Pechlaner and Sexl 1966, Fujii 1975, Sanders 1984, Assis 1992) have speculated that G may decline exponentially with distance. In the dynamical model, the nature of the variation of G will emerge directly from the analysis. It will also be seen that there are no adjustable free parameters in the model. However, to complete the relation quantitatively, $f(\theta)$ and $f(\phi)$ will be assigned a definite functional form satisfying the conditions set out above. Of course, at this stage the only important thing to note is that $f(\theta)$ and $f(\phi)$ have the same functional form.

Before proceeding further, it should be pointed out that the inclusion of a velocity dependent term in the dynamic model of gravitational interaction is more than just a small modification of Sciama's model. In fact, it introduces some profound changes in the nature of the problem, leading to some significant major results, as will be seen later. Furthermore, the occurrence of a velocity dependent term is also found in the work of Assis (1989), who has utilized a force law equivalent to Weber's law to obtain a quantitative model of *Mach's Principle*. However, in Assis' work, the force is a conservative one, and the velocity dependent term does not appear as a *drag* effect, as in our model. Because (4) represents a force law which is not conservative, an elegant analysis based on the concept of potential is unfortunately not possible.

The key assumptions which will be used in developing the model are as follows:

1. Even a single body can establish a field of influence, so that a three dimensional space is characterized. Thus, when another particle moves, it can be assigned a velocity and acceleration which need not be $\dot{r}$ and $\ddot{r}$. In other words, it is not meaningless to conceive anything other than $\dot{r}$ and $\ddot{r}$ in an isolated 2-particle system.
2. Gravitons (which can be assumed to carry the gravitational effect) emitted from one body can interact with those from other bodies. Furthermore gravitons are assumed to move at the speed of light, while G is assumed to be proportional to the energy of gravitons from the active body at the location of the passive object.

3. The universe is infinite, homogeneous, isotropic and quasistatic; i.e. there are statistical local fluctuating motions, but there is no universal motion. Most material bodies are also assumed to move at nonrelativistic speeds, save for some exceptional cases. A mean rest-frame of the universe exists.

Next, the interactive force between a body A of mass m (moving with a velocity **v** and acceleration **a** with respect to the mean rest-frame of the universe) with the matter in the rest of the universe will be determined. The universe can be considered to be composed of thin concentric spherical shells, with A as the centre. Figures 2(a) and (b) show the velocity and acceleration dependent interactive forces between A and an elemental ring of a thin spherical shell of radius r and thickness dr. When summed over the whole universe, the contribution from the first term will be zero because of symmetry, and the resultant force will be the sum of the velocity- and acceleration-dependent dynamic gravitational forces, as follows:

$$F = -2\int_0^\infty \int_0^{\pi/2} \mathbf{u}_v G \frac{2\pi r^2 \rho \sin\theta \cdot v^2 mf(\theta)\cos\theta \, d\theta \, dr}{c^2 r^2}$$

$$-2\int_0^\infty \int_0^{\pi/2} \mathbf{u}_a G \frac{2\pi r^2 \rho \sin\phi \cdot a \cdot mf(\phi)\cos\phi \, d\phi \, dr}{c^2 r} \qquad (5)$$

$$= -\mathbf{u}_v \frac{mv^2}{c} \int_0^\infty \frac{\chi G\rho}{c} dr - \mathbf{u}_a \frac{ma}{c^2} \int_0^\infty \chi G r\rho \, dr$$

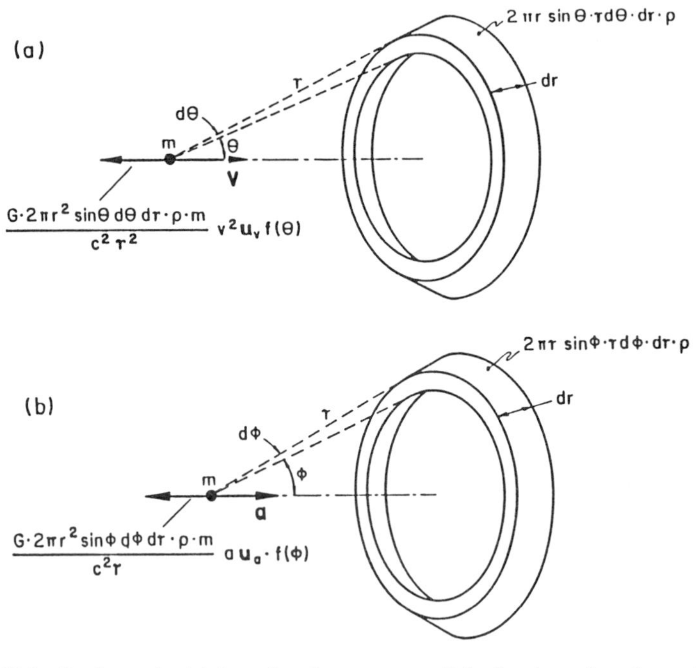

Figure 2. (a) Velocity-dependent interactive force on a particle due to a ring element of the universe.
(b) Acceleration-dependent interactive force on a particle due to a ring element of the universe.

where

$$\chi = 4\pi \int_0^{\pi/2} \sin\theta\cos\theta\, f(\theta)d\theta = 4\pi \int_0^{\pi/2} \sin\phi\cos\phi\, f(\phi)d\phi$$

and ρ is the average density of the universe.

To proceed further, it is essential to know G as a function of r. The problem is solved in the following manner. First, we write

$$\int_0^\infty \frac{\chi G \rho}{c} dr = k \qquad (6)$$

The expression for the force on the moving body a can then be written as follows:

$$\mathbf{F} = -\mathbf{u}_v k \frac{mv^2}{c} - \mathbf{u}_a \frac{ma}{c^2} \int_0^\infty \chi G \rho\, dr \qquad (7)$$

Equation 7 implies that if a particle moves—even with a *constant velocity* $\mathbf{v}$ ($=\mathbf{u}_v v$)—with respect to the mean rest-frame of the universe, it will be subjected to a *cosmic drag*

$$-\mathbf{u}_v \frac{kmv^2}{c}$$

and will loose both momentum and energy. This lost momentum and energy go to the rest of the universe. Everything moving, even the gravitons themselves, is subjected to this drag. Consequently, if at any instant, a graviton has an energy E (i.e. a mass E/c^2) it will be subjected to a drag

$$\frac{kE}{c}$$

because the speed of a graviton is equal to c. The energy change when it moves a distance dr is

$$dE = -k\frac{E}{c} \cdot dr$$

If the initial energy (i.e. at $r = 0$) is E_o, the solution of the above equation yields

$$E = E_o \exp\left(-\frac{k}{c}r\right)$$

Hence

$$G = G_o \exp\left(-\frac{k}{c}r\right) \qquad (8)$$

Thus, we have found that G is an exponentially decreasing function of r, as suggested by earlier workers on an *ad hoc* basis. Furthermore, we have identified the coefficient of the exponent as well. The value of G_o is just the local value of G, which is equal to the constant of gravitation in Newton's law (= 6.67×10^{-11} m³/ kg s²). Substituting the expression for G from (8) in (6) and (7) we obtain

$$\frac{\chi G_o \rho}{k} = k \qquad (9)$$

and

$$\mathbf{F} = \mathbf{u}_v \frac{mv^2}{c} \cdot k - \mathbf{u}_a ma \cdot \frac{\chi G_o \rho}{k^2} \qquad (10)$$

Solving (9) we have

$$k = (\chi G_o \rho)^{\frac{1}{2}} \qquad (11)$$

and using this in (10) we obtain

$$\mathbf{F} = -\frac{k}{c} mv^2 \cdot \mathbf{u}_v - m\mathbf{a} \qquad (12)$$

Thus, the acceleration-dependent part of the inertial force (due to dynamic gravitational interaction) is identically equal to

$$-m\mathbf{a}$$

where m is the gravitational mass of the moving particle. This means that the exact equivalence of inertial and gravitational masses is established, and Newton's second law can be derived from the model of dynamic gravitational interaction. The velocity-dependent part is equal to

$$-\frac{k}{c} mv^2 \mathbf{u}_v$$

and to determine it a knowledge of χ is essential. For this purpose, the inclination effects represented by $f(\theta)$ and $f(\phi)$ have to be given a definite form, as we noted earlier. Let us assume[1]

$$f(\theta) = \cos\theta |\cos\theta| \text{ and } f(\phi) = \cos\phi |\cos\phi|$$

With the above function, we find $\chi = \pi$, and taking $\rho = 7 \times 10^{-27}$ kg m⁻³, we get the following numerical value for k: $k = 1.21 \times 10^{-18}$ s⁻¹.

It is obvious that the force due to the velocity-dependent term is extremely small and cannot be detected by laboratory experiments. Perhaps this is why the existence of a velocity drag has not been suspected before. Even though the order of magnitude is extremely small, it has some profound implications; for example, photons moving through the universe are subjected to this drag, and a non-Doppler

---

[1] Other forms can also be assumed, satisfying the required conditions. This may be considered to be the only assumed parameter in the model. Another possibility is $f(\theta) = \cos\theta$

origin of the cosmological redshift becomes possible. This velocity-dependent inertial induction (VDII) can also act as a mechanism for the transfer of angular momentum. In Newtonian mechanics, there is no mechanism other than tidal friction for transferring angular momentum from rotating bodies.

Before presenting the various astrophysical and cosmological consequences, it is desirable to have some idea about the relative contributions of the inertial induction terms when one considers interactions with the Earth, the Sun, the galaxy and the whole universe. Table 1 shows the orders of magnitude of the interactive forces in the above mentioned cases. It is interesting to note that the magnitude of local velocity-dependent inertial induction in the vicinity of massive bodies predominates over the interaction with the whole universe. On the other hand, in case of acceleration-dependent inertial induction, the interaction with the whole universe is dominant. This is not unexpected, since the velocity-dependent inertial induction term in (4) is inversely proportional to r^2, whereas the acceleration-dependent interaction falls off as $1/r$, making it a long range force.

Table 1. Comparative magnitudes of inertial induction terms

Interacting system	Velocity-dependent inertial induction	Acceleration-dependent inertial induction
Earth (near its surface)	$\sim 10 \dfrac{mv^2}{c^2}$	$\sim 0.75 \times 10^{-9}\ ma$
Sun (near its surface)	$\sim 275 \dfrac{mv^2}{c^2}$	$\sim 1.5 \times 10^{-7}\ ma$
Milky Way Galaxy (near Sun)	$\sim 200 \dfrac{mv^2}{c^2}$	$\sim 10^{-6}\ ma$
Universe	$3.6 \times 10^{-10} \dfrac{mv^2}{c^2}$	ma

Cosmological Redshift in a Stationary Universe

In the previous section we showed that, in the proposed model of dynamic gravitational interaction (*i.e.* inertial induction) any object moving with a constant velocity **v** experiences a cosmic drag due to velocity-dependent inertial induction. The magnitude of the drag is

$$-\frac{k}{c}mv^2$$

Let us consider a photon starting from a source at a distance x from the earth. At any instant, the magnitude of the drag on the photon is

$$-kc \cdot \frac{E}{c^2}$$

where E is the energy of the photon at that instant. When the photon travels a distance $d\xi$, the decrease in energy is

$$dE = -kc\frac{E}{c^2}d\xi$$

Since $E = h\nu$, where h is the Planck's constant and ν is the frequency, the above equation reduces to

$$d\nu = -\frac{k\nu}{c}d\xi$$

Using the initial condition $\nu = \nu_0$ when $\xi = 0$, the solution of the above equation yields the following expression for the frequency of the photon upon arriving at the earth

$$\frac{\nu}{\nu_0} = \exp\left(-\frac{k}{c}x\right) \tag{13}$$

When $(k/c)x \ll 1$, the right hand side of (13) can be approximately linearized in the following form:

$$\frac{\nu}{\nu_0} \approx 1 - \frac{k}{c}x$$

or,

$$\frac{\nu}{\nu_0} = \frac{\nu - \nu_0}{\nu_0} \approx -\left(\frac{k}{c}\right)x$$

Using the relation (with λ representing the wavelength)

$$\nu\lambda = \nu_0\lambda_0$$

in the above relation we obtain the following redshift equation

$$z = \frac{\Delta\lambda}{\lambda_0} \approx \frac{k}{c}x \tag{14}$$

If the redshift is *assumed* to be due to a recessional velocity of the source, then

$$V = cz \approx kx \tag{15}$$

Thus we have identified k ($= \sqrt{\chi G_0 \rho}$) as the Hubble constant H! It should be noted that the derived magnitude of k ($= 1.2\times10^{-18}$ s^{-1} with $\chi = \pi$) is extremely close to the present estimate of H_0 ($= 1.6\times10^{-18}$ s^{-1})! In the short range, the redshift-distance relation is linear. However, (13) clearly indicates that in the very long range it deviates from the linear relation and becomes exponential, as has been observed. Figure 3 shows a comparison of a plot of (13) and observed redshift.

Unlike most other tired-light mechanisms, velocity-dependent inertial induction can be tested in other situations, and it will be shown in the next section that it yields results in excellent agreement with observations in all cases. According to the velocity-dependent inertial induction mechanism of the cosmological redshift, it is expected that such shifts will be more pronounced when a photon travels through volumes of space where galaxies are most concentrated. The observations indeed support this conclusion (Karoji and Nottale 1976, Vigier et al. 1990).

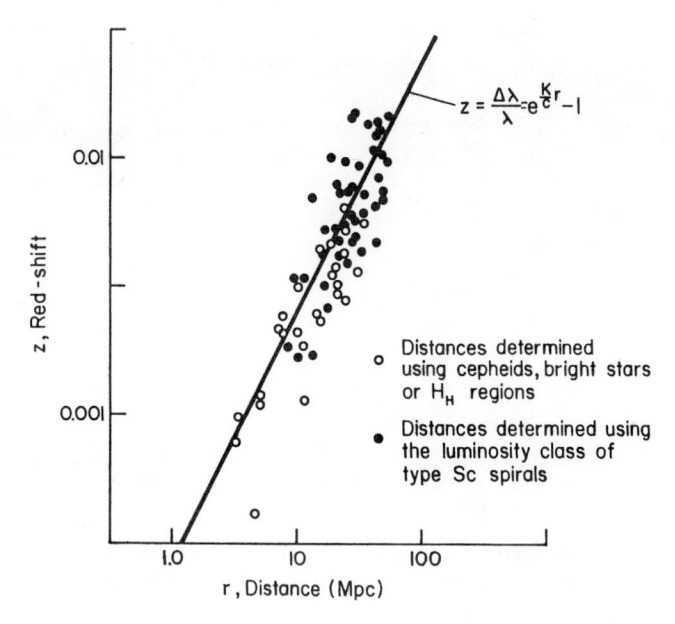

[Figure: plot of z vs r with equation $z = \frac{\Delta\lambda}{\lambda} = e^{\frac{K}{c}r} - 1$]

Figure 3. Cosmological redshift in a stationary universe due to velocity-dependent inertial induction.

Consequences of Local Velocity-Dependent Inertial Induction

From Table 1 it is quite obvious that effects of local velocity-dependent inertial induction are much stronger compared to the universal interaction. Furthermore, the universe is lumpy, and such interactions can lead to observable effects. In almost all cases of local interaction the effects of acceleration-dependent inertial induction will be comparatively small, and can generally be neglected.

Local interactions can be divided into two large groups: (A) interaction of photons with matter and (B) interaction of matter with matter. A major source of data is the solar system, as accurate observation is possible in many cases. It will be interesting to note that many of the effects that have been observed remained unexplained mysteries until now.

Interaction of Photons with Matter

When a photon moves near a massive object (*viz.* a star) the photon is subjected to two forces according to the proposed model (4). The contribution of the first term is the usual effect, resulting in gravitational redshift (or blueshift depending on the direction of motion). However, the contribution of the second term will result in a comparable effect (as the speed is equal to c), which should be observable.

Excess Redshift in Solar Spectrum. Figure 4a shows a photon originating at a point on the surface of the sun and moving towards earth. The total force on the photon at any instant will be due to the first two terms of (4), since photons move with constant speed. Hence, the photon will loose momentum and energy due to

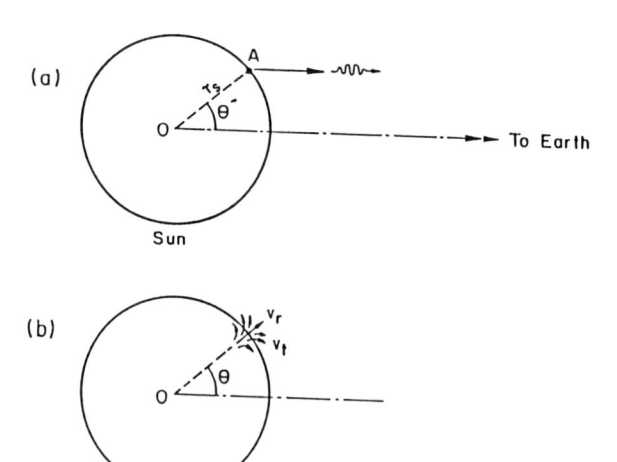

Figure 4. (a) A photon emitted by the sun and moving towards the earth. (b) Motion of photosphere material due to solar granulation.

the gravitational pull and velocity-dependent inertial drag. Assuming the mass of the sun to be concentrated at its centre (which is not at all unjustified, as the density increases very rapidly towards the centre) and the distance of the earth to be much larger than the radius of the sun, the fractional redshift of the photon when it arrives at the earth can be expressed (Ghosh 1986b) as follows:

$$z = \frac{\Delta\lambda}{\lambda} \approx \frac{G_o M_s}{c^2 r_s}\left(2 - \tfrac{1}{3}\sin^2\theta\right) \tag{16}$$

where M_s and r_s are the mass and the radius of the sun, respectively, and θ is the angle shown in Figure 4a. However, the sources emitting the photons at the surface of the sun are not stationary. Due to the solar granulation effect, the material of the photosphere possesses radial and transverse motion, as indicated in Figure 4b. The resultant redshift can be expressed in terms of an "equivalent velocity of recession" in the following form:

$$v_{eq}(\theta) \approx \frac{G_o M_s}{c^2 r_s}\left(1 - \tfrac{1}{3}\sin^2\theta\right) - v_r \cos\theta - v_t \sin\theta \tag{17}$$

where v_r and v_t are the mean radial and transverse speeds of photosphere matter due to the granulation phenomenon. Substituting the known and observed values of G_o, M_s, c, r_s, v_r and v_t, the above relation takes the following final form:

$$v_{eq}(\theta) \approx 0.636\left(2 - \tfrac{1}{3}\sin^2\theta\right) - \cos\theta - 0.2\sin\theta \text{ km s}^{-1} \tag{18}$$

Figure 5 shows a plot of the above function as well as the measured values of the redshift of the solar spectrum for different values of θ. It has not been possible to account for the observed excess redshift at the limb when $\theta = 90°$. However, taking velocity-dependent inertial drag into account, the observed and predicted values are found to be in good agreement. It must be pointed out that the values of v_r and v_t are measured in other experiments conducted to study the granulation phenomenon and cannot be treated as free parameters to adjust the value of v_{eq}.

Even if the effects of granulation phenomenon are ignored, the measured intrinsic redshift of a star should be more than the conventional expectation, which is $(G_o M/c^2 r)$. The effect is bound to be more pronounced for denser objects, where $(G_o M/c^2 r)$ is high. The next section shows that this is precisely the observation which has remained as a mystery.

White Dwarfs. White dwarf degenerate stars are much denser than common main-sequence stars, and consequently the order of magnitude of the expected relativistic redshift $(G_o M/c^2 r)$ is much higher than for normal stars. If the proposed velocity-dependent inertial drag is present, then the total intrinsic redshift (the effect of granulation can be ignored, because in the very high gravity atmosphere of white dwarfs granulation effects are absent) should be substantially greater than $(G_o M/c^2 r)$. Conversely, if we calculate the mass m from the measured intrinsic redshift, the obtained value will be substantially higher than the true value. The "apparent" relativistic mass, ignoring the velocity drag effect is given by

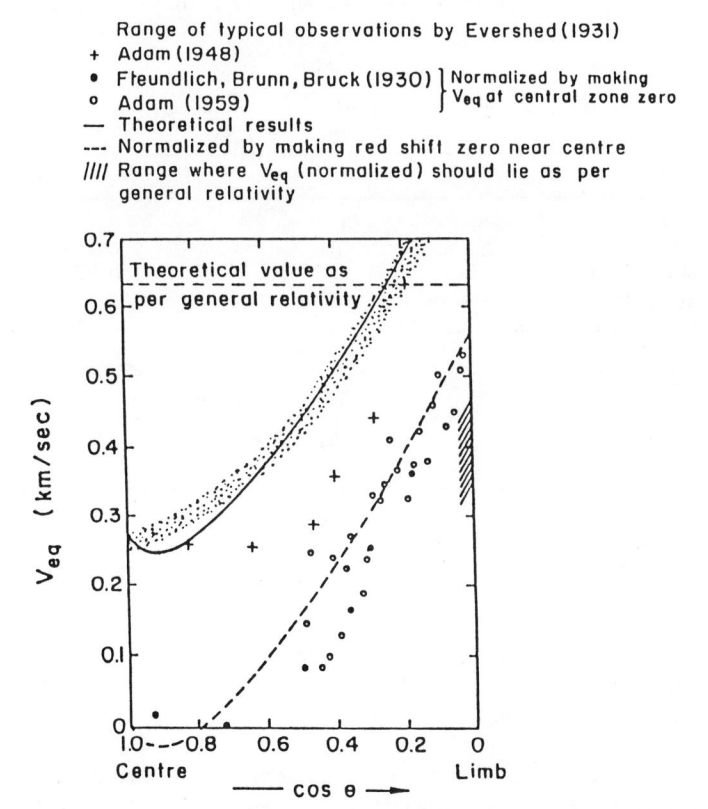

Figure 5. Variation of redshift in solar spectrum-theory and observation.

$$M_r = \frac{zc^2 r}{G_o} \tag{19}$$

where z is the intrinsic redshift of the star. On the other hand, if velocity-dependent drag is also present, then

$$M'_r \sim \frac{zc^2 r}{1.67 G_o} \tag{20}$$

taking the average effect of $(2 - \frac{1}{3}\sin\theta)$ to be approximately represented by 1.67. However, the mass of a white dwarf can be determined by other methods: the value obtained is termed the "astrophysical" mass (M_a) and should be equal to the value of the "relativistic" mass determined from the redshift.

Since 1967 it has been observed by a number of astronomers (Greenstein 1967, Trimble and Greenstein 1972, Moffett et al. 1978, Shipman and Sass 1980, Grabowski et al. 1987) that the "relativistic" masses (without considering velocity-dependent inertial drag effects) of the white dwarf stars are significantly greater than the "astrophysical" masses. Various attempts to explain the discrepancy have been made, but the problem is still unsolved. Table 2 shows a considerable amount of data compiled by Shipman and Sass (1980) and the values of M_a and M_r, as well as M'_r. It is clear that M'_r shows good agreement with M_a, but M_r is significantly larger than M_a.

Table 2. "Astrophysical" and "relativistic" masses of white dwarfs

Method	No. of Stars	Mean Mass
Photometry	110	0.55 M_s
Photometry	31	0.60 M_s
Binary stars	7	0.73 M_s
Two-colour Diagram	40	0.60 M_s
Two-colour Diagram	35	0.45 M_s
H-line Profiles	17	0.55 M_s
All together	240	Average M_a = 0.55 M_s
Gravitational Redshift (Conventional)	83	Average M_r = 0.80 M_s
Gravitational Redshift (proposed)	83	Average $M'_r \sim$ 0.50 M_s

As the values of M_a and M_r were expected to be same, a considerable amount of cooking was done in an attempt to bring M_a and M_r closer. However, more data (without any bias) taken using recent advanced techniques can throw more light on this subject.

The reason why the effect of the velocity-dependent inertial induction cannot be seen in the terrestrial experiments (viz. Rebka, and Pound, Vessot et al.) has been discussed by Ghosh (1986b). It is primarily because all these experiments are two way experiments and the VDII effect, being a drag, cancels out.

Redshift of Photons Grazing a Massive Object: According to conventional theory, when a photon grazes a massive body no resultant redshift is expected. The blue shift caused during the approach is canceled during its recession. However, because the velocity-dependent inertial induction is a drag, this effect will cause a photon to undergo a resultant redshift when it grazes past a massive object. If the mass of the object is assumed to be concentrated at its centre and r is the perpendicular distance of the centre from the path of the photon, the fractional redshift of the photon can be expressed as follows:

$$z \approx \frac{\Delta\lambda}{\lambda} \approx \exp\frac{4G_oM}{3c^2r} - 1 \qquad (21)$$

where M is the mass of the body. The maximum possible value of z will occur when r is equal to the radius of the object. The orders of magnitude of redshift for different classes of objects are given in Table 3.

Table 3. Redshift of electromagnetic waves grazing massive objects

Type of Object	M	r	z
Jupiter	$10^{-3} M_s$	$r_s/10$	$\sim 2.7\times10^{-8}$
Sun	M_s	r_s	$\sim 2.7\times10^{-6}$
White Dwarf	M_s	$r_s/80$	$\sim 2.2\times10^{-4}$
Neutron Star	$2 M_s$	10 km	~ 0.5
Black Hole	—	Schwarzchild Radius	~ 1

Evidently, the redshift caused by Jupiter is too small to be measured using the currently available techniques. However, once the method of laser heterodyne spectroscopy is perfected, it may be possible to detect redshifts of the order of 10^{-8}. Then

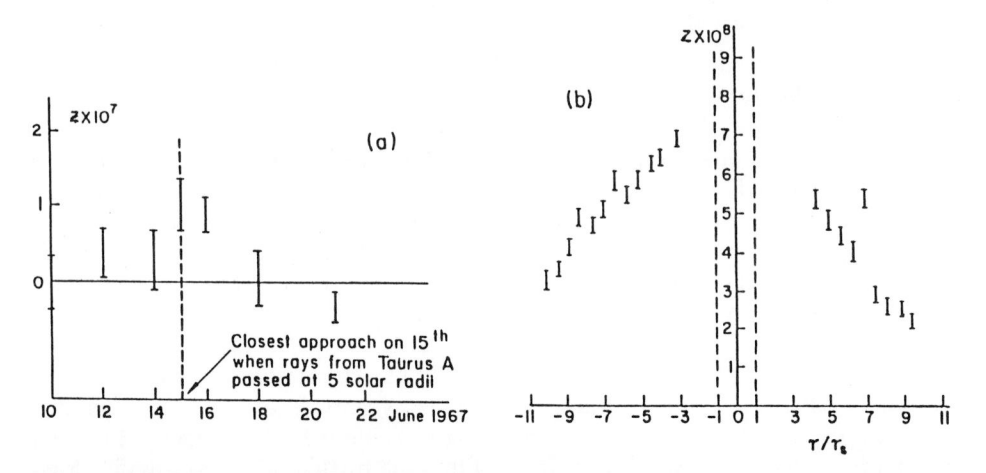

Figure 6. (a) Variation in excess redshift in the 21 cm signal from Tauras A at near occultation position by the sun.
(b) Variation in the ecxess redshift in the 2292 MHz signal from Pioneer 6 during occultation by the sun.

the redshift during occultation by Jupiter can be detected. However, the redshift produced by the sun is measurable, and the first report of such an unexplained redshift was made by Sadeh (Sadeh et al. 1968). It was reported that the 21 cm signal from Taurus A at a near occultation position by the sun suffered a redshift of 150 Hz at a distance of 5 solar radii. Figure 6a shows the results, which agree with the estimated values using (21) so far as the order of magnitude is concerned. The 2292 MHz signal from Pioneer-6 was also found to be subjected to an unexplained redshift when it passed behind the sun (Merat et al. 1974). Figure 6b shows the variation of the redshift with distance from the sun's centre. The results are in reasonable agreement with the expected redshift.

Grazing experiments can be conducted with the objective to determine the existence of any redshift. A considerable amount of data are available for eclipsing binaries, and the introduction of the velocity-dependent inertial drag may lead to many interesting results. An analysis of binary pulsar signals should also be conducted to investigate a possible VDII.

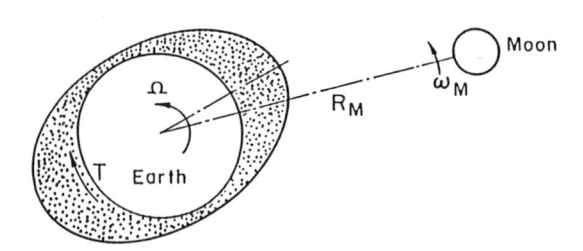

Figure 7. Tidal friction on the spinning earth.

Interaction of Matter with Matter

Of the many observed phenomena described in the previous section some have already been discussed proposing some photon-photon interaction. However, thus far no proposals for matter-matter interactions have been made apart from the standard gravitational interaction. In this section the proposal of a velocity-dependent inertial drag will be examined in connection with objects in the solar system for which accurate observational data exist.

Secular Retardation of Earth's Rotation. Of all the celestial systems, the most accurate data are available for the earth-moon system. Thus we will focus our attention on this system first, and investigate possible velocity-dependent inertial induction effects. It is now well established that the earth's spin is gradually slowing down. The conventional explanation for this has been tidal friction, as indicated in Figure 7. The torque due to tidal friction, T, cannot be calculated directly, but it can be derived from the estimated value of $\dot{\Omega}$, which has been determined from various observational data (Ω = spin rate of the earth). Investigations have been done to determine whether the derived value of T is within the feasible range. The difficulty with this explanation comes from another direction. To conserve the total angular momentum of the earth-moon system, the orbital angular momentum of the moon increases causing it to recede as per Kepler's law. With R_M as the orbital radius of the moon, the observational data yield the following values:

$$\dot{\Omega} = -6 \times 10^{-22} \text{ rad s}^{-2}$$

$$\dot{R}_M = 1.3 \times 10^{-9} \text{ m s}^{-1}$$

According to this rate of recession, some 1300 million years ago the moon must have been so close to the earth that both would have been destroyed because of the mutual gravitational pull. In fact, this should have happened much more recently, since the intensity of tides varies as R_M^{-3} and the torque due to tidal friction, $T \propto R_M^{-6}$. In the past, the tide also should have been much more intense. Surprisingly the geological evidence does not indicate any such disaster or strong tide, even though there is evidence of the presence of the tidal phenomenon during the last 3500 million years. This is a stumbling block, and there has been no satisfactory solution until now. Some have tried to reason that tidal friction was lower in the distant past, because the configuration of the land masses was different due to continental drift. However, no quantitative analysis has been done along these lines.

Analysis of the sun-earth-moon system shows (Ghosh 1986a) that velocity-dependent inertial induction produces a torque of 4.75 $\times 10^{16}$ N-m resisting the earth's spin. This torque results in a spindown rate of $\dot{\Omega} \approx -5.5 \times 10^{-22}$ rad s^{-2}! Only a small fraction (0.5×10^{-22} rad s^{-2}) is left to be taken care of by tidal friction. Using this model it is found that R_M is presently *decreasing* at a very small rate (0.15×10^{-9} m s^{-1}), and the difficulty of the moon's close approach is resolved. The almost exact amount of $\dot{\Omega}$ derived from velocity-dependent inertial induction without using any observational data on secular motion can hardly be pure chance.

Secular Acceleration of Phobos. Another system for which reasonably accurate observation has been possible is the Mars-Phobos system. It has been found that Phobos is spiraling down with acceleration of 10^{-3} deg yr^{-2} (Pollack 1977, Sinclair 1989). It is very difficult to satisfactorily explain such a large secular acceleration. The ability of Mars to provide enough tidal effect to produce the observed result is highly doubtful, even if a molten core is present. Applying velocity-dependent inertial induction to the Sun-Mars-Phobos system in a very approximate model yields a secular acceleration of 1.5×10^{-3} deg yr^{-2} (Ghosh 1986a)—surprisingly close to the observed value. The slight discrepancy could be due to the assumption that the density variation of Mars is the same as the earth's.

Greenberg (1977) has observed that tidal dissipation does not explain the large number of circulating near-commensurabilities among planets and satellites. Velocity-dependent inertial induction can explain the prevalence of these near-commensurabilities. This mechanism may also eliminate the need for an uncomfortably small Q for tidal effects on Saturn caused by Titan.

Transfer of Angular Momentum. Leaving the interaction between planets and satellites, let us now take up a more fundamental problem related to the cosmogony of the solar system. The present scientific community has unanimously accepted the nebular hypothesis for the origin of the solar system. One of the most serious problems in this model is to account for the anomalous distribution of angular momentum. Although the sun possesses about 99.9% of the total mass of the solar system, it contains only about 0.5% of the total angular momentum. Thus, if the whole system has evolved through the condensation of a rotating cloud, the only way the present distribution of angular momentum can be achieved is by transferring most of the angular momentum from the central body (the sun) to the planetary system. All the conventional mechanisms for angular momentum transfer proposed so far could have been active during the pre-main-sequence period only (which had a duration of about 2×10^7 yrs). However, in most cases it is very doubtful if the required intensities of the mechanisms are at all feasible.

Now, velocity-dependent inertial induction is capable of transferring momentum without any physical contact between material bodies. It has been shown (Ghosh 1988) that the angular momentum of the sun 4.7×10^9 years after the formation of the protoplanetary disc is equal to 1.4×10^{41} kg m^2 s^{-1} if the angular momentum of the original cloud is taken as 10^{44} kg m^2 s^{-1}, and about 2% of the central body is detached in forming the protoplanetary disc. Both values have been suggested by researchers in the field after investigating the various properties of the solar system. This value is very close to the actual value of the angular momentum of the sun. Thus, it is clear that the mechanism based on velocity-dependent inertial induction can transfer the requisite amount of angular momentum in the available time, 4.7×10^9 yrs. There is a major difference from other mechanisms: velocity-dependent inertial induction is active during the whole period and, therefore, the major part of the transfer takes place during the long main-sequence period. This is in agreement with the observation that new born stars are fast rotators because a comparatively small portion of the original angular momentum has been transferred in the preceding short pre-main-sequence period.

This mechanism also explains similar situations in many planetary satellite systems where the majority of the system's angular momentum is contained in the orbital motion of the satellites rather than the central body. The mechanism also yields the correct value of the orbital radius of the planet nearest to the central body.

Radial Matter Distribution in Spiral Galaxies. A very important feature of the spiral galaxies is that the stars rotate around the galactic centre with an almost constant speed except near the centre (Figure 8). The flat rotation curves (almost a universal feature of all spiral galaxies) have given rise to the theory of missing mass, though a few (*e.g.* Milgrom) have tried to explain the flatness by modifying Newton's law of gravitation. Whatever the reason, it is not possible to have a flat rotation curve unless the matter is distributed in a unique manner in the galactic disc. Thus, there must be a servomechanism to distribute the matter in a galaxy in the required way. No such servomechanism has been reported using the conventional mechanics.

When the model of velocity-dependent inertial induction, combined with Newton's law of gravitation, is applied to a self gravitating, rotating disc-like system (*viz.* spiral galaxies) it is found that a stable equilibrium configuration is achieved when a star's motion is such that its inward centripetal acceleration is equal to the total inward gravitational pull divided by its mass and the forward tangential pull due to the velocity-dependent inertial induction from the matter inside the orbit is

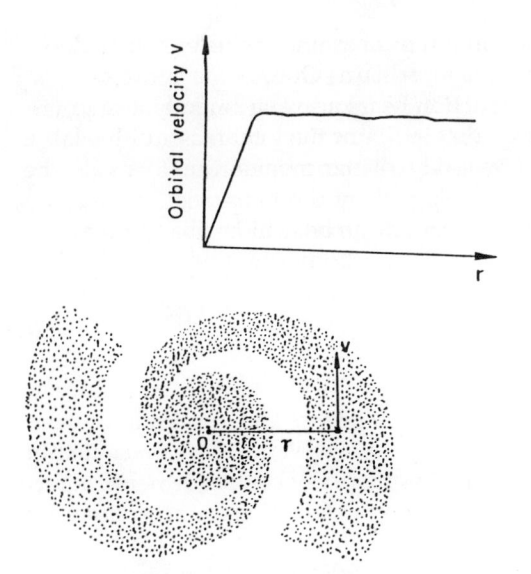

Figure 8. Typical rotation curves for spiral galaxies.

Figure 9. Stable equilibrium of rotating material in spiral galaxies due to velocity-dependent inertial induction.

balanced by the backward drag caused by the stars outside the orbit (Figure 9). It has been shown (Ghosh *et al.* 1988) that the mass distribution corresponding to this equilibrium configuration invariably results in an almost flat rotation curve. However, velocity-dependent inertial induction cannot resolve the problem of the matter-luminosity ratio, and finer details of the static term may yield a solution.

Of course, the missing mass problem can be partially taken care of by velocity-dependent inertial induction, since as many redshifts which are presently assumed to be of Doppler origin may be caused by the interactive mechanism proposed here. If the velocities are found not to exist, the need for a substantial amount of missing mass will also evaporate.

Miscellaneous Cases

There are a number of other problems which appear to be solvable if the proposed model of inertial induction is used. One such case is the formation time of globular clusters. The stellar drag on a star due to gravitational interaction with the rest of the cluster results in a relaxation time which is too large (Harwit 1973). In fact, the formation time becomes a few orders of magnitude larger than the estimated age of the universe in the standard model. Of course, there have been attempts to resolve the difficulty by considering the interaction of the globular clusters with the dense matter in the galactic nucleus. However, in the model proposed here, velocity-dependent inertial induction can act as the mechanism of momentum transfer, and the resulting relaxation time of a star in a typical globular cluster comes down to the order of magnitude of 10^{17} s, which is a realistic value.

Since velocity-dependent inertial induction bears a strong resemblance to viscous drag, an attempt can be made to explain the filamentary character of the large scale structure of the universe. The "large scale streaming of galaxies" can also be eliminated if the excess redshifts can be explained by the proposed mechanism. Thus, the elusive search for the so called "Great Attractor" can be abandoned.

Rationalization of Terminology

During the past few years quite a few research papers and articles have been published which have either directly or indirectly dealt with the ideas presented in this paper. It is now desirable to rationalize the terminology before further proliferation takes place. A few suggestions will be made in this section for acceptance by the researchers in the field.

To distinguish the proposed phenomenological model from that proposed by Newton (4), we propose the expression *dynamic gravitational interaction*. The first term in the right hand side of (4) representing the static interaction may be termed *gravitational attraction*. This is logical, since it is always an attraction and the proposed terminology has been used for the last three centuries. Thus, no confusion will be created. The third term represents a quantification of *Mach's Principle* (MP), and including the second term may be called an extension of *Mach's principle*. The entire model may be called *Extended Mach's Principle* (EMP). Sciama has coined an appropriate term for the dynamic gravitational interaction depending on acceleration: *inertial induction*. However, to distinguish between the second and third terms, we may call the second term *velocity-dependent inertial induction* (VDII) and the third term *acceleration-dependent inertial induction* (ADII). The first term on the right hand side of (12) may be called *cosmic drag*.

Concluding Remarks

Many important conclusions can be drawn on the basis of the results of the application of the proposed phenomenological model of dynamic gravitational interaction. In fact there are so many important and interesting results that it may be helpful to present the whole model in tabular form (Table 4).

The reader should note that the model does not have *any adjustable free parameters*. The only quantity which can be changed is the form of $f(\theta)$. However, it is quite obvious that no major adjustment of the order of magnitudes of the values is possible via $f(\theta)$. The author's opinion is that so many excellent results in widely differing circumstances cannot be pure accident. Thus, there are good reasons to design and perform further tests of the hypothesis. Until now, the model has remained purely phenomenological; theorists may wish to look into the possibility of providing a physical basis for the proposed interaction. Further work on the different problems within the solar system may be attempted by using Equation (4) in place of the Newtonian gravitation law.

Table 4. Consequences of velocity-dependent inertial induction

Particle–Photon Universal Interaction

1. G varies with distance: $G = G_o \exp[-(k/c)r]$ where $k = \sqrt{\chi G_o \rho}$ with $\chi \approx 0(\pi)$
2. Gravitational mass $\equiv$ Inertial mass
3. $\mathbf{F} = -\dfrac{k}{c} m v^2 \mathbf{u}_v - m\mathbf{a}$ (So when a particle moves, the force required is $m\mathbf{a}$ + small drag proportional to v^2.
4. In a steady state stationary universe, photons are subjected to a cosmological redshift as follows:

$$z = \exp\left(\frac{k}{c} \cdot x\right) - 1$$

$$\approx \frac{k}{c} \cdot x \text{ when } kx/c \ll 1$$

Therefore, $V \approx kx$; i.e. k is the so called Hubble constant H_o
5. If the local inhomogeneity of matter distribution in the universe is taken into account then k (or, H_o) will be larger when the photon travels through a comparatively denser region. This may appear as an extra velocity of the source when the redshift is assumed to be of Doppler origin.

Photon–Matter Local Interaction

1. The redshift from a source (viz. a star) is more than the conventional value ($GM/c^2 r$).
   (a) Explains the excess redshift at the solar limb quantitatively.
   (b) Eliminates the long-standing discrepancy between the "astrophysical" and "relativistic" mass of white dwarfs.
2. When photons graze massive objects they undergo a redshift which can be very substantial in the case of degenerate stars.
   (a) Explains the observed excess redshift of Taurus A at near occultation position with the sun.
   (b) Explains the observed excess redshift of the signal from Pioneer 6 during occultation by the sun.
   (c) Can explain some discordant redshifts

Matter-Matter Local Interaction

1. Produces a secular retardation of the earth's spin at a rate of -5.5×10^{-22} rad s^{-2} in comparison to the observed rate of -6×10^{-22} rad s^{-2}. Only a small fraction is due to tidal friction. Completely eliminates the problem of moon's close approach in the past.
2. Produces a secular acceleration of about 1.5×10^{-3} deg yr^{-2} of Phobos as compared to the observed value of 1×10^{-3} deg yr^{-2}. Eliminates the need for a small Q and large tidal dissipation in Mars.
3. Can explain the many near-commensurabilities in the satellite motions in the solar system. Eliminates the need for uncomfortably small Q. May explain the secular changes in the motion of the satellite *LAGEOS* without invoking theories such as elastic rebound of the earth's mantle due to gradual disappearance of polar ice sheets. May be responsible for the spiraling down of artificial satellites earlier than expected.
4. A feasible mechanism for transfer of angular momentum of the central body to the orbiting objects. Transfer of solar angular momentum matches the estimated values very well.
5. Automatically distributes matter in all spiral galaxies in a unique way which results in flat rotation curves, as observed.
6. Reduces the relaxation time of a star in a globular cluster to about 10^{17} s, which removes the age problem.

References

Assis, A.K.T., 1989, *Found. Phys. Lett.* 2:301.
Assis, A.K.T., 1992, *Apeiron* 13:3.
Cook, R.J., 1976, *Nuovo Cimento* 36:25.
Fujii, Y., 1975, *Gen. Rel. Grav.* 6:29.
Ghosh, A., 1984, *Pramana—Jr. of Phys.* 23:L671.
Ghosh, A., 1986a, *Pramana—Jr. of Phys.* 26:1.
Ghosh, A., 1986b, *Pramana—Jr. of Phys.* 27:725.
Ghosh, A., Rai, S. and Gupta, A., 1988, *Astrophysics and Space Science* 141:1.
Ghosh, A., 1988, *Earth, Moon and Planets*, 42:169.
Ghosh, A., 1991, *Apeiron*, 9-10:35.
Grabowski, B., Madej, J. and Helenka, J., 1987, *Astrophysical Jr.* 313:750.
Greenberg, R., 1977, Orbit-Orbit Resonances Among Natural Satellites, in: *Planetary Satellites* (ed.) J.A. Burns, University of Arizona Press, 157.
Greenstein, J.L. and Trimble, V.L., 1967, *Astrophysical. Jr.* 149:283.
Harwit, M., 1973, *Astrophysical Concepts*, John Wiley, 94.
Karoji, H. and Nottale, L., 1976, *Nature*, 259:259.
Kuhn, J.R. and Kruglyak, L., 1987, *Astrophysical Jr.* 313:1.
Laplace, P.S., 1980, Traité de Mécanique Céleste, in: *Oeuvres de Laplace*, Vol.5, Book 16, Chap. 4.
Lindley, D., 1992, *Nature* 359:583.
Lizhi, F. and Yaoquan, C., 1987, *From Newton's Laws to Einstein's Theory of Relativity*, Scientific Press and World Scientific, 93.
Merat, P., Pecker, J-C. and Vigier, J-P., 1974, *Astronomy and Astrophysics* 174:168.
Milgrom, M., 1983. *Astrophysical Jr.* 270:365.
Moffett, T.J., Barnes, T.G. and Evans, D.S., 1978, *Astronomical Jr.* 83:820.
Pechlaner, E. and Sexl, R., 1966, *Commun. Math. Phys.* 2:165.
Pollack, J.B., 1977, Phobos and Deimos, in: *Planetary Satellites* (ed.) J.A. Burns, University of Arizona Press.
Sadeh, D., Knowles, S.H. and Yaplee, B.S., 1968, *Science* 159:307.
Sciama, D.W., 1953, *Mon. Not. R. Astron, Soc.* 113:34.
Sciama, D.W., 1961, *The Unity of the Universe*, Doubleday.
Sciama, D.W., 1969, *The Physical Foundations of the General Theory of Relativity*, Doubleday.
Selak, S., 1984, *Astrophysics and Space Science* 107:409.
Shipman, H.L. and Sass, C.A., 1980, *Astrophysical Jr.* 235:177.
Sinclair, A.T., 1989, *Astronomy and Astrophysics* 220:321.
Vigier, J-P., Pecker, J-C. and Jaakkola, T., 1990, *Apeiron* 6:5.

A New Interpretation of Cosmological Redshifts: Variable Light Velocity

Eugene I. Shtyrkov

Kazan Physical-Technical Institute
Sibirsky Tr. 10/7
420029 Kazan, Russia

At present there are two alternative points of view to explain non-intrinsic redshifts in galaxy spectra which were discovered experimentally early in our century. One of them is based on the assumption of the Doppler effect and results in the Big Bang model of the Universe. The alternative approach is the notion of a step-by-step energy loss by a photon while it is traveling through space from a galaxy to a terrestrial observer (tired-light model). Though differing in essence, both assume that astronomical spectral experiments provide us with information about the frequency of the light being examined.

No experiments have ever directly measured either the frequency of light arriving from distant galaxies or its velocity. Information in the galaxy spectra is usually obtained with a diffraction spectrometer in values of the length of waves (λ) but not in frequency (ω). Indeed, the diffraction angle from spectral gratings is a result of interference of the waves shifted on multiples of λ, i.e. the angles depend on a wavelength. Though neither idea has been proved directly by experiment, the former has admittedly become the official version.

Nevertheless, the tired-light mechanism has a number of adherents as well. In my opinion, the main barrier to developing this idea is the fact that a light wave with variable frequency cannot be a solution of the wave equation with the stationary boundary conditions which is currently used in quantum electronics and optics to study propagating light in any medium. In effect, stepwise energy loss by a photon on its way from a star to a terrestrial observer is usually identified with a change of frequency. However, dependence of frequency on distance is not in accord with the stationary boundary conditions of the wave equation. A wave with variable frequency requires a new assumption: a source has to be unstable itself, *i.e.* the frequency of its irradiation should change with time.

This work presents a new point of view on the problem of cosmological redshift. While remaining in the framework of the steady state theory, one can obtain a result satisfying stationary boundary conditions and the experimental data on observed redshifts. This approach is based on a solution of the modified wave equation taking into account the real influence of the vacuum on a light wave while it is traveling through space.

The vacuum has been experimentally established to be not a void, but some material medium with definite, though not yet investigated, features. This is confirmed by several vacuum effects, for example, zero point oscillations and polarization of vacuum, generating the particles in the vacuum due to electromagnetic interaction. Therefore, it is reasonable to assume that this real matter can posses internal friction due to its small but a real viscosity. Hence, the vacuum can affect the light wave because of its resistance. Similarly, in the equation for a damped simple oscillator, such a resistance can be taken into account by formally introducing a term with some constant vacuum coefficient γ.

$$\frac{\partial^2 E}{\partial x^2} + \gamma \frac{\partial E}{\partial x} - \frac{k^2}{\omega_o^2} \cdot \frac{\partial^2 E}{\partial t^2} = 0 \tag{1}$$

This wave equation is written for the plane polarized monochromatic wave with a constant frequency ω_o which is traveling along the OX-axis. As was mentioned above, in the redshift experiments usually the wavelength is measured, not the frequency. The wavelength was found to depend on the distance (Hubble 1929). Thus in equation (1) the wave number (k) should be considered as a function of x. This equation can be solved under the following boundary conditions: at the point $x = 0$ the light source (any spectral component of stellar radiation) emits the stationary field

$$E(0, t) = E_o \exp(i\omega_o t) \tag{2}$$

with the initial wavelength $\lambda_o = 2\pi c/\omega_o$ and the wave number for the radiated wave $k(0) = k_o = 2\pi/\lambda_o$.

Let us seek a solution of (1) in the form of a periodic function with the same frequency but variable phase

$$E(x, t) = E_o \exp i\phi(x, t) \tag{3}$$

where $\partial \phi / \partial t = \omega_o$ is constant.

Performing differentiation of (3) with respect to x and t and inserting into (1), we have

$$-\left[\frac{\partial \phi}{\partial x}\right]^2 + i\frac{\partial^2 \phi}{\partial x^2} + i\gamma \frac{\partial \phi}{\partial x} + \left[\left(\frac{k}{\omega_o}\right)\frac{\partial \phi}{\partial t}\right]^2 = 0 \tag{4}$$

After separating the real and imaginary parts we obtain simple equations

$$\frac{\partial^2 \phi}{\partial x^2} = -\gamma \frac{\partial \phi}{\partial x} \quad \text{and} \quad \frac{\partial \phi}{\partial x} = \left(\frac{k}{\omega_o}\right)\frac{\partial \phi}{\partial t} \tag{5}$$

Since a time derivative of a phase is a frequency ($\partial \phi / \partial t = \omega_o$) the latter term in (5) results in $\partial \phi / \partial x = k$ as a function of x. Taking into account the boundary conditions we obtain the expressions for the wave number, phase and field strength

$$k(x) = k_o \exp(-\gamma x) \tag{6}$$

$$\phi(x,t) = \omega_o t - \frac{k_o}{\gamma}\left(1 - e^{-\gamma x}\right) \tag{7}$$

$$E(x,t) = E_o \exp i\left[\omega_o t - \frac{k_o}{\gamma}\left(1 - e^{-\gamma x}\right)\right] \tag{8}$$

Since $k(x) = 2\pi/\lambda(x)$, Equation 6 gives directly

$$\lambda(x) = \lambda_o \exp(\gamma x) \tag{9}$$

This implies a gradual increase of the wavelength at a constant frequency while the wave propagates through the vacuum. To compare this dependence with one well-known in cosmology, we write from (9) the result for the relative spectral shift

$$\frac{\Delta\lambda}{\lambda_o} = \exp(\gamma x) - 1 \tag{10}$$

where $\Delta\lambda = \lambda(x) - \lambda_o$.

Let us expand the exponential in terms of (γx)

$$\frac{\Delta\lambda}{\lambda_o} = (\gamma x) + \frac{(\gamma x)^2}{2!} + \cdots \tag{11}$$

and compare this with the well-known expression (Zelmanov 1962)

$$c_o \frac{\Delta\lambda}{\lambda_o} = Hx + \frac{(1+q)(Hx)^2}{2c_o} \tag{12}$$

where c_o is the electrodynamical constant, H is the Hubble coefficient, and the dimensionless parameter $(q = -R\ddot{R}/\dot{R}^2)$ determines a relative acceleration of Metagalactic expansion. For a stationary universe $q = 0$. Comparing (11) with (12) leads to

$$H = c_o \gamma \tag{13}$$

It is interesting to note that the value of the vacuum constant $\gamma = H/c_o$ derived from (13) is identical with the cosmological parameter Λ introduced by A. Einstein into the gravitational potential equation (Zelmanov 1962). According to current data, the Hubble constant is in the interval 60–140 km s^{-1} Mpc^{-1}. To estimate γ we use the mean of H =100 km s^{-1} Mpc^{-1}. As one would expect, the vacuum constant is very small—about 10^{-28} cm^{-1}. At such a small value of γ, gigantic distances are required for appreciable changes in light characteristics. Thus, it follows directly from this that the phase velocity of light in vacuum does not remain constant while a wave is traveling. This conclusion is certainly not in agreement with the generally accepted point of view. However, several additional considerations can be discussed in favor of it.

Firstly, there is no direct experimental evidence that light velocity does not depend on distance traveled. No such test has been carried out. Moreover, in terrestrial experiments it has not been possible to find any influence of such an effect even occasionally because of its infinitesimal value. Even over distances comparable with the size of our galaxy (30 kps) this effect cannot be detected by modern apparatus.

Secondly, phase velocity is known to depend on the permittivity (ε) and permeability (μ) of a medium and the electrodynamical constant (c). Since the vacuum is real matter, all of these characteristics are real parameters of the interaction of light with this real matter. In their turn, these interaction parameters depend on conditions both of the medium and of the light field itself. During the travel of light through the vacuum, some of its characteristics may be changed due to vacuum resistance, i.e. the interaction parameters can depend on x. As for ε and μ, the situation with variable permittivity and permeability was discussed in (Shtyrkov 1992) where a wave equation with the term $\partial E/\partial x$ was derived from Maxwell's equations. Variable ε and μ were shown in principle to result in stepwise redshifts of λ with conservation of frequency. However, in the linear approximation, ε and μ of the vacuum are actually constant. Therefore, another solution is more promising as an explanation of the step-by-step mechanism of the redshifts, namely, treating the electrodynamical parameter c as an interaction parameter which can depend on distance covered.

In effect, from Maxwell's equations one can see that the process of electromagnetic wave generation is consistent according to the principle of short-range interaction: the change of E causes the electrical induction to change, then a generated magnetic field curl gives rise to magnetic induction and a new curl of E at the next point in space. This process reproduces itself, and the excitation propagates with a definite velocity depending on the state of the vacuum. In this case the curl can be considered as an excited state of the vacuum. Therefore, it is not ruled out that the change of such a state due to vacuum resistance during a cycle $E \to H \to E$ may lead to a significant change in the further excitation rate. The electrodynamical constant, then, is a parameter which links electrical and magnetic phenomena. Hence, a change of rate of the unit $E \to H \to E$ cycle is determined by any change in the parameter. At large distances, these changes are accumulated and may lead to the effect detected experimentally.

We consider the problem in more detail by writing Maxwell's equations

$$\nabla \times \mathbf{E} = -\frac{1}{c}\frac{\partial \mathbf{B}}{\partial t}$$

$$\nabla \times \mathbf{H} = \frac{1}{c}\frac{\partial \mathbf{D}}{\partial t}, \qquad \mathbf{D} = \varepsilon \mathbf{E} \qquad (14)$$

$$\nabla \cdot \mathbf{D} = 4\pi\rho, \; \nabla \cdot \mathbf{B} = 0 \qquad \mathbf{B} = \mu \mathbf{H}$$

With linear polarization of light wave in vacuum and $\rho = 0$, $\varepsilon = \mu = 1$, parameter $c = c(x)$, it follows from (14) that

$$\frac{\partial E}{\partial x} = -\frac{1}{c(x)}\frac{\partial H}{\partial t}$$
$$\frac{\partial H}{\partial x} = -\frac{1}{c(x)}\frac{\partial E}{\partial t} \qquad (15)$$

Differentiating the former with respect to x and substituting $\partial H/\partial x$ from the latter and $\partial H/\partial t$ from the former we obtain

$$\frac{\partial^2 E}{\partial x^2} + \frac{1}{c}\frac{dc}{dx}\frac{\partial E}{\partial x} - \frac{1}{c^2}\frac{\partial^2 E}{\partial t^2} = 0 \qquad (16)$$

Comparing this equation with (1) yields

$$\gamma = \frac{d(\ln c)}{dx} \quad \text{and} \quad k(x) = \frac{\omega_o}{c(x)}$$

After integrating the former at the previous boundary conditions $k(0) = k_o$ and $c(0) = c_o = \omega_o/k_o$ we derive

$$c(x) = c_o \exp(\gamma x) \qquad (17)$$

and obtain the same dependence of wavelength on distance as in (8). Thus a change in the electrodynamical parameter can be a reason for variable phase velocity of light *in vacuo*.

At present the idea that a traveling electromagnetic wave is created according to the Carman wave principle (Azjukovskii 1990) is being widely discussed. On this hypothesis a photon is a stable formation (a particle) made up of vacuum curls with periodic distances between them equal to λ. It is reasonable to assume that an interaction with the vacuum due to resistance causes this formation to come to a new dynamic equilibrium state. This leads to the gradual rearrangement of the system, *i.e.* to a change of λ. With this assumption the photon gradually propagates as much as it is allowed to by its stability range. When a definite value of λ is achieved, the structure leaves the stability region and the photon collapses. This conclusion may be in agreement with experimental data on redshift limits. If the maximum value of $\Delta\lambda$ detected for galaxies does not exceed $0.5\,\lambda_o$, we may assume that this value corresponds to the critical collapse distance.

The estimate obtained from (10) would give the value for limiting distance of about 4×10^{27} cm, *i.e.* 1000 Mpc. This value may be increased by a factor of 9–10 if quasar redshifts are not of gravitational origin, but of the same cosmological nature

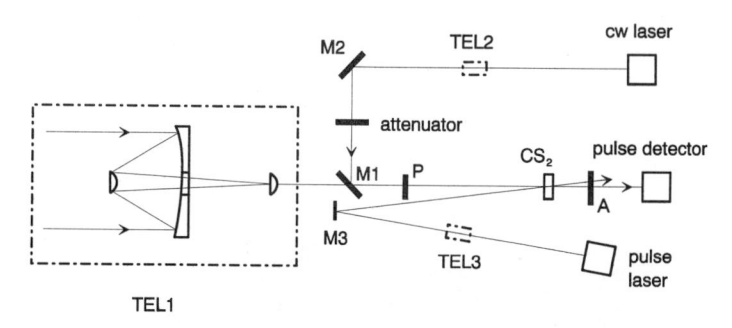

Figure 1. Experiment to test variable light velocity hypothesis

(the limiting value of $\Delta\lambda/\lambda_o$ for quasars does not exceed 5). Thus the light from galaxy associations situated beyond this distance may not reach the Earth due to a total collapse of the photon. Consequently, the photometric Olbers' paradox has a natural explanation.

The concept considered in this work certainly requires additional experimental examination. At least two types of experiments can be suggested: 1) direct measurements of either the velocity of light from far galaxies or the frequency of this light (measuring its wavelength redshift value as well). For instance, one possible experimental setup to test whether light velocity depends on x is illustrated in Figure 1. To examine redshifted light, a telescope (Tel. 1) and pulsed optical shutter (POS: P-polarizer, A-analyzer, CS_2 cell) are used. The POS is opened by a short light pulse from the pulsed laser. Another light from the reference source (cw-laser in the figure or some star without redshift) is passed through POS as well. By comparing the difference of arrival times for both signals, one can detect a difference of light velocities, if it is measurable. Employing a POS with an exposure of about one picosecond time scale, for instance, it is possible confidently to measure $\Delta c/c$ of about 0.01 if the distance from the CS_2 cell to the detector is of the order of only one meter.

In order to detect light frequency, either the heterodyne method or a resonant absorption medium could be used directly. In the latter case, instead of the CS_2 cell a medium with some resonant absorption should be inserted. Choosing an appropriate spectral line of a specific atomic element in a redshifted spectrum as well as a relevant absorption line for the same element in the medium, one should be able to measure some absorption, if the frequency is, in fact, not shifted. Measuring a frequency and/or light velocity could yield real data concerning a photon at its source location and, hence, advance our understanding of the real origin of the cosmological redshifts.

References

Azjukovskii, V.A., 1990, *Obshaja efirodinamika*, Energoatomizdat (in Russian).
Hubble, E.P., 1929, *Proc. Nat. Acad. Sci.* 15,3:168.
Shtyrkov, E.I., 1992, *Galilean Electrodynamics* 3:4.
Zelmanov, A.L., 1962, Kosmologiya, in: *Fizicheskii Enciklopedicheskii Slovar*, (in Russian).

Quantized Vacuum Energy and the Hierarchy of Matter

Henrik Broberg

Skirnervägen 1b
18263 Djürsholm, Sweden

The cosmological redshift is treated as the loss of units of quantized energy by the photon travelling through a nonzero energy vacuum, and the quantum characteristics of the Metagalaxy are deduced. Particles are created from the vacuum, appearing as confined waves of quantized energy in a system with quantum characteristics. A model of energy circulation in an equilibrium universe is discussed, and hypotheses to explain the stability and energy production of neutron stars and other astrophysical objects are advanced.

1. Introduction

One of the major problems facing modern physics is a general understanding of the phenomenon of particle rest mass. Previous attempts, by de Broglie and Einstein in particular, to explain the rest mass of elementary particles using singularities in space-time met with little success. Present efforts aimed at a unification of the forces in the universe are intimately related to this problem, and the most common approach is to use a rather arbitrary notion of rescaling during the expansion of the universe since the big bang.

A great deal is known about relations between elementary particles, their energy levels and their forces. However, the fundamental problem persists: Why do particles have the observed scale of masses? Why is the mass of the neutron on the order of 10^{-27} kg, and not, say one kilogram? How do particles confine their energy? What is the basic material of which particles are made?

In this paper, we begin with the postulate of a minimum quantum of energy, which we deduce from the phenomenon of the cosmological redshift (for a detailed treatment, see Broberg 1981, 1982, 1984). This quantum corresponds to a standing wave across the gravitational diameter of the universe (or Metagalaxy), which we treat as a large black hole. The entire energy of the Metagalaxy would be made up of 10^{120} such quanta, each having energy hH_o, i.e. the product of the Planck and Hubble constants. The gravitational force constant can thus be directly linked to the mass of the Metagalaxy, conceived as a quantum system.

From the above understanding of the Metagalaxy, we apply the notion of a quantum system to the microcosm, where we analyze elementary particles as "miniature" universes, each confined within an event horizon. The energy confined within such a system forms its own quantum conditions, with the minimal energy quantum corresponding to a wave across the gravitational diameter and an internal vibration, while each system has a "gravitational" force constant of its own, depending on its mass. From these assumptions, we may calculate the quantum numbers for each level of energy condensation. Naturally, we find that the smaller the mass of the quantum system, the larger the constituent quantum. The resulting mathematical series is brought back to the starting point when one elementary quantum provides all the energy of the smallest local "universe". This case generates the mass of the lightest of the elementary particles subject to the strong nuclear force, or the π-meson group (the value of Hubble's constant used for the numerical calculations is consistent with a Hubble time of 12 billion years).

As noted above, the clue to a description of elementary particles as singularities was found by analyzing the cosmological redshift of photons. One conclusion we will draw from this is that photons move in space with a group velocity just below c, while the theoretical velocity c is only applicable to an ideal case of energy transfer, due to the fact that vacuum space has a certain energy density. The ideal energy transfer applies to a spiral along the photon pattern. The elementary quantum can therefore be seen as the limiting case of a standing wave across the universe: it has zero group velocity. The energy of a photon is relativistic energy with the elementary quantum as the rest state, and therefore depends on the photon group velocity. In the photon's internal system, its energy can be described as a rotating system, a disk with peripheral velocity equal to the group velocity. In the system of an observer, the circumference of the disk will appear shrunk by the relativistic factor, giving the periphery of the photon's spiraling information front. When the energy is high enough, the circumference shrinks back to that of a black hole containing the photon energy. This happens for a distinct energy, which is found to be below the energy of the π-meson above the energy of the μ-meson. Hence, this case separates the particles subject to the strong nuclear force from the lighter particles. Finally, the magnitude of the force at the particle level is obtained from a generalized expression for the force coefficient, as a function of Newton's constant G_o and the constituent mass of the quantum system. Thus, the force coefficient is determined by the number of elementary quanta in the particle system.

2. The Cosmological Redshift as a Probe of Space

The original Hubble law was an empirical relation between redshift and magnitude, or distance to the observed source. This relation is generally written

$$z = \frac{H_o}{c} \cdot r \qquad (1)$$

where r is the distance to the object and H_o is Hubble's constant. The time taken for the light to travel from the source to the observer is $\Delta t = r/c$, and with this notation the above formula can be rewritten as

$$\frac{\Delta \lambda}{\lambda} = H_o \Delta t \qquad (2)$$

This is almost the same relation as the one used in the standard "big-bang" model, with the only difference that H_o is replaced by $H(t)$ in the latter.

However, contrary to the standard model, we will assume that Hubble's constant is a true constant, and that its significance is to be found in an analysis of the interaction between the photon and the energy of vacuum space. In fact, this assumption is supported by a comprehensive survey by Jaakkola *et al.* (1979), from which the authors conclude that the redshift is most probably of a non-Doppler origin, as well as contributions by Arp, Napier, Jaakkola, Lerner and others to these proceedings.

Hence, using formula (2) the photon frequency loss per cycle is

$$\Delta \nu(\tau) = -H_o \qquad (3)$$

This energy is independent of the energy of the individual photons.

We know that angular momentum is quantized in multiples of $\hbar/2$. Could another quantized property in the universe lurk behind the above derived unitary loss of frequency by all photons on the cosmic time-scale? The immediate answer is that angular frequency is quantized as well. This will mean that energy is also quantized. Therefore, the basic property of energy itself seems to be natural candidate for a universal law, and the following postulate is made:

The energy of the universe is quantized in integer numbers of a system-invariant elementary energy quantum.

We add to this a hypothesis on the energy of the elementary quantum:

The elementary energy quantum in the universe is equal to the product of Planck's and Hubble's constants.

The elementary quantum is therefore treated as an isotropic two-dimensional quantum oscillator (Levich et al. 1973) with ground level energy

$$E_{eq} = \hbar \omega_{eq} = h H_o \qquad (4)$$

The photon as a system of N elementary quanta thus has the energy

$$E_q = N h H_o \qquad (5)$$

where N is the integer number of elementary quanta that make up the photon energy.

It should be recalled that the vacuum in the lowest energy state is characterized by Maxwell's equations, and that there is no formulation of quantum electrodynamics by which the mean energy density becomes zero in the vacuum (Zeldovich and Novikov 1971). Consequently, the vacuum is assumed to have a certain positive energy density $\rho_o c^2$.

The loss of one elementary unit of angular frequency ω_{eq} from the photon during each cycle will be treated as an absorption of one elementary oscillator into vacuum space. A cross section σ_q is defined as a characteristic interaction surface for the interaction between the photon and the vacuum space it encounters. In the following, we will describe the interaction between the photon and the vacuum space globally in terms of the following basic concepts: the elementary energy quantum, the vacuum density and the cross section. A more comprehensive discussion is given later, in part 4.2.

The energy flow from the photon to the vacuum space during one photon cycle is given, in equivalent mass units, by

$$\begin{cases} \Delta m_q(\tau) = -\sigma_q \rho_o \lambda_q \\ \Delta m_q(\tau) = -E_{eq}/c^2 \end{cases}$$

Hence, the cross section is

$$\sigma_q = \frac{hH_o}{\rho_o c^2} \frac{1}{\lambda} = \frac{H_o}{\rho_o c} m_q = A m_q \qquad (6)$$

The equivalent mass of the photon is m_q, and with the above assumptions, the cross section becomes proportional to m_q by the constant A

$$A = \frac{H_o}{\rho_o c} \qquad (7)$$

A quantum volume can now be defined as the volume covered in vacuum space by the photon cross section during one cycle

$$V_o = \sigma_q \lambda_q = \frac{hH_o}{\rho_o c^2} = \frac{Ah}{c} \qquad (8)$$

This quantity is invariant of the photon energy and corresponds directly to the volume of space with density ρ_o that contains one elementary quantum of energy.

We can as well define a cross section σ_s, characteristic for the vacuum space, to describe the absorption of photon energy. In this case, the absorption during one cycle is described by

$$\begin{cases} \text{absorbed energy}: & \sigma_s(\rho_q c^2)\lambda_q = hH_o \\ \text{energy density}: & \rho_q c^2 = E_q/V_o \end{cases}$$

The density of the photon ρ_q is defined using the above found quantum volume V_o. The cross section characteristic for a quantum volume of vacuum space becomes

$$\sigma_s = A\frac{hH_o}{c^2} \qquad (9)$$

Hence, an interaction surface can be defined for vacuum space as well, using a constant A for the relation between the surface and the equivalent mass of vacuum energy.

The above concept of a cross section is equivalent to the application of the term in nuclear physics, and it is a measure of the likelihood that a wave of energy will interact with the vacuum space it penetrates.

The expression (9), however, is independent of the photon energy, and can, therefore, also be assumed to remain valid for interactions with the lowest energy state, or without the vacuum itself. Compared to the interaction surface of the photon, which is singly directed in space, the total interaction surface of a closed quantum volume in comoving space will be six times larger, representing energy flows in all the $\pm X, \pm Y, \pm Z$ directions, which gives an interaction constant equal to $6A$.

For the interactions within the vacuum itself, it can be assumed that two interacting quantum volumes share a common surface, each having the interaction constant $3A$; this case will be applied to the static comoving space in our surrounding.

Any object that can be described as a packet of energy waves can be characterized by a certain cross section for its interaction with the vacuum. We will show that particle rest-mass energy can be described as a system of energy waves in a microscopic vacuum state, isolated from the surrounding universe by its interaction surface, and therefore the particle will be characterized by the interaction constant $6A$. For any physical object, the interaction surface σ_u is given by the constant A, the dimensionality L_u and the mass M_u

$$\sigma_u = L_u A M_u \tag{10}$$

3. The Comoving Quantum Universe

We will now discuss a local universe of vacuum energy with density ρ_o which is comoving with the observer. The idea is first to analyze such a universe in terms of the quantum parameters defined in the preceding section, and then to see how it can be generalized and rescaled to particle dimensions.

The static Gaussian curvature of the three dimensional universal space in a comoving frame is (Zeldovich and Novikov 1971)

$$\Lambda = \frac{1}{R_o^2} = \frac{8}{3}\pi \frac{G_o \rho_o}{c^2} \tag{11}$$

where G_o is Newton's constant. For this space, we will use $L_u = L_o = 3$ in equation (10) for the interaction constant, although we will retain the symbol L_o in the calculation in order to allow us to later use our results in the case of particles.

Combining equations (7) and (11) to eliminate ρ_o gives

$$R_o^2 = \frac{3Ac^3}{8\pi G_o H_o} \tag{12}$$

The gravitational radius of the comoving vacuum space is

$$R_g = \frac{2G_o M_o}{c^2} \tag{13}$$

where M_o is the equivalent mass of the comoving vacuum space. It is easily shown that R_g is equivalent to R_o from equation (11) by setting M_o equal to the mass of a Euclidean sphere with density ρ_o and radius R_g.

The surface of our comoving space in geometrized units in the Kerr metric becomes (Stephani 1982)

$$\sigma_o = 4\pi \left(\frac{R_g^2}{2} - Q^2 + \left(\frac{R_g^4}{4} - R_g^2 Q - P^2 \right)^{1/2} \right) \tag{14}$$

By assuming that the charge Q and the angular momentum P are negligible compared to the mass, here represented by R_g, we can describe the universal surface σ_o with a spherically symmetrical surface in three dimensional space

$$\sigma_o = 4\pi R_g^2 \qquad (15)$$

This surface is the characteristic surface of a black hole, and we will now make the assumption that it is identical to the surface given by equation (10), namely, the total interaction surface for all transfer of energy (and information) from the exterior to the interior vacuum space. Combining equations (10) and (15) yields

$$\sigma_o = L_o A M_o = 4\pi R_g^2 \qquad (16)$$

We can now eliminate M_o from equations (13) and (16). Using $R_g = R_o$ and equation (12), we obtain an expression linking Newton's constant directly to the Hubble constant

$$G_o = \frac{L_o^2}{24\pi} H_o A c \qquad (17)$$

The radius of the comoving universal vacuum space is thus, using equations (12) and (17)

$$R_o = \frac{3}{L_o} \frac{c}{H_o} \qquad (18)$$

And, when equation (18) is combined with equation (16), we obtain the equivalent mass

$$M_o = \frac{36\pi}{L_o^3 A} \left(\frac{c}{H_o}\right)^2 \qquad (19)$$

We are now in a position to develop a direct relationship between the force constant and the equivalent mass of the vacuum universe by eliminating H_o from (17) and (19). We therefore obtain

$$G_o = \left(\frac{L_o A c^4}{16\pi M_o}\right)^{1/2} \qquad (20)$$

With the observed value of Newton's constant G_o, we may compute the values of the mass, radius and other characteristics of the large scale vacuum universe. We use a reasonably safe mid-range value of the Hubble constant $H_o = 2.64 \times 10^{-18}$ s^{-1}, equivalent to $(12 \times 10^9 \text{ years})^{-1}$ (Weinberg 1972). The Metagalaxy is assumed to be characterized by the observed values of Newton's and Hubble's constants. With these constants, the basic features of the comoving vacuum universe may be obtained as follows (with $L_o = 3$):
From equation (17), the value of the interaction constant is

$$A = \frac{8\pi G_o}{3 H_o c} = 0.70 \ [\text{m}^2/\text{kg}] \qquad (21)$$

Equation (18) gives the radius of the Metagalaxy as

$$R_o = \frac{c}{H_o} = 1.1 \times 10^{26} \ [\text{m}] \qquad (22)$$

Equation (19) provides an evaluation of the mass of the comoving system

$$M_o = \frac{4\pi}{3A}\left(\frac{c}{H_o}\right)^2 = 8 \times 10^{52} \ [\text{kg}] \qquad (23)$$

Combining equations (4) and (23) yields an estimate of the number of elementary quanta of energy in the system

$$N_o = \frac{M_o c^2}{h H_o} = 4 \times 10^{120} \ [\text{elementary quanta}] \qquad (24)$$

And finally, for the energy density, equations (11) and (18) yield

$$\rho_o = \frac{3H_o^2}{8\pi G_o} = 1.2 \times 10^{-26} [\text{kg/m}^3] \qquad (25)$$

The values obtained from the above computation are on the proper order of magnitudes for the observed universal dimensions, although they apply only to the comoving vacuum space. The constant A calculated above is assumed to be a universal constant, valid in the microcosm as well. The number N gives the quantum number of our comoving universe, and is equivalent to the number of elementary quanta that make up the energy of the total vacuum space of any observer.

The density of the vacuum ρ_o is equivalent to the equilibrium density in the Friedmann models (Weinberg 1972).

4. Local Confinement of Energy in Particles

4.1 The Rescaled Comoving Vacuum Universe

We now return to formula (21), which gave the gravitational force constant as a function of the equivalent mass of the comoving vacuum universe with the dimensionality $L_u = L_o$ as a parameter. The "surface-to-mass" constant A in the formula is assumed to be valid generally, in both the microcosm and the macrocosm, following the examples of h and c. We will now generalize equation (20), and let each L_u represent a class of vacuum universes with different values of the force constant, generally written G_u, for different values of M_u

$$G_u = \left(\frac{L_u A c^4}{16\pi M_u}\right)^{1/2} \qquad (26)$$

Of course, each such universe will have its own typical value of an elementary energy hH_u and vacuum density ρ_u.

We substitute now the proper numerical value for A from formula (21) and our observed values for Newton's constant G_o and Hubble's constant H_o from our own universe with the dimensionality L_o, which gives the following expression for the force constant as a function of the mass for any universe belonging to a certain class, defined by the parameter L_u

$$G_u = \left(\frac{3L_u G_o c^3}{2L_o^2 H_o M_u} \right)^{1/2} \tag{27}$$

In the same sense we can now also develop a formula for the mass of any comoving vacuum universe belonging to the same class of universes, defined by L_u, as a function of a quantum number K giving the number of elementary energy quanta that make up the total equivalent mass of the universe. In our surrounding, the frequency of the elementary energy quantum is H_o from (3), and the relation between that frequency and the gravitational radius of the universe is given by (18). In the generalized case, we use the latter formula with L_o replaced by L_u and H_o by H_u, while G_o is replaced by G_u and M_o by M_u in the expression for the gravitational radius (13).

We arrive then at the following equation system:

$$\begin{cases} M_u = \dfrac{KhH_u}{c^2} \\ R_u = \dfrac{3c}{L_u H_u} = \dfrac{2G_u M_u}{c^2} \\ G_u = \left(\dfrac{L_u A c^4}{16\pi M_u} \right)^{1/2} \end{cases}$$

Eliminating H_u and G_u, the following expression is found for the mass

$$M_u = \left(\frac{36\pi K^2 h^2}{L_u^3 A c^2} \right)^{1/3} \tag{28}$$

Hence, we have all the masses of the same class of the universes (common L_u) expressed in the constants A, h and c and a quantum number K.

Using equation (17) we can eliminate A, and reinstate the mass formula with the parameters G_o and H_o instead

$$M_u = \left(\frac{3L_o^2 K^2 h^2 H_o}{2L_u^3 G_o c} \right)^{1/3} \tag{29}$$

In the same way the radius becomes

$$R_u = \left(\frac{L_u A M_u}{4\pi} \right)^{1/2} = \left(\frac{2L_u G_o M_u}{3H_o c} \right)^{1/2} \tag{30}$$

We will now evaluate the quantities applications to the family of comoving universes for which the dimension parameter is $L_u = 6$. As we already discussed in part 3, this case may be thought of as applicable to the total interaction surface of isolated volumes in vacuum space.

To distinguish the following case ($L_u = 6$) from one of part 3 above ($L_u = 3$), the symbol K is used instead of N for the quantum number.

The physical characteristics of this class ($L_u = 6$) of isolated universes become

from (21):
$$A = \frac{8}{3}\pi \frac{G_o}{H_o c}$$

from (29):
$$M_u = \left(\frac{K^2 \hbar^2 H_o}{16 G_o c}\right)^{1/3} \qquad (31)$$

from (27):
$$G_u = \left(\frac{G_o c^3}{H_o M_u}\right)^{1/2} \qquad (32)$$

from (30):
$$R_u = \left(\frac{4 G_o M_u}{H_o c}\right)^{1/2} \qquad (33)$$

The above formulae are evaluated for the ground state and resonance masses of the hadronic particles (the pion and the nucleon) in the following table. The same value of Hubble's constant is used here as earlier in part 3.

Table 1. Experimental values of baryon and meson masses compared with predicted results.

	Experimental data			Estimated quantum number	Quantum mass
Particle group		Mass (GeV)	K_{est}	K	(GeV)
Baryons	p; n	0.938; 0.940	35.94	36	0.94
	Λ	1.116	46.56	46	1.11
	Σ	1.189; 1.192; 1.197	51.42	52	1.20
	Δ_r	1.236	54.26	54	1.23
	Ξ	1.315; 1.321	59.76	60	1.32
	Σ_r	1.379	63.96	64	1.38
Mesons	$\pi^+; \pi^\circ; \pi^-$	0.140; 0.135; 0.140	2.04	2	0.14
	K	0.494; 0.498	13.8	14	0.50
	η_r	0.548	16.02	16	0.55
	ω_r	0.783	27.36	28	0.79
	φ_r	1.019	40.62	40	1.01

From the table, we can easily see that the ground state particle masses (proton, neutron and pions) fall directly on integer quantum numbers, *i.e.* with the numerical value used for Hubble's constant. Furthermore, the ratios between the calculated force constants from equation (32), and Newton's constant $G_p/G_o \sim 10^{40}$ agree with the known ratio between the strong nuclear force and the gravitational force, while the calculated particle radii calculated from equation (33) are compatible with the reach of the strong nuclear force, and apply to known particle dimensions. Using a more refined approach, the author has calculated a fuller series of meson and baryon masses (Broberg 1991).

4.2 The Interaction Between the Photon Energy and Vacuum Space

The physical significance of the above rescaling of the comoving vacuum universe can be understood through a discussion of the interaction between the photon and vacuum space.

The velocity c will be assumed to represent an idealized case of information transfer in space, and we will use the Lorentz-Minkowski geometry with reference to idealized information cones instead of light cones.

From the point of view of a stationary observer who looks at the system of a passing photon, the internal oscillator with frequency v in the photon's rest system cannot have its frequency reduced (by relativistic "time dilation") to less than a minimum value v_{min}, which is identical to the equivalent frequency of the minimal energy quantum H_o postulated above. Therefore, while the proper frequency v and wavelength λ apply in the photon frame, the coordinate time is assumed to be $\Delta t = 1/H_o$ in the system of the observer, in which the photon moves a distance Δx during the interval Δt. The minimum frequency corresponds to a maximum wavelength λ_{max}, such that

$$\begin{cases} \Delta x = (\lambda_{max}^2 - \lambda^2)^{\frac{1}{2}} \\ \Delta t = \dfrac{\lambda_{max}}{c} \end{cases}$$

The velocity of energy transport in the observer system therefore becomes

$$v_g = \frac{\Delta x}{\Delta t} = c\left(1 - \left(\frac{\lambda}{\lambda_{max}}\right)^2\right)^{\frac{1}{2}} \tag{34}$$

which is just the group velocity of photon energy. The relation between λ and λ_{max} can be expressed as

$$\gamma = \frac{\lambda_{max}}{\lambda} = \frac{E_q}{E_{min}} \tag{35}$$

Therefore, γ will play the same role as the usual relativistic factor for the elementary energy quantum

$$\gamma = \left(1 - \left(\frac{v_g}{c}\right)^2\right)^{-\frac{1}{2}} \tag{36}$$

If $\lambda_{max} = c/H_o$ corresponds to the energy hH_o of the elementary energy quantum from equation (4), using equation (27) we find that all "normal" photons with, say, $\lambda \leq 10^6$ m, have $\gamma \geq 10^{20}$. We may also note that, in accordance with de Broglie's theory of the photon (de Broglie 1926), the difference between the ideal velocity of information (c) and the actual group velocity of the electromagnetic wave in vacuum is real, though extremely small

$$\begin{cases} \left(1 - \dfrac{v_g}{c}\right) \leq 10^{-40} \\ \lambda \leq 10^6 \ [\text{m}] \end{cases}$$

Hence, "normal" light is a very good approximation of ideal information, although this approximation will cease to apply when we discuss particle rest mass.

At each cycle of the photon's internal oscillator τ, the photon's energy will be contained in a new quantum volume V_o, from equation (8), which is defined in the observer system as the product of the photon cross section and the wavelength λ. Conversely, from the photon system, the wavelength is contracted and this volume may be expressed as V_o/γ. When the contracted wavelength reaches the dimension of the gravitational diameter of the photon's internal energy, a stationary, self-confining particle energy system can be created in its ground form. The photon energy is then trapped in a quantum volume, and we have the following equation system:

$$\begin{cases} \lambda' = \dfrac{\lambda}{\gamma} \\ \lambda' = 2R_g = \dfrac{4G_o}{c^2}\dfrac{h}{c\lambda} \\ \gamma = \dfrac{E_q}{E_{eq}} = \dfrac{1}{\lambda}\dfrac{c}{H_o} \end{cases}$$

Eliminating λ' and γ gives the equivalent mass

$$M_{\lim} = \dfrac{h}{c\lambda} = \left(\dfrac{H_o h}{4G_o c}\right)^{1/3} = 137 \ \text{MeV} \qquad (37)$$

This energy limit is almost identical to the mass of the lightest hadron $m_{\pi^o} = 135 \ \text{MeV}$), and the mass formula is also equivalent to the mass generated by equation (31) for the case $K = 2$.

In order to recover the ground form of equation (31), we need to treat the case of an ideal optical resonator, where the wavelength is twice the distance between the "mirrors". We must also adopt the parameter $L_u = 6$, which reduces the radius of the reference universe to $R = \frac{1}{2}c/H_o$, according to equation (18). The "reference resonator" will then have the diameter c/H_o and the "reference resonant wave" will have the wavelength $2c/H_o$. With these adjustments, our equation system becomes

$$\begin{cases} \lambda' = \dfrac{\lambda}{\gamma} \\ \lambda' = 4R_g = \dfrac{8G_o}{c^2}\dfrac{h}{c\lambda} \\ \gamma = \dfrac{2}{\lambda}\dfrac{c}{H_o} \end{cases}$$

Solving this equation system yields

$$M = \frac{h}{c\lambda} = \left(\frac{H_o h^2}{16 G_o c}\right)^{1/3} \tag{38}$$

which is the ground state ($K=1$) of equation (31).

However, equation (37) gave the lowest limit for capturing photon-energy locally in space, corresponding to quantum number $K=2$ in equation (31). The ground state expressed by equation (38) cannot, therefore, actually exist. Instead, the above established relationship for the confinement of one wave within a gravitational surface must be extrapolated to a system made up of an integer number of identical waves. The total mass of this system has a gravitational diameter which we set equal to half the wavelength of the constituent elementary waves. This procedure completely restores equation (31), with the number of waves corresponding to the quantum number (K) of that equation. The elementary wave thus has the wavelength $K\lambda$, and we obtain the equation system

$$\begin{cases} \lambda' = \dfrac{\lambda}{\gamma} \\ K\lambda' = 4R_g = \dfrac{8G_o}{c^2}\dfrac{h}{c\lambda} \\ \gamma = \dfrac{2}{K\lambda}\dfrac{c}{H_o} \end{cases}$$

Solving this system then brings back equation (31). It thus follows that one captured photon is transformed into a pair (or pairs) of waves in the particle system.

5. Implications for Astrophysics and Cosmology

The elementary quantum hypothesis affords new insights into a number of astrophysical phenomena and processes. Here we will discuss only a few of the model's predictions, and only in the most general terms.

The cornerstone of the cosmological model must be the existence of a universal balance of energy as elementary quanta pass between different states (mass particles, electromagnetic energy, free vacuum quanta). We established above that the energy loss by photons to the vacuum, by virtue of the fact that their velocity approaches c, was given by

$$\frac{dm}{dt} = -H_o m$$

Conversely, the transfer of energy from the vacuum to mass particles must therefore take place at the rate

$$\frac{dm_p}{dt'} = H_o m_p$$

where t' is the time in the particle system. When the particle translates at a certain velocity, it would have to give up energy to conserve its velocity, just as a photon does. Therefore, in relation to its kinetic energy, we propose that it has to give up energy (or mass) to space according to

$$\frac{dm_k}{dt'} = -H_o \frac{E_k}{c^2} \equiv -H_o m_p (\gamma - 1)$$

Assuming that the above relations are valid in the system of the particle, the resulting flow of mass to the particle in its own time system would be

$$\frac{dm}{dt'} = H_o m_p - H_o m_p (\gamma - 1) \equiv H_o m_p (2 - \gamma)$$

However, the above relation must also apply to the particle's internal system and the background space. In the particle system, the internal frequency provides the time standard, while in background space the frequency of a stationary particle could be chosen. Therefore, the relation between the time in the particle system and the time in background space is just

$$\frac{dt'}{dt} = \frac{\tau'(v)}{\tau} \equiv \gamma^{-1}(v)$$

where t is the time in background space and τ is the background cycle time of the particle at rest. Transferring the above relation for the mass flow to the system of background space thus gives

$$\frac{dm}{dt} = \frac{dm}{dt'} \cdot \frac{dt'}{dt} \equiv H_o m_p \frac{2-\gamma}{\gamma} \equiv H_o m_p \alpha \tag{39}$$

Now, if we suppose that the density of particles in the universe ρ_{mp} can be divided into a ground state portion, ρ_{mo}, corresponding to the mass of particles at rest relative to a fundamental frame, and another purely kinetic (or relativistic) portion ρ_{mk}, corresponding to kinetic energy, we obtain

$$\rho_m = \rho_{mo} = \rho_{mo} + \rho_{mk}$$

We may therefore express equation (39) in terms of densities, as follows

$$\begin{cases} \dfrac{d\rho_m}{dt} = \alpha H_o \rho_m \\ \alpha = \dfrac{2-\gamma}{\gamma} \end{cases}$$

Combining the expressions for ρ_m and α above yields

$$\alpha = \frac{(+1) \cdot \rho_{mo} + (-1) \cdot \rho_{mk}}{\rho_{mo} + \rho_{mk}}$$

This latter expression can be understood as the weighted average of the factor α for the distribution of mass, or its equivalent, between rest mass and kinetic energy, the rest mass being associated with a flow $H_o \rho_{mo}$ to particles from background space, and the kinetic mass associated with a flow $H_o \rho_{mk}$ in the opposite direction.

Applying the above defined concept of mass flow vis-à-vis the inertial and kinetic mass components, and with the rate of flow of energy from "free" electromagnetic quanta to background space as given above, we obtain a cycle of matter in the universe along the lines of Figure 1. Rather than give absolute values of the mass quantities in the different states, the densities have been used, such that the figure relates to any sufficiently large volume of space (*e.g.* $V \sim (c/H_o)^3$). The transfer rates indicated in the figure should be understood as the time derivatives of the quantities concerned. For example

$$\frac{d\rho_r}{dt} = \lambda_r \rho_{mo} - H_o \rho_r$$

The dotted line in the figure indicates the possibility of creation of matter directly from the energy of background space, and this would be necessary for a dynamic analysis starting from all energy in the ρ_o state. We will not undertake this analysis here, since we assume that λ_{cr} is several orders of magnitude smaller than the other quantities in the process.

In simple terms, then, we have flows of elementary quanta between states of:

1. Rest mass, or mass at rest relative to background space, corresponding to the ground states of the elementary particles.
2. Kinetic mass, corresponding to kinetic energy, or the equivalent excited part of the inertial mass content.
3. Radiation, or "free" electromagnetic quanta in space.
4. Background space, or "free" elementary quanta in space.

We may, for example, assume that the universe is in a state of equilibrium, and derive relations for the time derivative of energy exchange and for the kinetic mass density. The following data are used in the numerical calculation:

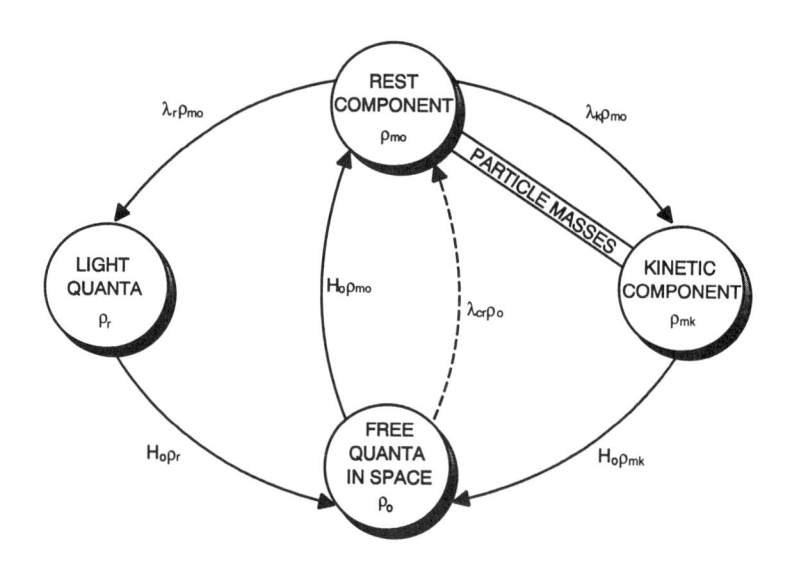

Figure 1. Circulation of matter through different different states in a universe in equilibrium.

a) Particle matter in the rest (ground) state:
The observed density of galactic mass from dynamical analysis gives (Weinberg 1972)

$$\rho_{mo} \approx 3 \times 10^{-28} \text{ kg m}^{-3}$$

b) Radiation
The total spectrum corresponding to the 2.7° K background radiation gives the equivalent mass density from the Stefan-Boltzmann laws as:

$$\rho_r = 4.4 \times 10^{-31} \text{ kg m}^{-3}$$

c) Background space
The density of "free" elementary quanta in background space from equations (7) and (21) is

$$\rho_o = \frac{3 H_o^2}{8 \pi G} \approx 10^{-26} \text{ kg m}^{-3}$$

With the above values, the following relations may be derived:

$$\begin{cases} \lambda_r = H_o \dfrac{\rho_r}{\rho_{mo}} \approx 4 \times 10^{-21} \text{ s}^{-1} \\ \lambda_k = \dfrac{H_o(\rho_{mo} - \rho_r)}{\rho_{mo}} \approx 3 \times 10^{-18} \text{ s}^{-1} \\ \rho_{mk} = \rho_{mo} - \rho_r \approx 3 \times 10^{-28} \text{ kg m}^{-3} \end{cases}$$

From the above expression for the density of elementary quanta in background space, we obtain a total density of

$$\rho_{tot} = \rho_o + \rho_{co} + \rho_{mo} + \rho_{mk} + \rho_r \equiv \frac{3}{2} \rho_o + 2 \rho_{mo}$$

or simply

$$\rho_{tot} = \frac{3}{8} \frac{H_o^2}{\pi G} + 2 \rho_{mo}$$

If we compare this result with General relativity (Friedmann models), the first term can be identified as the critical density, or the equilibrium density between an open and closed universe. In a universe dominated by relativistic matter, this is also the expected density from the Einstein equations (Weinberg 1972). The second term is a factor of 0.05 smaller, accounting for the non-relativistic matter; therefore, the elementary quantum concept appears to lead to the same density as General Relativity, though the "missing mass" is contained in elementary quanta in a free (or possibly compound) state in space.

Interactions between mass particles (fusion, fission, etc.) produce relativistic energy in the form of photons or neutrinos, and this contributes to the entropy of the Universe. The transfer of energy back to the vacuum when elementary quanta are lost by relativistic particles counterbalances the increasing entropy, and this balance between entropy creation and destruction gives rise to a constant entropy density on the large scale.

It can be argued that much of the electromagnetic radiation from stars is generated by the same process that is responsible for the cosmic background radiation. In other words, the energy absorbed by the particles that make up a star might be re-radiated into space through the surface of the star. This mechanism, if real, would account for the lack of neutrinos in solar radiation, and might provide the energy required to fuse the particles comprising a normal star into a neutron star, *i.e.* bypassing the heavy elements. It is thus conceivable that fusion is only responsible for part of a star's energy output, mainly during the early phases of the stellar cycle. Radiation produced in the later stages of the cycle could arise through the release of residual energy once the fusion process has absorbed part of the energy provided by the interaction between the star's constituent particles and the vacuum energy.

The vacuum energy absorption hypothesis may likewise be applied to the problem of the energy supply to neutron stars and X-ray burst sources. If we suppose that the mass of a stable neutron star is roughly one solar mass, the rates at which energy is absorbed and radiated by the star must approximately balance. A neutron star of mass M_o will, therefore, absorb energy according to

$$\frac{dM}{dt} = H_o M_o \approx 6 \cdot 10^{12} \text{ kg s}^{-1} \approx 5.4 \cdot 10^{29} \text{ W}$$

and the black body temperature of the radiation is given by

$$\begin{cases} \frac{dE}{dt} = H_o M_o c^2 \\ \frac{dE}{dt} = \sigma T^4 \cdot 4\pi R^2, \quad \sigma = 5.67 \cdot 10^{-8} \text{ W m}^{-2} \text{ T}^{-4} \end{cases}$$

leading to a temperature of

$$T_n = \sqrt[4]{\frac{H_o M_o c^2}{4\pi R^2 \sigma}} \approx 9.3 \cdot 10^{6 \circ} \text{ K}$$

in good agreement with observations of a steady state temperature for neutron stars on the order of $10^{10 \circ}$ K or less (Lewin 1981). Photon energy levels are characteristically in the range 2-10 KeV (Marshall and Millit 1981), which agrees well with the 4 KeV maximum average temperature that would correspond to a maximum energy density at $\lambda_{max} \sim 3$Å. In other cases, the intensity decreases from 3 KeV upward, though the spectra are flatter than a black body spectrum, and would appear to suggest Compton scattering of optical or UV photons in a very hot plasma. The range of energy emission and spectral temperature due to the energy absorption mechanism proposed here is evidently compatible with observational data, although the black body spectrum may not be entirely adequate as a model.

In an equilibrium situation, we may assume that the elementary particles in the star absorb very long-wavelength energy (very light particles) from space in proportion to their rest masses, *i.e.* the first term in the general formula, while they return energy to background space in relation to their kinetic energy, given by the second term in the general formula. When the net energy exchange is zero, the relativistic factor becomes 2, *i.e.* the kinetic energy is equal to the rest energy.

The energy released from normal particles would take the form of electromagnetic energy radiated directly back into space, but in the case of a very dense neutron star, many processes may occur before the released energy reaches the surrounding space. It may be interesting to consider the following possibilities:

a) neutrons decay into protons and electrons (lifting the collapsed electron shells to higher energy levels) and the subsequent collapse into neutrons again, accompanied by radiation.
b) spontaneous creation of particles in the dense flow of radiation surrounding the star, leading to thermonuclear reactions at intervals when sufficient amounts of new matter have been deposited on the star.
c) buildup of thermal and dynamical instabilities, while the star increases its mass above the stable maximum, leading to violent energy bursts to reset equilibrium.
d) total destruction of the star due to instabilities when its interior material composition has developed to a certain stage, where energy bursts are no longer sufficient to reset the equilibrium.

The general formula for mass exchange utilized here also provides an alternative explanation regarding the source of energy that powers X-ray bursts on the order of 10^{32} Joule emanating from objects in the galactic nucleus and from globular clusters. On the conventional view, of course, this energy is furnished by a companion star in a binary system, which gives up hydrogen to a dense object, normally assumed to be a black hole, or more likely a neutron star. In many cases, however, the binary companion cannot be accounted for. Energy absorption by neutron stars directly from space could provide the missing energy for X-ray burst sources. Indeed, if this hypothesis is confirmed, and if the neutron star is the final steady state of stellar evolution, the masses of these objects being subject to a certain limit, then it must be conceded that there is very little prospect of finding objects denser than neutron stars, *i.e.* black holes. The same model may ultimately prove successful in explaining the extremely high redshifts, energy output and spectral variations of quasars and active galactic nuclei.

6. Discussion

The hypothesis of a nonzero vacuum energy provides an affirmative answer to the question whether there exist definite relationships between the microcosm and the macrocosm. To illustrate the significance of this claim, we consider the case of the mass formula devised by Weinberg (1972) by means of an arbitrary combination of fundamental constants

$$\left(\frac{h^2 H_o}{Gc}\right)^{1/3} \sim m_\pi$$

which yields a value m_π approximately equal to the mass of the pion. According to the method adopted here, the pion mass arises when vacuum energy becomes confined within a volume equal to the volume defined by the cross-section radius of the mass equivalent

$$m_\pi = \left(\frac{h^2 H_o}{4G_o c}\right)^{\frac{1}{3}} \sim 137 \text{ MeV}$$

In a sense then, the Weinberg mass (pion) represents the boundary between the microcosm and the macrocosm. From the above analysis, we may conclude that the Weinberg mass formula is merely a special case of a general mass formula incorporating a parameter that establishes the scale of the quantum system, extending from elementary particles up to the Metagalaxy. Within this series, the force coefficient is rescaled for each system. For low values of the quantum parameter, the force corresponds in magnitude to the strong force, while at large values, it represents the gravitational force, with a proportionality of approximately 10^{40} between the two extremes. The derivation of this figure, which specifies the relationship between the large and small scale Universe, from a general model of quantum systems, provides a theoretical foundation for the Dirac large numbers.

Elsewhere (Broberg 1991, 1993) I have suggested a relationship between the spiral structure of galaxies and the toroidal geometry which, I have found, appears to characterize elementary particle masses. As noted above, especially in the discussion of the Tifft effect, galaxies, like fundamental particles, must absorb elementary quanta from the vacuum and re-radiate energy at shorter wavelengths. Presumably, stars would also conform to the toroidal geometry. This analogy leads to some interesting consequences. First, we note that the gravitational radius of an average galaxy is on the same order as the cross-section, or toroid radius of an average star. Moreover, the sum of the star toroid radii

$$r_r = \left(\frac{Am_\odot}{4\pi}\right)^{\frac{1}{2}} \approx 10^{15} \text{ m}$$

in a galaxy yields a distance on the order of the Hubble length c/H_o. Thus, nature appears to be ordered in a hierarchical sequence, with definite relations between structures at different scales.

Certainly, more research will be required to elucidate all the steps and nuances in this sequence. However, in view of the foregoing, we can assert with some degree of confidence that the hierarchical order extends from elementary particles to the Metagalaxy.

References

Broberg, H., 1981, The Elementary Quantum—Some Consequences in Physics and Astrophysics of a Minimal Energy Quantum, *ESA STM-223*.
Broberg, H., 1982, Energy, matter, and gravitation in an unlimited, renewable Universe, *ESA Journal* 6:207.
Broberg, H., 1984, The Basic Concept of Particle Rest-Mass in a Quantized, Relativistic Universe, *ESA STM-233*.
Broberg, H., 1991, Mass, energy, space, *Apeiron* 9-10:62.
Broberg, H., 1992, *On the kinetic origin of mass*, Apeiron 15:1.
de Broglie, L., 1926, *Ondes et Mouvements*, Gauthier-Villars.

Jaakkola, T., Moles, M. and Vigier, J-P, 1979, Empirical status in cosmology and the problem of the nature of redshifts, *Astron. Nach.* 300:5.
Levich, B.G., Udovin, Y.A. and Myamlin, V.A., 1973, *Theoretical Physics, Part 3. Quantum Mechanics*, North Holland.
Lewin, W., 1981, The sources of celestial X-ray bursts, *Scientific American* May 1981.
Marshall, N. and Millit, J.M., 1981. Evidence for an X-ray period in the Sco-like source 250614+091, *Nature* 293:1.
Stephani, H., 1982, *General Relativity*, Cambridge University Press, p. 30.
Weinberg, S., 1972, *Gravitation and Cosmology*, S. Wiley & Sons.
Zeldovich Ya. B. and Novikov I. D., 1971, *Relativistic astrophysics*, vol. 1, University of Chicago Press.

Index

Symbols

0957+561, 221
1038+528, 221
1146+111, 221
1525+227, 217
1635+267, 221
2016+112, 221
2237+030, 221
2345+007, 221
3C 120, 220
3C 273, 21, 217, 219, 232, 264–265
3C 279, 260, 265–267
3C 345, 219
3C 66B, 218
3C 75, 221
4C 29.45, 217

A

Absolute motion, 187
Absorption. *See* Energy: absorption
Abundance of elements, 89, 143–144
   heavy elements, 92
   light elements, 90–93, 103, 126–127, 144
Accelerators. *See* High-energy accelerators
Accretion. *See* Neutron star: accretion
Action-at-a-distance, 288, 289, 299
Active galactic nuclei, 259–268
   acceleration mechanism, 206
   blue-UV bump, 231
   broad emission lines, 207, 228–229
     variability, 230, 260–261
   Cerenkov radiation. *See also* Cerenkov effect
   gamma-radiation
     source, 260
     spectrum, 260, 261–263
   source
     dimension, 207, 231–232
   spiral arm production, 188
   synchrotron radiation, 206–233
   variability, 217–219

Aether, 187, 285–286
   and relativity, 197, 269–270
   cosmological redshift, 188, 196, 201
   dynamic, 87
   matter-antimatter creation, 196, 198
   transmission of radiation, 188, 197, 202
Anaxagoras, 137
Anaximander, 137
Anaximenes, 137
Annular channel, 208
AO 235+164, 216, 217
Apeiron, 137
Apocalypse, 192–194
Arecibo radio telescope, 4
Arp, H., 7
Astration, 91
Astronomy and Astrophysics, 11
Astrophysical Journal, 62
Atoms, 137
Aurora. *See* Galactic aurora
Axial vortex, 220

B

B2 1308+326, 218
Balmer lines
   anomalous decrements, 229–230
Baryonic matter, 97, 137–138
BATSE, 250, 252, 254, 255, 257
Bell, John, 279
Bending of light, 183
Berkeley, George, 307
Big Bang. *See also* Cosmological models: Big Bang
   nucleosynthesis, 90, 97
     inhomogeneity, 93
   observational evidence, viii, 177
Bipolar outflows, 220, 226
Birkeland current, 243–246
BL Lac objects, 206, 228, 259
   synchrotron emission, 216–220
Black hole, 259

Blazars. *See* BL Lac objects
Born, Max, 164
Bow shock wave, 220
Burbidge, E.M., 7
Burbidge, G., 7

C

C-field theory (Hoyle & Narlikar), 46
C-gravitons, 182
Campbell, W.W., 10
Carman wave principle, 331
Cascade model, 260–268
   adjoint cascade equation, 261
   cascade unit, 261
   development region, 267
   differential energetic spectrum, 261
   electron-photon cascade, 260
Catastrophism, 189, 192
CBR. *See* Cosmic background radiation
Centaurus A, 217, 219
Cerenkov continuum, 231
Cerenkov effect, 206, 221–225, 226–232
   polarization, 231
   resonance line, 221
Cerenkov instability, 246
Cerenkov pumping, 207, 228, 229–230
Cesium clocks, 283
Chiron, 191, 194
Clock retardation, 182, 199, 282–283, 342
Clouds
   intergalactic, 67–69
   Okroy, 68–69
Clusters of galaxies, v, 21, 43–44, 49, 57, 117
   Coma cluster, 29
   Fornax, 34
   Local Supercluster, 17, 27, 30, 37, 44
   UMa, 34
Coalescence, 54
COBE, 94, 99
Cognition
   and physical world, 171
Cold dark matter. *See* Dark matter: cold
Collisions. *See* Inelastic collision: photon-atom (molecule)
Comets, 188–189, 201
   Encke, 194
   giant, 189, 191
   impact cratering, 190
   short-period, 189, 194
Compact groups, 49–65
   age, 53
   interlopers, 61, 62–64
   selection criteria, 51–53
   Southern sky catalog, 52
   surface brightness, 52
   surface density, 50, 55
   velocity dispersion, 58, 60

Compton Gamma-ray Observatory, 250
Compton scattering, 215, 348
   inverse, 260
      cross-section, 262
Computer simulations, 103, 237–248
Conduction current, 210
Connection coefficients, 294
Conservation principles, 144
   angular momentum, 279
   baryon number, 100
   energy, 270
   momentum, 287, 273
Convection velocity, 285–286
Cosmic background gravitation, 130
Cosmic background radiation, 24–25, 98–100, 125–129, 164
   anisotropy, 128, 178
   energy density, 27, 126, 127
   fluctuations, 24, 99–100, 117, 135
   isotropy, 125
   origin, 125
   spectrum, 90, 97, 98, 103, 110, 117, 125, 347
   temperature, 25, 117, 126, 164, 178
Cosmic drag, 311
Cosmic plasma, 102
Cosmic rays, 138, 213–215
   physics, 283
Cosmic time, 118
Cosmological constant, 96–97, 98
   and CBR, 99–100
Cosmological cycle, 138
Cosmological horizon, 169
Cosmological models, 169–175, 27, 257
   Big Bang, 1, 188. *See also* Big Bang
      empirical inconsistency, 24, 89–104, 112, 114, 147, 158–159
      logical disproof, 82
      theoretical foundations, 147
   Equilibrium, 173, 111–149
      empirical tests, 113–118, 145–147
      theoretical foundations, 118–120
   Meta, 177–186
   Plasma, 102–104
   Steady-state, 87, 106, 119, 148, 153–166
Cosmological paradox. *See* Paradoxes: cosmological
Cosmological principles, 169–175, 178, 180
   Ancient Greek, 174
   Ancient Hindu, 175
   Anthropic, 175
   Copernican, 171–172
   Fully Perfect, 173
   Generalized Ancient Greek, 174
   Generalized Perfect, 173
   Lucretian, 173
   Perfect, 172, 112, 118–119, 153

Critical collapse distance. *See* Carman wave principle
Critical density, 347
Current filaments, 209
Current sheet, 209
Cyclotron lines, 213, 215
Cygnus A, 220

D

Dark matter, 135, 188
 cold, 94
 in intergalactic space, 67
 inflationary, 98–99
Deceleration parameter, 114
Democritus, 137
Density parameter, 90–91
deSitter. *See* Redshift: deSitter
Deuterium
 primordial, 90–93. *See also* Abundance of elements: light elements
Diffraction spectrometer
 and wavelength measurement, 327
Dimensionality, 180
Dirac large numbers, 350
Discordant redshift. *See* Redshift: discordant
Disk. *See* Galactic disk
Doppler blueshifts, 226
Doppler effect, 199
Doppler redshift
 transverse, 215
Double layer, 206
Drift velocity
 of free charges, 210–211
Dust (intergalactic)
 in voids, 67

E

Earth
 rotation
  secular retardation, 321–322
Eddington, A.S., 25
Eddington limit, 254
Effective charge, 277
Einstein Cross, 23, 24
Ejection. *See* Galaxies: spiral: ejection model
Electrical charge
 potentials and fields, 276–279
Electrogravitational coupling, 119–122, 131
 field equation, 132
 tests, 145
Electron beams, 210
Electron orbits
 velocity compression, 182
Elementary particles
 hadrons, 341
 masses, 333, 337

Elementary quantum, 333–334, 335–337, 342
 movement between states, 346
Elements. *See* Abundance of elements
Emden isothermal gas sphere, 134
Empedokles, 137
Energy
 absorption, 344, 348
Entropy, 348
 equilibrium, 185
Eotvös experiment, 299
Equation of state, 134
Equations of motion
 parameter free, 287, 296–297, 301–302
Equivalence principle, 288, 292, 312
Eschatological record, 189, 192, 201
Ether. *See* Aether
Euclidean space, 292
Euler-Lagrange equations, 293, 294, 296
Evolution effects, 21
 non-existence, 115
Exploring sphere, 278

F

Fields. *See* Magnetic fields: intergalactic
Fine structure constant, 46
Finger of God, 20–21
Fireball
 excess, 190–191
Fisher formula, 42
Fitzgerald contraction, 198, 269, 279–281, 284
Force, 180. *See also* Gravitation: force
 strong (nuclear), 341
Friedmann, A., 14
Fundamental frame, 269–284
 equivalent to laboratory, 275

G

Galactic aurora, 125
Galactic disk, 239–240
Galactic halo, 239
Galaxies
 age, 146, 165
 binary, 29
 blue
  excess, 184
 compact, v
 companions, 1, 3
  excess redshift, 3–5, 13, 21
  hot young stars, 7
 dissolution, 140
 distribution, 27, 45–46, 94, 117, 135, 325
  uniform, 165
 dominant, 1–3, 4, 7, 18
 dwarfs
  low surface brightness, 6

Galaxies (cont'd)
    elliptical, 24, 142
    field, 40
    formation, 139–140
    morphology, 140–142, 146–147
    nearby, 4, 12, 17
    nests, 140
    nuclei, 143, 195
        outflows, 135
        redshift, 113
    parent, 140
    radio, 21, 178
    rotation rate, 142, 323–324
    rotation velocity, 56, 84–85, 103, 243
        dispersion, 242–243
        vs. blue magnitude, 85
    spiral, 29, 40, 142, 187–188, 201, 246–247
        ejection
            model, 6, 59, 139, 196, 201, 247
Galaxy formation era, 97
Galilean. See also Gravitation: Galilean
    group, 286, 294
    manifold, 287
    scalar function, 292
Gamma-ray
    bursts, 214–215, 249–257
        absorption lines, 254
        annihilation lines, 253
        energy spectrum, 250, 252–254
        isotropic distribution, 251, 255–256
        source, 214, 251, 257
        time history, 252
    continuum from AGNs, 259–260
Gaussian curvature, 337
GB 191178, 254
GB 780325, 254
GB 790305, 253
GBS 05236, 214
General Relativity, 129, 147, 183
    field equations, 14
GINGA, 250, 254
Globular clusters
    age, 98, 325
Gravitation. See also Mock gravity
    constant
        distance-dependence, 306, 309, 312
        mass-dependence, 334, 339–341
    force, 143
        dynamical terms, 306, 307, 312, 316
        quantization, 299
    Galilean, 285–303
    law
        inverse linear, 185
        inverse square, 185
        Seeliger-Neumann, 155

Gravitation (cont'd)
    Newtonian, 154–157, 195, 197, 286–287, 289
        potential, 290–291
        pressure induced, 129–130, 180
        tests, 145
        propagation velocity, 184
        shielding, 130, 139, 184–185
Gravitational cosmology, 129–136
Gravitational lensing, 23–24, 221
Gravitational N-body experiments, 237–238
Gravitational paradox, 154–155
Gravitons, 310. See also C-gravitons
Gravity
    absorption, 155
GRB. See Gamma-ray: bursts
Great Attractor, 128, 139, 325
Greek philosophers, 137
GRO, 265
Group crossing time, 53
Guthrie, B., 9
Gyroradius, 208

H

$h + \chi$ Persei, 10
Halo. See Galactic halo
He4, 90–93
HEAO A-4, 254
HEAO-1, 254
Heliocentric model. See Cosmological principles: Generalized Ancient Greek
Helios, 255
Helium (3,4), 91, 92–93
    primordial, 90
Heraclitus, 137
Hertzsprung-Russell diagram, 184
HI emission, 59, 60, 61
Hidden hypotheses
    beauty, vii
    origin, vii
    reduction of phenomena, vii
    simplicity, vii
High-energy accelerators, 283
HII regions, 59
HO II, 3
Homoiomers, 137
Hoyle, F., 25
Hubble age, 90
Hubble constant, 90, 98, 329, 338
    calculated from superluminal knots, 220
    derivation, 16, 315
    variation with direction, 101
    variation with distance, 17, 165
Hubble diagram, 114
    for QSOs, 124
Hubble, Edwin, 82

Hubble sequence, 141–142
Hubble Space Telescope, 23, 90–91, 218
Hubble-Tolman test. See Redshift: vs. surface brightness
Hubble's law, 122, 123, 328
 and Copernican Cosmological Principle, 171
 and Perfect Cosmological Principle, 172
Humason, Milton, vi
Hydrogen atom, 283
Hydrogen Balmer lines, 7

I

IC 2574, 3
IC 3, 255
IC 342, 1, 9
Inelastic collision
 photon-atom (molecule), 163
Inelastic processes
 threshold, 273–276
Inertial frame, 197
Inertial induction
 acceleration, 308
 velocity, 308–313
Inertial interaction, 286–288
Infinite dimension, 185
Infinite divisibility, 180
Inflation, 46, 98–99
Interaction cross-section, 336–337
Interaction manifold, 292
 metric tensor, 293
Intergalactic medium
 and cosmic background radiation, 25
 hot, 95
 plasma, 103
Inverse Landau damping, 247
IRAS, 94
Iwanowska lines, 71

J

Jaakkola
 universe model. See Cosmological models: equilibrium
Jeans collapse, 190, 195
Jeans instability, 239
Jet-counterjet systems, 220
Jets. See also Galaxies: spiral: ejection model
 filamentary character, 220
 hot spots, 217
 knots, 219
 superluminal expansion, 219–220
 viewing angle, 220

K

K effect, 10
Karlsson, K.G., 7
Kelvin-Helmholtz instability, 246

Kepler's Laws, 289
Kerr metric, 337
Kinetic energy, 345
KONUS, 249, 252, 254, 255

L

Lagrange equation. See Euler-Lagrange equations
Lagrangian density, 293, 294, 296
Large Magellanic Cloud, 9–10, 113
Large scale structure, 100
Large-scale structure, 94–98, 102, 132–133
Length contraction, 182
Lensing. See Gravitational lensing
Leucippus, 137
Light clock, 282
Light element abundances. See Abundance of elements: light elements
Light velocity. See also Velocity of light: as limit
 experimental measurement, 332
 resonant absorption medium, 332
 in a moving frame, 282
Light-carrying medium, 182–183
 and curvature of space-time, 183
Lightbending. See Bending of light
Local Group, 1, 27, 49, 71–79
 bipolar jets, 72–73, 77
 M31 group, 72, 77–78
 Milky Way group, 72, 74–75
 redshift groupings, 73–74, 75–76
 velocity dispersion, 71–72
Local standard of rest (l.s.r.), 201
Local Supercluster, 113
Lorentz contraction. See Fitzgerald contraction
Lorentz group, 286
Lorentz transformation, 182, 198, 199
 in Galilean space and time, 300
Lorentz-Dicke theory, 195
Luminosity
 radio, 103
Luminous envelope
 around compact groups, 56
Lyman alpha forest
 absorption lines, 226–227

M

M31, 1, 12
M32, 3
M33, 3, 12
M51, 3
M81, 3
M82, 3, 7
Mach, Ernst, 307
Machian force, 130–131, 132
Mach's principle, 170–171, 155, 307
 extended, 325
Macmillan, William, 148

Magnetic fields
  intergalactic, 96
  lines
    reconnection, 212
Magnetic vortex tube (MVT), 207–209, 214–215
  active galactic nuclei, 218
  and synchrotron jets, 220
  displacement current, 209–213
  in galaxies, 215–216, 220–221
  origin of cosmic rays, 213
  pairing, 221
  precession, 220
Mann-Whitney test, 42
Markarian 205, 23
Markarian 421, 260
Mars, 322
Mass
  density, 138
  hypermassive states, 196
  inertial, 289
    Galilean invariant, 299
  missing, 347
  photon, 46
  rest, 333, 345
  variable
    in subatomic particles, 14, 46
Mass loss wind, 10, 12
Mass-energy equivalence, 271–272, 275
Matter flows, 138, 143
Maxwell's equations, 285–286, 280, 330–331, 335
Mercury. See Perihelion advance (Mercury)
Mergers, 140
Metagalaxy, 333, 338
  equilibrium state, 118
  homogeneous distribution, 117, 119
Metric tensor, 295
Michelson-Morley experiment, 280
Microwave background. See Cosmic background radiation: temperature
Milky Way, 10, 27
Missing mass. See Mass: missing
Mock gravity, 140
Models, cosmological. See Cosmological models
Momentum
  as quantity of motion, 271
Monte Carlo simulation, 40
Morphological method. See Zwicky, Fritz: morphological method

N

Napier, W., 9
Natural frame of reference, 294
Nature, 3
Neutral hydrogen, 59, 142
  deficiency, 56

Neutrino
  tau-, 93
Neutron, 341
  lifetime, 91
Neutron star. See also Accretion: neutron star
  accretion, 254
  as gamma-ray burst source, 214–215, 250, 252, 254
  radiation temperature, 348–349
Newtonian binary interaction, 290
Newtonian dynamics, 297
Newton's constant, 130, 338
Newton's Law, 182, 289, 270, 306
  derivation, 130
Newton's Laws of motion, 155–156
  Second law, 312
  Third law, 288, 290, 291
NGC 185, 3
NGC 205, 3
NGC 3077, 3
NGC 404, 7
NGC 4051, 217
NGC 4151, 217, 231, 264
NGC 4319, 23
NGC 5068, 3
NGC 5102, 3
NGC 5128, 3
NGC 5195, 3
NGC 5236, 3
NGC 5253, 3
NGC 6814, 217, 232
NGC 7317, 57
NGC 7318ab, 59
NGC 7318B, 59, 60
NGC 7319, 59
NGC 7320, 58–61
NGC 7331, 59, 60
NHC 3384, 221
Nuclear processing, 91

O

Olbers-de Cheseaux paradox. See Paradoxes: Olbers-de Cheseaux
Oort cloud, 189, 191, 195
OQ 172, 227
Origin, of universe, vii

P

Pair production, 215, 344
Palomar Sky Survey, 51, 62
Paradoxes
  cosmological, 118
  Olbers-de
    Cheseaux, 112, 119, 129, 137, 332
  Seeliger-Neumann, 112, 119
  Zeno's, 179
Particle-in-cell approach, 243

Particles. *See* Elementary particles
Pencil beam surveys, 8
Perihelion advance (Mercury), 183, 286
PG 1115+080, 221
PHL 5200, 227
PHL 938, 227
PHL 957, 226
Phobos
  secular acceleration, 322
Photoelectric effect, 199
Photon
  energy loss, 335–337
Photon-baryon number ratio, 137–138
Physical laws, 306–307
Pinching, 208
Pion, 341, 350
Pioneer 6, 113, 321
PKS 0208-512, 215, 217
PKS 0237-23, 226
PKS 0521-36, 228
Planetary orbits
  predicted from Galilean gravitation, 297–299, 302–303
Planetesimals, 188, 191, 194
Plasma
  degenerate, 215
  focus, 206, 209
  nonthermalized
    in gamma-ray bursts, 254
  outflows, 189, 195
  simulations, 243–246
  vortices, 247
Plato, 137
Poisson equation, 295–296
Polarization position angle, 217–218
Positronium, 215
Power spectrum analysis, 31–33
Preferred frame, 280
PROGNOZ 9, 252
Proton, 341
Proton-antiproton pairs, 100
Pulsed optical shutter, 332
PVO, 255
Pythagoras, 137

Q

Q 0307-195, 221
Q 1246-057, 227
QQ 1145-071, 221
Quantization. *See* Redshift: periodicity (quantization)
Quantum mechanics, 197
  and cosmology, 25–26
Quantum system, 334
Quasars/QSOs, 259
  ejection, 23
  gravitational field, 125

Quasars/QSOs (*cont'd*)
  high redshifts, 109, 122, 124, 184
    galaxy associations, 23–24
  synchrotron emission, 215, 216–220
  X-ray bright, 7

R

Rabi frequency, 225, 228
Radial velocity, 58
Radio arc, 59, 60
Radio lobes, 220
Radio sources, 59, 61
  double
    jets, 246
  steep spectrum, 59, 60
Radio telescopes, 39
Raman emission, 207
  stimulated, 224–225, 227–228, 230
Re-radiation, 348
Reasoning
  inductive, 178, 179
Redshift, 1
  absorption, 206
  cosmological, vi, 105–110, 157–163, 187, 197, 199, 201, 306, 314–315, 327–332, 333, 334–337
  deSitter, 105–110
  differential, 3, 5, 43
  discordant, 3, 47, 57
    statistical analysis, 61–65
  electrogravitational coupling, 123–125
  gravitational, 196
  grazing experiment, 320
  HI, 33, 39
  interaction effect, 113, 122–125
  intrinsic, 1–3, 4, 21, 113, 158, 162, 207
    in stars, 9–13, 215
    of Raman emission, 225, 228–229
    related to distance, 15
    time-variability, 122, 124, 145
  mean, 18–20
  negative, 19
  occultation by Sun, 120
  peaks, 34–35, 38–40, 41–43, 44
  periodicity (quantization), 7–9, 25–26, 29–47, 74, 122, 124, 178, 200
    and Local Group, 8, 78–79
    in quasars, 7–8
  quadratic Doppler, 200, 201
  related to galaxy age, 7, 13, 17, 21
    mechanism, 14–15, 124
  related to size of system, 113–114
  Solar limb, 160–162, 316–318
  tired-light mechanism, 69, 83, 158–160, 184, 315, 327
    problems, 13, 160–161
  vs. angular diameter, 100–101, 124

Redshift (cont'd)
  vs. apparent magnitude, 21
  vs. rotation rate, 85–86
  vs. surface brightness, 82, 114–115
Rees, Martin, 7
Reference resonator, 343
Reference scalar, 292–293, 294
  potential structure, 295–296
Relative kinetic energy density, 294, 295, 302
Relativistic energy
  and momentum, 270–272
Relativistic invariance
  of scalar products, 275
Relativistic spacetime, 173
Relativity principle, 198, 200
Retarded field, 276
Retarded local action, 288
Retarded potential, 277
Revised Shapley Ames Catalog, 19, 20
Ricci condition, 295
Roemer, Olaf, 83
ROSAT, 23
Rotation
  of a receding object, 83–84
RS 23, 227

S

Sandage, Allan, 19, 21
Saunders, W., 94
Seeliger-Neumann. *See* Gravitation: law: Seeliger-Neumann
Seeliger-Neumann paradox. *See* Paradoxes: Seeliger-Neumann
Selection effects, 43
Sensitivity technique, 262–263
Seyfert galaxies, 116, 228, 259
  emission lines, 230
  synchrotron emission, 216–220
Seyfert's Sextet, 50, 61
Shapley Ames Catalog. *See* Revised Shapley Ames Catalog
Shock waves, 217. *See also* Bow shock wave
SIGMA, 265
SIGNE, 250, 252
Small Magellanic Cloud, 9–10, 113
Solar limb shift, 113. *See also* Redshift: Solar limb
Solar Maximum Mission, 249, 252
Solar motion, 29
Solar nebula, 189, 190
Solar system
  transfer of angular momentum, 322–323
Solar vector
  correction, 30, 34
  galactocentric, 34, 35, 37, 44
Space. *See* Euclidean space

Spearman rank correlation coefficients, 35, 41
Special Relativity, 183, 197
  foundations, 269–284
Spectral displacement. *See* redshift
Spiral arms, 240. *See also* Galaxies: spiral
  radial matter distribution, 141
  rotation curve, 242
SSC. *See* Synchrotron radiation: self-Compton
Stability of matter, 134
Stahl, Otmar, 12
Stark effect, 207, 225, 228
Stars
  Ap type, 226
  supergiants, 9–13
    blue, 12
    in nearby galaxies, 12
  supermassive, 195
  white dwarfs
    mass discrepancy, 318–319
  young, 7, 23
Statistical testing, 30–31
  ranking procedure, 33
  signal strength, 37–38, 42–43
Stellar population, 142
Stephan's Quintet, 50, 57–61
Streaming of galaxies, 178
Strong force. *See* Force: strong (nuclear)
Substance, 179
Sulentic, J., 7
Supergalaxies, 133
Supernova, v, 59
Synchrotron radiation, 215
  beaming, 217
  non-uniform field model, 216–217
  self-Compton, 260

T

Taurid stream, 189, 191, 192, 194
Taurus A, 113, 321
Texture models
  explosive mechanisms, 96
Thales, 137
The Monist, 270
Thermodynamics, 144
Tifft effect. *See* Redshift: periodicity (quantization)
Tifft, W.G., 7
Time dilation. *See* Clock retardation
Tolman, Richard, 82
Toomre's analytical limit, 238–239
Toroidal geometry, 350
Transformation of radiation to matter, 137
Tully, R. Brent, 4, 94
Tully-Fisher
  distance, 17
  relation, 85

Tunguska, 190, 192, 194

U

U 12810, 85
Ultrarelativistic
   electrons, 230
   particles, 206
Uniformitarianism, 192
Unit vector, 277, 279
Universal balance of energy, 344
Universe
   age of, 90, 94–98
   comoving vacuum, 337–339
   effective radius of, 132
   expansion, 1
   Galilean model, 291, 294
   mean rest frame, 310

V

Vacuum
   and wave propagation, 187, 328, 342–344
   energy, 46, 201, 335
      density, 339
   permeability, 330
   permittivity, 330
Vector potential, 279
Velocity of light
   as limit, 272–273, 334
Velocity space, 293
VENERA, 255
Virgo Cluster, 17–21
   velocity dispersion
      ellipticals, 71
Virgocentric infall, 139
VLA, 60

Voids, 67–69
   South Coma, 69
Voronoy Foam, 170
Vortex current filaments, 102
VV 172, 50

W

Walls of galaxies, 185
Wave equation
   phase, 328
   phase velocity, 329–330
   stationary boundary condition, 327
Wave transmitting medium. *See* Light-carrying medium
Weber's law, 155, 309
Wedge diagram, 20
Weinberg mass, 349–350
Wickramasinghe, C., 25

X

X-ray
   background, 117
   bursts, 214–215, 349
   extensions, 23
   pulsars, 213
   sources
      diffuse, 56, 60, 61
   spectrum, 261, 263

Z

Zeno's paradox. *See* Paradoxes: Zeno's
Zero-mass hypersurfaces, 15
Zwicky, Fritz, v–vi
   morphological method, v